S0-BTB-571

Interdisciplinary Research

Fourth Edition

To our spouses, children, and grandchildren—
"The greatest force in the world is an idea whose time has come."

Sara Miller McCune founded SAGE Publishing in 1965 to support the dissemination of usable knowledge and educate a global community. SAGE publishes more than 1000 journals and over 600 new books each year, spanning a wide range of subject areas. Our growing selection of library products includes archives, data, case studies and video. SAGE remains majority owned by our founder and after her lifetime will become owned by a charitable trust that secures the company's continued independence.

Los Angeles | London | New Delhi | Singapore | Washington DC | Melbourne

Interdisciplinary Research

Process and Theory

Fourth Edition

Allen F. Repko
The University of Texas at Arlington (Retired)

Rick Szostak
University of Alberta, Canada

Los Angeles | London | New Delhi
Singapore | Washington DC | Melbourne

FOR INFORMATION:

SAGE Publications, Inc.
2455 Teller Road
Thousand Oaks, California 91320
E-mail: order@sagepub.com

SAGE Publications Ltd.
1 Oliver's Yard
55 City Road
London EC1Y 1SP
United Kingdom

SAGE Publications India Pvt. Ltd.
B 1/I 1 Mohan Cooperative Industrial Area
Mathura Road, New Delhi 110 044
India

SAGE Publications Asia-Pacific Pte. Ltd.
18 Cross Street #10-10/11/12
China Square Central
Singapore 048423

Acquisitions Editor: Helen Salmon
Editorial Assistant: Megan O'Heffernan
Production Editor: Bennie Clark Allen
Copy Editor: Lana Arndt
Typesetter: C&M Digitals (P) Ltd.
Proofreader: Molly Hall
Indexer: Alison Syring
Cover Designer: Candice Harman
Marketing Manager: Shari Countryman

Copyright © 2021 by SAGE Publications, Inc.

All rights reserved. Except as permitted by U.S. copyright law, no part of this work may be reproduced or distributed in any form or by any means, or stored in a database or retrieval system, without permission in writing from the publisher.

All third party trademarks referenced or depicted herein are included solely for the purpose of illustration and are the property of their respective owners. Reference to these trademarks in no way indicates any relationship with, or endorsement by, the trademark owner.

Printed in the United States of America

Library of Congress Cataloging-in-Publication Data

Names: Repko, Allen F., author. | Szostak, Rick, 1959- author.

Title: Interdisciplinary research : process and theory / Allen F. Repko, University of Texas at Arlington (Retired), Rick Szostak, University of Alberta, Canada.

Description: Fourth edition. | Los Angeles : SAGE, [2021] | Includes bibliographical references and index.

Identifiers: LCCN 2019037613 | ISBN 9781544398600 (paperback) | ISBN 9781544398594 (epub) | ISBN 9781544398587 (epub) | ISBN 9781544398570 (pdf)

Subjects: LCSH: Interdisciplinary research.

Classification: LCC Q180.55.I48 R47 2021 | DDC 001.4—dc23
LC record available at https://lccn.loc.gov/2019037613

This book is printed on acid-free paper.

20 21 22 23 24 10 9 8 7 6 5 4 3 2 1

BRIEF CONTENTS

DETAILED CONTENTS

PART II • DRAWING ON DISCIPLINARY INSIGHTS

PREFACE

THE BOOK

The purpose of the fourth edition of *Interdisciplinary Research: Process and Theory* is to reflect the substantial research on all aspects of interdisciplinarity that has been published since the appearance of the third edition in 2017. The literature on interdisciplinary research continues to expand; we have drawn in this revision upon many works from Europe and Australia, as well as North America. This book also reflects feedback from faculty and students who have used the third edition. Our goal in this edition is to provide a comprehensive and systematic presentation of the interdisciplinary research process and the theory that informs it, not only for students, but also for professionals and interdisciplinary teams. The book emphasizes the relationships among theory, research, and practice in an orderly framework so that the reader can more easily understand the nature of the interdisciplinary research process.

NEW IN THE FOURTH EDITION

The fourth edition incorporates the following revisions:

- We have added a list of guiding questions to the start of each chapter.

- We have expanded our discussion of creativity within the interdisciplinary research process, especially in Chapters 3, 4, 12, and 13.

- We have added more detail on the strategies associated with several STEPS.

- We have incorporated insights from dozens of recent publications.

- We have extended our discussion in the Preface on the importance of students performing independent research while reading this book.

- We have expanded and revised our discussion of epistemology in Chapters 1 and 2.

- We address confirmation bias and social media in Chapter 1.

- We discuss how interdisciplinary causal linkages may destabilize disciplinary systems of stability in Chapters 2 and 4.

- We give extended advice on how to choose a research question in Chapter 3.

- We note in Chapter 4 that many of the strategies outlined in this book have applicability beyond the academy.

- We give specific advice to graduate students on literature search in Chapter 5.

- We emphasize in Chapters 6 and 7 that disciplinary adequacy involves both evaluating insights in terms of their disciplinary perspective and pursuing debates in the discipline regarding the research question.

- We stress that interdisciplinary researchers bring new questions to the evaluation of disciplinary theories and methods in Chapter 7.

- We address the philosophy of integration in Chapter 8. We also added an example of the technique of organization.

- We discuss the role of scholarly disagreements in Chapter 9.

- We added discussions of student work patterns, mapping, finding common ground in different situations, and ethical conflicts in Chapter 10.

- We provide some clarification of the nature of both philosophical theory and models, and clarify how to address situations where disciplinary insights are complementary in Chapter 11.

- We clarify the different types of integration in Chapter 12.

- We add discussions of job interviews, policy side-effects, metacognition, understanding scholarship as a conversation, and the importance of persuasion in Chapter 13.

- We also deleted material in several chapters that was tangential to our main purpose.

The new edition continues using features that students and instructors have said they find helpful. These include the easy-to-follow step-by-step approach to describe the research process, tables and figures to illustrate aspects of each STEP, and a variety of examples oriented toward students working in the natural sciences, the social sciences, and the humanities. The additions and changes reflect the concerns and developments that have surfaced in the field since the publication of the third edition. From the constructive criticism offered by instructors and students, we have refined the prose to make it more readable, made key concepts and processes more accessible to students, and reduced the use of in-text quotations except where it is preferable to read the author's own words. Sources are cited in the text to demonstrate best scholarly practice.

THE NEED FOR THIS BOOK

This book is needed for four reasons. First, interdisciplinarity is an emerging paradigm of knowledge formation whose spreading influence can no longer be denied, discounted, or ignored. The reason is explicit: "Interdisciplinarity is associated with bold advances in knowledge, solutions to urgent societal problems, an edge in technological innovation, and a more integrative educational experience" (Klein, 2010, p. 2).

Second, this book is a corrective to those who argue that interdisciplinarity is too hard to do, who reject the notion that the field should aspire to its own methodology, or who worry that if the field becomes "disciplined it cannot offer the peculiar kind of insights that our times require" (Frodeman, Klein, Mitcham, & Holbrook, 2010, p. xxxi). It is also a corrective to those who argue that interdisciplinarity is easy in the sense that it can be done without reflection. Moreover, it helps to correct several mistaken understandings of interdisciplinarity, such as that interdisciplinarity is hostile to disciplines (see Szostak, 2019).

Third, those involved in interdisciplinary education have requested this book. As noted by Carol Geary Schneider (2010), president of the Association of American Colleges and Universities, "Interdisciplinarity is now prevalent throughout American colleges and universities" (p. xvi). Faculty are concerned that students learn how to do interdisciplinary research and writing. This is one of the important findings reported in the 2003 volume of *Issues in Integrative Studies* titled "Future Directions for Interdisciplinary Effectiveness in Higher Education: A Delphi Study." The study posed this question to its participants, all of whom are leading interdisciplinary practitioners: "What changes in interdisciplinary studies programs need to take place over the next decade in order to better serve the needs of students whose academic goals are not adequately addressed by traditional discipline-based programs?" Under "Curriculum," the participants recommended a textbook that provides an overview of disciplinary perspectives, theories, and methodologies, and especially integrative techniques, along with concrete examples (Welch, 2003, p. 185). Further evidence that the topic is neglected comes from Klein (2005a), who in *Humanities, Culture, and Interdisciplinarity: The Changing American Academy* criticizes the tendency of scholars to "hover at the level of theory with little or no attention to what is happening on the ground of practice" (p. 7). This book is a response to these concerns. It attempts to apply theory to the "ground of practice" and make the interdisciplinary research process comprehensible and achievable for students.

Fourth, the book enables students to differentiate between interdisciplinary research and disciplinary research. An important contribution of the book is that it surveys the dozen or so research methodologies used by the disciplines and explains how these are foundational to, but different from, the interdisciplinary research process. The book also reflects an emerging consensus about the meanings and operations of important inter-disciplinary theories and concepts.

THE INTENDED AUDIENCE

The book is aimed at four audiences: undergraduate students, graduate students, faculty, and members of interdisciplinary research teams. Through its extensive discussion of the disciplines and their defining elements, the book provides students not only with under-standing of the interdisciplinary research process, but also with useful discipline-specific information. This information on disciplinary perspectives, phenomena, epistemologies, assumptions, theories, and methods is as necessary for multidisciplinary research as it is for interdisciplinary research. Students in disciplinary majors may also find this infor-mation helpful to tie together courses they may take in different disciplines. Graduate students and faculty will appreciate the book's glossary of key terms, endnotes, extensive sources, various teaching aids, numerous examples that demonstrate best practices from professional work, and recommended readings organized by specialty from the field's extensive literature in the Appendix.

Most books on research methods can assume professional consensus about the principles of the field they present. Because the field of interdisciplinary studies has only just reached the point where there is sufficient potential for scholarly consensus on the principles of the field, this book has to point the reader toward a scholarly rationale in the literature for each principle, in addition to explaining the principle itself. In a sense, then, the book is aimed at faculty teaching an interdisciplinary course as much as at students taking that course. Undergraduate and graduate students can learn about interdisciplinary stud-ies from the rationale for each interdisciplinary principle, as well as from the principles themselves.

The book is intended as either a core text or a supplemental resource for undergraduate and graduate courses that are interdisciplinary. More specifically, the book is useful in a variety of academic contexts: intermediate-level courses that focus on interdisciplinary research and theory; upper-level topics, problems, or theme-based courses that involve working in two or more disciplines; integrative capstone and senior seminar courses that require an in-depth interdisciplinary research paper/project; keystone courses that integrate general education for upper-level undergraduate programs; graduate courses in interdisciplinary teaching and/or research; teaching assistant training/certificate courses in interdisciplinary learning, thinking, and research; first-semester master's-level research courses; and administrators and faculty who wish to develop interdisciplinary courses and programs at their institutions. The book, particularly its early chapters, may serve as a primary text for introductory interdisciplinary studies courses. For a text designed especially for those courses, see Repko, Szostak, and Buchberger (2020), *Introduction to Interdisciplinary Studies,* third edition. This book is also useful to multidisciplinary programs calling on students to cross several disciplinary domains, professionals, and interdisciplinary teams practicing interdisciplinary research.

THE APPROACH USED AND STYLE OF PRESENTATION

This book's approach to interdisciplinary research is distinctive in at least six respects. (1) It describes how to *actually do* interdisciplinary research using processes and techniques of demonstrated utility whether one is working in the natural sciences, the social sciences, the humanities, or applied fields. (2) It integrates and applies the body of theory that informs the field into the discussion of the interdisciplinary research process. (3) It presents an easy-to-follow, but not formulaic, decision-making process that makes integration and the goal of producing a more comprehensive understanding achievable. The term *process* is used rather than *method* because, definitionally, *process* allows for greater flexibility and reflexivity, particularly when working in the humanities, and it distinguishes interdisciplinary research from disciplinary methods. (4) The book highlights the foundational and complementary role of the disciplines in interdisciplinary work, the necessity of drawing on and integrating disciplinary insights, including insights derived from one's own basic research. (5) The book includes numerous examples of interdisciplinary work from the natural sciences, the social sciences, and the humanities to illustrate how integration is achieved and how an interdisciplinary understanding is constructed, reflected on, tested, and communicated. (6) This book is ideally suited for active learning and problem-based pedagogical approaches, as well as for team teaching and other more traditional strategies.

DESIGN FEATURES

The book aids student content comprehension by using current learning strategies that characterize the modern textbook. The book's self-contained yet interconnected chapters promote flexibility in structuring courses, depending on the individual instructor's needs and interests. Conceptual and organizational approaches include chapter objectives and learning outcomes, section headings and subheadings, boldfacing of key concepts, italicizing of key statements, graphics to illustrate key concepts, tables to present content in a concise and coherent way, notes to readers, chapter summaries, a glossary of key concepts with chapter and page references, an author index, and a detailed subject index. Faculty can profitably use chapter components that correspond to their own approach to interdisciplinary research while omitting others.

HOW TO USE THIS BOOK

Students will best master the material in this book by applying it to their own research. Only then can students fully appreciate the value of the advice and information provided in this book. One strategy of proven utility is for students to work (solo or in small groups) on a semester-long project in three or four segments that correspond to various STEPS in the interdisciplinary research process (IRP). For example, the first segment might include the first two STEPS. Providing feedback on each segment of the project in a timely manner will aid students to make improvements in their work before they proceed to the next segment.

Both authors have in our own courses required students to perform research projects of their own choice. We have then discussed how students address each STEP in class. We have found that students enjoy aiding each other, and come to appreciate the value of the STEPS and strategies for performing these when they see these usefully applied to a student research project (especially their own!). We recognize, though, that some instructors and students may wish to deviate from this approach. They may prefer some sort of video or live presentation of results along with or instead of the traditional research paper. They may prefer team research: This may be especially valuable, for it forces students to appreciate the perspectives of other team members. They may pursue community service learning, where the student volunteers with some community organization during the term. At the end of the term, they write a reflective essay on their experience and the advice that they have for the organization. They should explore how techniques of interdisciplinary analysis can be applied to the work of the organization. Some instructors may wish to ask students to produce an ePortfolio or intellectual autobiography in which students are guided to reflect on various questions—about themselves and about the course—as the course proceeds. Students can then observe how their understanding of interdisciplinarity, and their ability to perform interdisciplinarity analysis, has been enhanced by this course (see Repko, Szostak, & Buchberger [2020] for advice on community service learning, ePortfolios, and intellectual biographies).

End-of-chapter exercises can be used in two ways: to stimulate class discussion and to facilitate deep learning of critical components of the IRP. Students should be encouraged to discuss both the challenges and successes they experience in performing each STEP with their group and/or with the class as a whole. Importantly, instructors should candidly share their experience with interdisciplinary research and explain how they overcame certain challenges. Students who struggle with a particular STEP, and then learn that a certain strategy allows them to move forward, will long remember the strategy.

Instructors new to interdisciplinary studies should be aware that some STEPS take longer for students to understand and perform than others. For example, they often stumble at STEP 1: identifying a research topic. The criteria for selecting a topic or problem suitable for interdisciplinary inquiry and justifying using an interdisciplinary approach is not explained until Chapter 3. Nevertheless, instructors are well advised at the outset of the term to ask students to reflect on the sorts of problems or topics they care about and that could be researched using information from more than one discipline.

For best results, instructors should encourage students to *do research while learning* about the various STEPS. They should urge students to start looking for relevant literature early on even though they have not yet mastered the intricacies of conducting the full-scale literature search discussed in Chapter 5 (STEP 4). Instructors who do this report that students are often more appreciative of learning the various literature search techniques after they have tried searching on their own.

Some STEPS are more time intensive than others. Such is the case for the critical STEPS of creating common ground (STEP 8) and producing a more comprehensive understanding of the problem (STEP 9) near the end of the IRP. Instructors should make certain that students have ample time to understand this material and complete these critical STEPS before the research project is due.

By carefully pacing coverage of the information presented in this book, instructors will enable students to produce a high-quality product and develop a deep appreciation for interdisciplinary studies.

CONTENTS

The book is divided into three sections, each organized around a theme that addresses a central issue of the field of interdisciplinary studies. Part I, consisting of two chapters, defines interdisciplinary studies, explains the intellectual essence of the field, and introduces the disciplines and their perspectives. Part II, consisting of five chapters, introduces the model of the IRP and explains how to draw on disciplinary insights and theories. Part III, consisting of six chapters, explains how to integrate conflicting disciplinary insights by creating common ground between them, construct a more comprehensive understanding of the problem, test it, and communicate it to an appropriate audience.

Part I: About Interdisciplinary Studies and Disciplines

Today, interdisciplinary learning at all academic levels is far more common, and there is greater understanding of what it is. Early definitions of interdisciplinary studies were quite general and disparate, but the range of meanings has narrowed dramatically over the last decade and these are integrated into the definition presented in this book. Interdisciplinary learning is more widespread because educators recognize that it is needed, that the disciplines though necessary are insufficient by themselves to address the complex problems that are demanding attention in today's world.

Chapter 1: Introducing Interdisciplinary Studies. Chapter 1 answers the question: *What is interdisciplinary studies?* The popularity of the term *interdisciplinarity* in the academy, the multiplication of interdisciplinary initiatives and programs, and the persistence of exaggerated claims and outdated suppositions about interdisciplinarity heighten the importance of achieving clarity about its meaning that is grounded in authoritative sources. The chapter defines interdisciplinary studies, describes the intellectual essence of interdisciplinarity, and distinguishes interdisciplinarity from disciplinarity, multidisciplinarity, transdisciplinarity, and integrative studies.

Chapter 2: Introducing the Disciplines and Their Perspectives. Interdisciplinary studies is based on the generally held assumption that the disciplines are foundational to interdisciplinarity. If so, then students should know how knowledge is typically structured in the modern academy and how it is reflected in its organization. They should also know how the disciplines usually associated with each major category—the natural sciences, the social sciences, the humanities, and the applied fields—engage in learning and produce new knowledge. There are unresolved questions and significant differences of opinion, however, over precisely what interdisciplinarians should use from the disciplines and how they should use it. The chapter attempts to bridge these differences by defining the term *disciplinary perspective* to mean those defining elements of a discipline—the phenomena, epistemology, assumptions, concepts, theories, and methods—that constitute its intellectual "center of gravity" and differentiate it from other disciplines. The chapter unpacks the meaning of these elements and explains how they are used in the interdisciplinary research process. The chapter also presents two approaches to ascertain the relevance of a particular discipline's perspective on the problem: the "perspectival approach," which calls for linking the problem to those disciplines whose perspectives embrace it, and the "classification approach," which calls for connecting the problem (at least initially) directly to the phenomena typically studied by disciplines. By focusing on phenomena, researchers can broaden their investigation without focusing prematurely on particular disciplines. Subsequent chapters draw heavily from the information provided in this chapter.

Part II: Drawing on Disciplinary Insights

Three more questions follow in the focus on interdisciplinary studies: *What is the inter-disciplinary research process? How is it achieved? What theory or body of theory informs it?* Part II introduces the IRP and describes how to select a problem or research question and justify using an interdisciplinary approach (Chapter 3), discusses how to identify disciplines relevant to the problem or research question (Chapter 4), explains how to conduct the literature search (Chapter 5), examines how to develop adequacy in relevant disciplines (Chapter 6), and demonstrates how to analyze the problem and evaluate insights (Chapter 7).

Chapter 3: Beginning the Research Process. Chapter 3 presents an integrated and step-based research model of the IRP. The chapter asserts that the process the model delineates is not linear, but a series of carefully considered decision points called "STEPS," which can lead to integration and a more comprehensive understanding of the problem. The process is heuristic and involves a good deal of reflexivity. This chapter begins the process of identifying decision points and research pathways and provides examples of them from published and student work. STEP 1 discusses how to develop a good research question and stresses the importance of framing the research or focus question in a way that is appropriate to interdisciplinary inquiry. STEP 2 urges students to justify using an interdisciplinary approach. The chapter addresses the need for greater transparency in interdisciplinary writing that will make these decision points and research pathways explicit.

Chapter 4: Identifying Relevant Disciplines. STEP 3 asks researchers to identify disciplines relevant to the problem, topic, or question. This chapter focuses on two decision points. The first is to decide which disciplines (including subdisciplines, interdisciplines, and schools of thought) are *potentially* interested in the problem. This decision should be made before conducting the full-scale literature search. Researchers are urged to map the problem to reveal its constituent disciplinary parts and connect each part to the discipline which studies that part. The second decision point is to reduce the number of "potentially relevant" disciplines to those that are "most relevant."

Chapter 5: Conducting the Literature Search. After defining *literature search* in the context of interdisciplinary studies, the chapter presents reasons for conducting a full-scale systematic literature search (STEP 4) and notes the special challenges confronting interdisciplinarians. The search process is divided into two phases: the initial search conducted at the outset of the project and the full-scale search conducted later. The chapter describes the organization and classification of books in research libraries and presents various search strategies, noting mistakes commonly made when beginning the literature search. The chapter then discusses how to conduct the full-scale search.

Chapter 6: Developing Adequacy in Relevant Disciplines. This chapter introduces how to develop adequacy in relevant disciplines, focusing on *how much* knowledge is required from each discipline and *what kind* of knowledge. Interdisciplinary researchers need to appreciate both disciplinary perspective and disciplinary debates regarding the issue being researched. The chapter discusses developing adequacy in disciplinary *theories*, explains the reason to understand theories and their concepts, and demonstrates how to proceed in identifying relevant theories. The chapter also discusses developing adequacy in disciplinary *methods*. Adequacy requires familiarity with the dozen or so methods used in the natural sciences, the social sciences, and the humanities, and their strengths and limitations. And adequacy involves understanding the interdisciplinary position on methods, how a discipline's preferred methods correlate to its preferred theories, and the importance of providing in-text evidence of disciplinary adequacy.

Chapter 7: Analyzing the Problem and Evaluating Insights. The chapter explains how to analyze the problem from each disciplinary perspective and evaluate its insights. This involves identifying the strengths and limitations of each author's perspective and theory, recognizing that each author's approach to the problem may be skewed and understanding the implications of this, recognizing that the data or evidence upon which the insight and theory are based may also be skewed, and recognizing that the methods used by authors may be skewed as well. Interdisciplinarians evaluate disciplinary theories and methods in a way that is different from, but complementary to, disciplinary evaluation. Analyzing the problem also necessitates reflecting on how one's personal or disciplinary bias may skew one's understanding of the problem, thus compromising the integrity of the interdisciplinary research process.

Part III: Integrating Insights

Engaging in the interdisciplinary research process raises further questions: How does one perform the integrative task? What, precisely, is being integrated? What is the understanding that is produced? How should it be tested? Integration, as presented here, is a process that involves making a series of decisions (Chapter 8). These involve identifying conflicts among insights and their sources (Chapter 9), creating common ground among concepts and/or assumptions (Chapter 10), creating common ground among theories (Chapter 11), constructing a more comprehensive understanding (Chapter 12), and reflecting on, testing, and communicating the understanding (Chapter 13).

Chapter 8: Understanding Integration. Integration is the key distinguishing characteristic of interdisciplinary work. After noting the controversy between generalists and integrationists over integration, the chapter establishes that integration should be the goal of interdisciplinary work, even though undergraduates and those working in the humanities may achieve only partial integration. The chapter identifies conditions necessary to perform integration, discusses the philosophy of integration, and describes the model of integration used in this book. The chapter discusses *what* the model of the IRP integrates, *how* it integrates, and what the *results* of integration look like. It answers three fundamental questions concerning integration.

Chapter 9: Identifying Conflicts Among Insights and Their Sources. This chapter focuses on the integrative part of the IRP, beginning with identifying conflicts among insights and their sources (STEP 7). These conflicts stand in the way of creating common ground and, thus, of achieving integration. Conflicts among insights are generally discovered when conducting the full-scale literature search. Possible sources of conflict among insights are their embedded concepts, theories, and the assumptions underlying them. The chapter concludes by discussing the importance of communicating one's research to the appropriate audience(s).

Chapter 10: Creating Common Ground Among Insights: Concepts and/or Assumptions. This chapter begins STEP 8 and is guided by the idea that disciplinary insights are potentially complementary if their concepts and theories and the assumptions underlying their concepts and theories are sufficiently modified. The theories of common ground and cognitive interdisciplinarity are the basis for collaborative communication and integration. The chapter defines common ground and presents six core ideas that form the basis for creating it. It explains how to create common ground among conflicting concepts or assumptions, identifies four techniques used to modify concepts and assumptions, and explains how to create common ground when ethical positions conflict.

Chapter 11: Creating Common Ground Among Insights: Theories. This chapter continues STEP 8, defining disciplinary theory. It describes the models, variables, concepts, and causal relationships that one typically encounters when working with disciplinary theories. Researchers working with a set of theories will have to create common ground by modifying them directly through their concepts or indirectly via their underlying assumptions. The chapter discusses the modification strategies commonly used.

Chapter 12: Constructing a More Comprehensive Understanding or Theory. After defining "a more comprehensive understanding or theory," the chapter explains how to construct the understanding from concepts and/or assumptions that were modified and from which common ground was created. STEP 9 lays out two pathways. The first pathway is applicable to the humanities, the fine and performing arts, and some applied fields where the focus of integration is directly on concepts and indirectly on their underlying assumptions. In these contexts, achieving full interdisciplinarity involves consciously choosing to construct an understanding that is comprehensive and nuanced.

The second pathway is applicable to the natural and social sciences, sometimes to the humanities, and to some applied and multidisciplinary fields where the focus of knowledge formation is on the development of theories to explain the phenomena of interest. The chapter identifies and illustrates five strategies demonstrated to achieve integration and construction of a more comprehensive theory.

Chapter 13: Reflecting on, Testing, and Communicating the Understanding or Theory. The final task of the IRP is STEP 10 to reflect on, test, and communicate the more comprehensive understanding or theory. The chapter discusses the four sorts of reflection that are called for in interdisciplinary work, including what has actually been learned from the project, what STEPS (of the IRP) were omitted, what were one's own biases, and what are the strengths and limitations of the insights and theories used, including the utility of the IRP itself. Students are invited to reflect on how they might describe their interdisciplinary education in a job interview. The chapter explains how to test the quality of one's work in a way that takes into account the literature on cognition and instruction, and then identifies four approaches to test the more comprehensive understanding.

Finally, the chapter stresses the importance of communicating the understanding *persuasively* in multiple ways to multiple audiences regardless of academic level. The activity of communicating the results of integrative work is, in fact, another way of testing its coherence, unity, and balance, and thus whether it constitutes partial or full interdisciplinarity.

The field of interdisciplinary studies is beginning to demonstrate its full potential and generate the volume and scope of new knowledge that its founders envisioned. The process of knowledge formation can be accelerated and find a wider audience as its practitioners produce more and better interdisciplinary work. To this end, we offer this fourth edition with its undoubted limitations to facilitate interdisciplinary education and research.

Editable, chapter-specific PowerPoint® slides and tables and figures pulled from the book are available on the instructor site at **study.sagepub.com/repko4e**.

Professor Allen F. Repko
Former Director
Interdisciplinary Studies Program
The University of Texas at Arlington

Professor Rick Szostak
Professor and Chair, Department of Economics
University of Alberta, Edmonton, Canada

ACKNOWLEDGMENTS

SAGE Publishing gratefully acknowledges input from the following reviewers:

Carol Barrett, Union Institute & University, and Saybrook University

Robert S. Garner, Salisbury University

Charles R. Hamilton, Texas A&M University, Central Texas

Kristin Picardo, St. John Fisher College

Dianna Rust, Middle Tennessee State University

Anissa J. Sorokin, Stevenson University

Marcus Tanner, Texas Tech University

Sabrina Wengier, Middle Georgia State University

ABOUT THE AUTHORS

 **Allen F. Repko**, PhD, is the former director of the interdisciplinary studies program in the School of Urban and Public Affairs at the University of Texas at Arlington, where he developed and taught the program's core courses for many years. Repko has written extensively on all aspects of interdisciplinary studies, has twice served as coeditor of the interdisciplinary journal *Issues in Integrative Studies*, and has served on the board of the Association for Interdisciplinary Studies (AIS). Though "retired," he continues to work on newer editions of his books.

 **Rick Szostak**, PhD, is a professor of economics at the University of Alberta, where he has taught for 31 years. He is the author of a dozen books and 50 articles, all of an interdisciplinary nature. Several of his publications address how to do interdisciplinary research, teach interdisciplinary courses, administer interdisciplinary programs, or organize information to facilitate interdisciplinarity. As an associate dean, he created the Office of Interdisciplinary Studies at the University of Alberta, the Science, Technology and Society program, an individualized major, and two courses about interdisciplinarity. He has twice served as coeditor of the interdisciplinary journal *Issues in Integrative Studies*. He was president of the Association for Interdisciplinary Studies (AIS) from 2011 to 2014. He can be contacted at rszostak@ualberta.ca.

ABOUT INTERDISCIPLINARY STUDIES AND DISCIPLINES

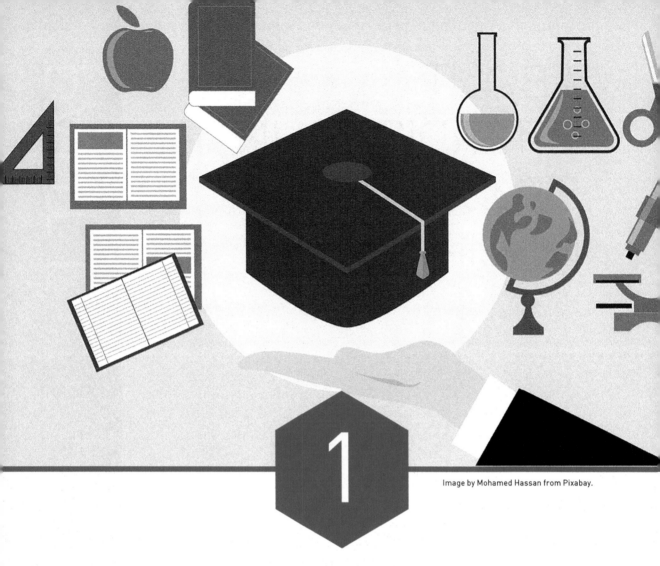

PART I

Image by Mohamed Hassan from Pixabay.

INTRODUCING INTERDISCIPLINARY STUDIES

LEARNING OUTCOMES

By the end of this chapter, you will be able to

- Define interdisciplinary studies
- Describe the intellectual essence of interdisciplinarity
- Distinguish interdisciplinarity from multidisciplinarity, transdisciplinarity, and integrative studies

GUIDING QUESTIONS

What is interdisciplinary studies?

What are the key characteristics of interdisciplinary studies?

How can we define interdisciplinarity, and carefully distinguish this from multidisciplinarity, transdisciplinarity, and integrative studies?

CHAPTER OBJECTIVES

In any university, whether physical or virtual, you will definitely encounter the disciplines. They are powerful and pervasive approaches to learning and knowledge production. They shape our perceptions of the world, our ability to address complexity, our understanding of others and ourselves—and usually the administrative structure of colleges and universities. Less than 200 years old in their modern form, the disciplines have come to dominate the ordering, production, and communication of knowledge. Today, however, disciplinary dominance is being challenged by interdisciplinarity.

This chapter introduces interdisciplinary studies as an academic field. We define interdisciplinary studies and present the intellectual essence of the field in terms of its assumptions, theories, and epistemology. We then distinguish interdisciplinarity from multidisciplinarity, transdisciplinarity, and integrative studies.

DEFINING INTERDISCIPLINARY STUDIES

Interdisciplinary studies refers to a diverse and growing academic field with its own literature, curricula, community of scholars, undergraduate majors, and graduate programs. Importantly, it uses a research process designed to produce new knowledge in the form of more comprehensive understandings of complex problems. The focus of this book is on this research process.

Before defining interdisciplinary studies, we unpack the meaning of its three parts: *inter, disciplinary,* and *studies.*

The "Inter" Part of Interdisciplinary Studies

The prefix *inter-* means "between, among, in the midst," or "derived from two or more." **Disciplinary** means "of or relating to a particular field of study" or specialization. Thus, a starting point for understanding the meaning of *interdisciplinary studies* is between two or more fields of study.

This "between" space is contested space—problems, issues, or questions that are the focus of several disciplines. For example, urban riots are an interdisciplinary problem because they are an economic problem *and* a racial problem *and* a public policy problem. The important point is that the *disciplines are not the focus of the interdisciplinarian's attention; the focus is the problem or issue or intellectual question that each discipline is addressing.* The disciplines are simply a means to that end.

The "Disciplinary" Part of Interdisciplinary Studies

Inside the academy, *discipline* refers to a particular branch of learning or body of knowledge such as physics, psychology, or history (Moran, 2010, p. 2). **Disciplines** are scholarly communities that specify which phenomena to study, advance certain central concepts and organizing theories, embrace certain methods of investigation, provide forums for sharing research and insights, and offer career paths for scholars. It is through their power over careers that disciplines are able to maintain these strong preferences: Disciplinary scholars generally gain a PhD within the discipline, get hired by a disciplinary department, and are granted tenure, promotions, and salary increases depending in large part on how that department judges their research and teaching. An **insight** is a scholarly contribution to the understanding of a problem based on research.

Each discipline has its own defining elements—phenomena, assumptions, philosophical outlook (i.e., epistemology), concepts, theories, and methods—that distinguish it from other disciplines (the subject of Chapter 2). For example, disciplines choose methods that are good at investigating their theories. All of these characteristics are interrelated and are included within a discipline's overall disciplinary perspective on reality.

History is an example of a discipline because it meets all of the above criteria. Its knowledge domain consists of an enormous body of *facts* (everything that has been recorded in human history). It studies an equally enormous number of *concepts or ideas* (colonialism, racism, freedom, and democracy). It generates *theories* about why things turned out the way they did (e.g., the great man theory argues that the American Civil War lasted so long and was so bloody because President Abraham Lincoln decided to issue the Emancipation Proclamation in 1862), although many historians strive to be atheoretical. Furthermore, it uses a research *method* that involves close reading and critical evaluation of primary sources (e.g., letters, diaries, official documents) and secondary sources (e.g., books and articles) to present a coherent picture of past events or persons within a particular time and place. **Close reading** is a method that calls for careful analysis of a text and close attention to individual words, syntax, potential biases, and the order in which sentences and ideas unfold.

Categories of Traditional Disciplines

There are three broad categories of traditional disciplines[1] (see Table 2.1 in Chapter 2):

- The natural sciences tell us what the world is made of, describe how what it is made of is structured into a complex network of interdependent systems, and explain the behavior of a given localized system.

- The social sciences seek to explain the human world and figure out how to predict and improve it.

- The humanities express human aspirations, interpret and evaluate human achievements and experience, and seek layers of meaning and richness of detail in written texts, artefacts, and cultural practices.

The Fine and Performing Arts

In addition to the traditional disciplines is the category of the fine and performing arts. These include art, dance, music, and theater. They rightly claim disciplinary status because their defining elements are very different from those of the humanities disciplines.

The Applied and Professional Fields

The **applied fields** also occupy a prominent place in the modern academy. These include business (and its many subfields such as finance, marketing, and management), communications (and its various subfields including advertising, speech, and journalism), criminal justice and criminology, education, engineering, law, medicine, nursing, and social work. (*Note:* Many of these applied and professional fields and schools claim disciplinary status.)

The Emergence of Interdisciplines

The line between the disciplines and interdisciplinarity has begun to blur in recent years with the emergence of **interdisciplines**. These are fields of study that cross traditional disciplinary boundaries and whose subject matter is taught by informal groups of scholars or by well-established research and teaching faculties. *Interdisciplines may or may not be interdisciplinary.* Frequently cited examples of interdisciplines are neuroscience, biochemistry, environmental science, ethnomusicology, cultural studies, women's studies, urban studies, American studies, and public health (National Academies, 2005, pp. 249–252). Some interdisciplines use a wide range of theories, methods, and phenomena, while others behave much like disciplines by focusing on a narrow set of these (see Fuchsman, 2012).

NOTE TO READER

The disciplines, applied fields, and interdisciplines are not rigid and unchanging but are evolving social and intellectual constructs. That is, they take on new theories, methods, and research questions over time, while shedding other theories, methods, or questions. They nevertheless retain their control over the careers of disciplinary scholars.

The "Studies" Part of Interdisciplinary Studies

The first fields to describe themselves using the word "**studies**" were those focused on particular sociocultural groups (including women, Hispanics, and African Americans). The word then became common in a host of contexts in the natural sciences and social sciences. In fact, "studies" programs are proliferating in the modern academy. In some cases, even the traditional disciplines (particularly in the humanities) are renaming themselves as studies, such as English studies and literary studies (Garber, 2001, pp. 77–79).

Why "Studies" Is an Integral Part of Interdisciplinary Studies

Studies programs in general represent fundamental challenges to the existing structure of knowledge. These new arrangements share with interdisciplinary studies (as described in this book) a broad dissatisfaction with traditional knowledge structures (i.e., the disciplines) and a recognition that the kinds of complex problems facing humanity demand that new ways be found to order knowledge and bridge different approaches to its creation and communication. Today, there are programs that include a core of explicitly interdisciplinary courses, established interdisciplinary fields such as area studies (e.g., Middle Eastern studies) and materials science, and highly integrated fields such as environmental studies, urban studies, sustainability studies, and cultural studies.

Comparing the Disciplines and Interdisciplinary Studies

The seven main characteristics of the established disciplines are compared and contrasted with those of interdisciplinary studies in Table 1.1. There are three differences (#1, #2, and #3) and four similarities (#4, #5, #6, and #7). The differences explain why the use of "studies" in interdisciplinary studies is appropriate:

- Interdisciplinary studies does not lay claim to a universally recognized core of knowledge as, say, physics does, but rather draws on existing disciplinary knowledge, while always transcending it via integration (#1).

- Interdisciplinary studies has a research process of its own (the subject of this book) to produce knowledge but freely borrows methods from the disciplines when appropriate (#2).

- Interdisciplinary studies, like the disciplines, seeks to produce new knowledge, but unlike them, it seeks to accomplish this via the process of integration (#3).

TABLE 1.1 ● Comparison of Established Disciplines to Interdisciplinary Studies

Established Disciplines	Interdisciplinary Studies
1. Claim a body of knowledge about certain subjects or objects	1. Claims a burgeoning professional literature of increasing sophistication, depth of analysis, breadth of coverage, and thus, utility. This literature includes subspecialties on interdisciplinary theory, program administration, curriculum design, research process, pedagogy, and assessment. Most important, a growing body of explicitly interdisciplinary research on real-world problems is emerging.
2. Have methods of acquiring knowledge and theories to order that knowledge	2. Makes use of disciplinary methods, but these are subsumed under an interdisciplinary research process that involves drawing on relevant disciplinary insights, concepts, theories, and methods to produce integrated knowledge
3. Seek to produce new knowledge, concepts, and theories within or related to their domains	3. Produces (via integration) new knowledge, more comprehensive understandings, new meanings, and cognitive advancements (We will define "more comprehensive understanding" and "cognitive advancement" in later chapters.)
4. Possess a recognized core of courses	4. Is beginning to form a core of explicitly interdisciplinary courses
5. Have their own community of experts	5. Is forming its own community of experts

(Continued)

TABLE 1.1 ⬢ (Continued)	
Established Disciplines	**Interdisciplinary Studies**
6. Are self-contained and seek to control their respective domains as they relate to each other	6. Draws on the disciplines for material but also on an interdisciplinary literature
7. Train future experts in their discipline-specific master's and doctoral programs	7. Is training future experts in older fields such as American studies and in newer fields such as cultural studies through its master's and doctoral programs and undergraduate majors. Though new and explicitly interdisciplinary PhD programs are emerging, interdisciplinary studies still typically hires those with disciplinary PhDs.

Source: Adapted from Vickers, J. (1998). Unframed in open, unmapped fields: Teaching the practice of interdisciplinarity. *Arachne: An Interdisciplinary Journal of the Humanities,* 4(2), 11–42.

Why "Studies" Is Plural

"Studies" is plural because of the idea of interaction between disciplines (Klein, 1996, p. 10). Imagine the world of knowledge wherein each discipline is like a box containing thousands of dots, each dot representing a bit of knowledge discovered by an expert in that discipline. Then imagine similar boxes representing other disciplines, each filled with dots of knowledge. Scholars interested in "studies" are excited by the prospect of examining a broad issue or complex question that requires looking inside as many disciplinary boxes as necessary to identify those dots of knowledge that have some bearing on the issue or question under investigation. "Studies" scholars, including those in interdisciplinary studies, are in the business of identifying and connecting dots of knowledge regardless of the disciplinary box in which they reside (Long, 2002, p. 14). Interdisciplinarians are interested not in merely rearranging these ever-changing dots of knowledge but in *integrating* them into a new and more comprehensive understanding that adds to knowledge.

Studies programs recognize that many research problems cannot easily be addressed from the confines of individual disciplines because they require the participation of many experts, each viewing the problem from its distinctive disciplinary perspective.

Critics of studies programs charge that they lack disciplinary "substance and good scholarship" (Salter & Hearn, 1996, p. 3). **Scholarship** is a contribution to knowledge that is "*public,* susceptible to *critical review and evaluation,* and accessible for *exchange and use* by other members of one's scholarly community" (Shulman, 1998, p. 5). "Substance" and "scholarship" are typically code words for disciplinary depth-intensive focus on a discipline or **subdiscipline**. By emphasizing a narrow set of theories, methods, and phenomena, disciplines are able to carefully police whether their theories and methods are correctly applied to appropriate phenomena.

A contrasting view is that a purely disciplinary focus sacrifices breadth, comprehensiveness, and realism for depth. An integrated view, which this book reflects, recognizes that there is a symbiosis between disciplinary and interdisciplinary research. By articulating the nature of the interdisciplinary research process, we can encourage comparable rigor in interdisciplinary analysis, while utilizing any relevant disciplinary theories and methods.

This is not to say that a "studies" program is superior to a disciplinary one. That would be a mistake because the purpose of each is different. *Both are needed*, particularly in a world characterized by increasing complexity, conflict, and fragmentation.

A Definition of Interdisciplinary Studies

It is possible to identify key elements that practitioners agree should form the basis of an integrated definition of interdisciplinary studies:

- The focus of interdisciplinary research extends beyond a single disciplinary perspective.

- A distinctive characteristic of interdisciplinary research is that it focuses on a problem or question that is complex. (*Note:* We provide a precise definition of complexity further in text.)

- Interdisciplinary research is characterized by an identifiable process or mode of inquiry.

- Interdisciplinary research draws explicitly on the disciplines.

- The disciplines provide insights about the specific substantive focus of particular interdisciplinary research projects.

- Interdisciplinary research has integration as its goal.

- The objective of the interdisciplinary research process is pragmatic: to produce a cognitive advancement in the form of a new understanding, a new product, or a new **meaning**. (*Note:* The term *meaning* is important in the humanities, where it is often equated with the intent of the author or artist or the effect on the audience [Bal, 2002, p. 27].)[2]

From these elements, it is possible to offer this integrated definition of interdisciplinary studies:

> **Interdisciplinary studies** is a process of answering a question, solving a problem, or addressing a topic that is too broad or complex to be dealt with adequately by a single discipline, and draws on the disciplines with the goal of integrating their insights to construct a more comprehensive understanding.

This definition includes four core concepts—process, disciplines, integration, and a more comprehensive understanding—which are the subjects of later chapters. Importantly, this definition has both a *what* and a *how* component. Typically, when defining an experiment, one almost unavoidably describes how to do it. Chapters 1 and 2 of this book explain the *what* part; the rest of the chapters, which deal with the interdisciplinary research process, explain the *how* part. (*Note:* More detail on the historical evolution of this definition is provided in Repko, Newell, & Szostak [2012].)

Rick Szostak (2015b) notes that some philosophers, aware of the ambiguity of language, urge what are termed "extensional" definitions—which list examples of a thing—as a complement to (or even a substitute for) the sort of "intensional" definition above, which attempts to capture the essence of a thing in a couple of sentences. His extensional definition—which he intends as a complement to the above intensional definition—necessarily focuses on the ways in which **interdisciplinarity**, the intellectual essence of the field of interdisciplinary studies, is performed: It seeks to integrate insights from multiple disciplines after evaluating these in the context of disciplinary perspective.

> Interdisciplinarity involves a set of practices: asking research questions that do not unnecessarily constrain theories, methods, or phenomena; drawing upon diverse theories and methods; drawing connections among diverse phenomena; evaluating the insights of scholars from different disciplines in the context of disciplinary perspective; and integrating the insights of those disciplinary scholars in order to achieve a holistic understanding. (Szostak, 2015b, p. 109)

Much of this book will be devoted to outlining these very practices that collectively constitute interdisciplinarity.

THE INTELLECTUAL ESSENCE OF INTERDISCIPLINARITY

There are two dominant forms of interdisciplinarity: instrumental and critical. **Instrumental interdisciplinarity** is problem driven. It is a pragmatic approach that focuses on research, borrowing from disciplines, and practical problem solving in response to the external demands of society. Borrowing alone, however, is not sufficient; it must be supplemented by integration. For instrumental interdisciplinarity, it is indispensable to achieve as much integration as possible given the insights currently available from the contributing disciplines.

Critical interdisciplinarity seeks to transform the nature of the academy. It "interrogates the dominant structure of knowledge and education with the aim of transforming them, while raising epistemological and political questions of value and purpose" (Klein, 2010, p. 30). This focus is silent in instrumental interdisciplinarity. Critical interdisciplinarians fault the instrumentalists for merely combining existing disciplinary approaches without advocating their transformation. Rather than building bridges across academic units for practical problem-solving purposes, critical interdisciplinarians seek to transform and dismantle the boundary between the literary and the political, treat cultural objects relationally, and advocate inclusion of marginalized cultures (Klein, 2005a, pp. 57–58).

These distinctions between instrumental and critical interdisciplinarity are not absolute or unbridgeable. Research on systemic and complex problems such as the environment and health care often reflects a combination of critique and problem-solving approaches. The integrated definition of interdisciplinary studies used in this book reflects an emerging consensus approach to the field: It is pragmatic, yet it leaves ample room for critique and interrogation of the disciplines, as well as economic, political, and social structures. This "both/and" approach is reflected in the definition of interdisciplinarity stated earlier: It refers to "answering a question, solving a problem, or addressing a topic," so it reflects an instrumentalist approach. But it also refers to "integrating [disciplinary] insights and theories to construct a more comprehensive understanding." Integrating disciplinary insights (i.e., their concepts and assumptions) or theories typically includes interrogating the disciplines. Similarly, constructing a more comprehensive understanding of a problem and communicating this understanding may involve raising philosophical and political questions or proposing transformative policies. Interdisciplinarity, then, "has developed from an idea into a complex set of claims, activities, and structures" (Klein, 1996, p. 209).

These two forms of interdisciplinarity share certain commonalities: assumptions, theories, and a commitment to **epistemological pluralism**. This refers to the diverse attitudes that disciplines have about how to know and describe reality. These commonalities constitute the intellectual essence of interdisciplinarity and provide coherence to this diverse field. We discuss them in turn below. (*Note:* This section draws heavily from Chapter 6 of Repko, Szostak, & Buchberger [2020], *Introduction to Interdisciplinary Studies,* third edition.)

Assumptions of Interdisciplinarity

All disciplines, interdisciplines, and fields of study are based on certain assumptions that provide cohesion to the field. In this regard, interdisciplinary studies is no different. There are at least four assumptions that anchor this diverse and rapidly evolving field, though the extent of agreement on each of them varies.

The Complex Reality Beyond the University Makes Interdisciplinarity Necessary

Broadly speaking, there are two categories of problems we face today: those that require a specialized disciplinary approach, and those that require a broader interdisciplinary approach. For example, a specialized disciplinary approach to the subject of freshwater scarcity could focus on depletion rates of freshwater aquifers (Earth science), the destruction of wetlands (biology), or types of pollutants (chemistry). But the same topic of freshwater scarcity would require an interdisciplinary approach if you wanted to learn about it as a complex whole. This would require drawing not only on these disciplines, but also on political science (to investigate existing or needed legislation), economics (to evaluate costs of stiffer environmental regulations), and interdisciplinary fields such as environmental science.

The Disciplines Are Foundational to Interdisciplinarity

The disciplines are foundational to the unique purpose of interdisciplinarity, though this notion is vigorously contested by some critical interdisciplinarians (see Box 1.1). The integrated definition of interdisciplinary studies presented earlier makes this assumption explicit: Interdisciplinary studies is a cognitive process by which individuals or groups draw on *disciplinary perspectives* and integrate *disciplinary insights and modes of thinking* to advance their understanding of a complex problem with the goal of applying it. Interdisciplinarity, particularly in its instrumental form, is not a rejection of the disciplines; it is firmly rooted in them, but offers a corrective to their dominance. We need specialization. But we also need interdisciplinarity to broaden our understanding of complex problems. This "both/and" position is reflected, for example, in the interdisciplinary fields of health sciences and health services. It is also the position of this book and reflects the majority opinion in interdisciplinary literature.

BOX 1.1

Some interdisciplinarians . . . share an **antidisciplinary** view, preferring a more "open" understanding of "knowledge" and "evidence" that would include "lived experience," testimonials, oral traditions, and interpretation of those traditions by elders (Vickers, 1998, pp. 23–26). However, there is a problem with this approach. Without some grounding in the disciplines relevant to the problem, borrowing risks becoming indiscriminate and the result rendered suspect. Moreover, those who reject the knowledge claims of the disciplines altogether may be uncertain how to make knowledge claims other than on arbitrary grounds of life experience. Transdisciplinarity and integrative studies integrate disciplinary insights and nonacademic insights of various sorts.

The Disciplines by Themselves Are Inadequate to Address Complexity Comprehensively

Disciplinary inadequacy is the view that the disciplines by themselves are inadequate to address complex problems. This inadequacy stems from several factors:

- The disciplines lack breadth of perspective.

- The disciplines are unwilling to assume responsibility for offering broad-based and comprehensive solutions to complex societal problems.

- The disciplines possess an unreasonable certainty that they provide all that is needed to make sense of the modern world.

- The disciplines do not have the cognitive or methodological tools to make sense of complex reality and provide us with a complete picture.

- Integrative strategies are needed to combine the best elements of disciplinary insights into a more comprehensive understanding.

Underlying the assumption of disciplinary inadequacy is the judgment that disciplinary approaches are "partial" and "biased." They are partial in that a discipline views a particular problem through the lens of its own unique and narrow perspective. Economists, for instance, are skeptical of research from other disciplines because they value their own theories and methods, and they tend to ignore insights generated by alternate theories and methods (Pieters & Baumgartner, 2002). Disciplinary approaches are biased in that they are interested in only those concepts, theories, and methods that the discipline embraces, while rejecting different concepts, theories, and methods preferred by other disciplines. For example, although power is a concept relevant to virtually all the social sciences, each discipline has its own definition of power, and each definition is undergirded by certain assumptions, methods, and so forth that are unique to it. To gain a more balanced and comprehensive understanding of power as it relates to a problem, we must first understand how each discipline understands the concept of power before attempting to create common ground between these varied and conflicting notions.

Disciplinary inadequacy as applied to the health sciences is the subject of a study by Terpstra, Best, Abrams, and Moor (2010). Their conclusion is summarized in Box 1.2.

BOX 1.2

Over the last century, there have been many lessons learned in the health field. A key lesson is that health is a complex phenomenon and the underlying causal pathways for disease and illness are more than just biological. . . . Health is a phenomenon deeply rooted within a social system, and health outcomes result from a dynamic interplay between factors across

(Continued)

(Continued)

the lifetime, originating from the cellular level, to the socio-political level. . . . As such, efforts to improve health must consider the multifactorial nature of the problem and integrate appropriate knowledge across disciplines and levels of analysis. . . . Health research has implicated a myriad of factors involved in HIV prevention. . . . Unfortunately, incidence rates continue to rise because the knowledge is not being applied in the unified manner necessary to address the complexity of the problem. . . .

Unfortunately, the majority of health research is conducted for the sake of science, and not for the sake of dissemination and implementation. Knowledge created for science's sake tends to be discipline specific and reductionist, producing results that are not easily applied to inform practice and policy decisions. The reality is that health and health service challenges cannot be handled well by any single discipline or social sector, and the traditional reductionist approach to science does not work well for the majority of health service problems. Disciplinary knowledge and levels of analysis are intertwined in health service problems, and as such, application requires integrative theoretical models and knowledge. As stated by Rosenfeld (1992), "to achieve the level of conceptual and practical progress needed to improve human health, collaborative research must transcend individual disciplinary perspectives and develop a new process of collaboration" (Terpstra et al., 2010, p. 1344).

Source: Terpstra, J. L., Best, A., Abrams, D., Moor, G. (2010). Interdisciplinary health sciences and health systems. In Julie Thompson Klein & Carl Mitcham (eds.), *The Oxford Handbook of Interdisciplinarity.* OUP, Oxford.

Interdisciplinarity Is Able to Integrate Insights From Relevant Disciplines

It is feasible to integrate insights concerning a complex problem from relevant disciplines. This bold assumption is based not on wishful thinking, but on a carefully constructed process to achieve integration that instrumental interdisciplinarians have developed, and applied successfully, in recent years.

Theories of Interdisciplinary Studies

Theory refers to a generalized scholarly explanation about some aspect of the natural or human world, how it works, and how specific facts are related, that is supported by data and research (Bailis, 2001, p. 39; Calhoun, 2002, p. 482; Novak, 1998, p. 84). An example is the "broken windows theory of crime," which communicates the idea that seemingly trivial acts of disorder such as a broken window in a vacant house tend to trigger more serious crime in the neighborhood.

Every discipline embraces certain theories that provide its intellectual core and give it coherence. This is true also of interdisciplinary studies that draws on a body of theory to justify using an interdisciplinary approach and inform the research process. This body of theory includes theories on complexity, perspective taking, common ground, and integration.

Complexity

What distinguishes phenomena and problems that are merely complicated from those that are complex is the nature of the relationships among the parts. **Complexity** refers to the parts of a phenomenon or problem that *interact* in surprising/unexpected ways. **Interdisciplinary complexity theory** states that interdisciplinary study is necessitated when the problem or question is multifaceted and functions as a "system" (see Box 1.3). (*Note:* As used here, "system" does not imply either that the system tends toward equilibrium or that it is closed—that is, isolated from other phenomena—because in reality, almost all phenomena influence almost all other phenomena somehow.)

BOX 1.3

What do acid rain, rapid population growth, and the legacy of *The Autobiography of Benjamin Franklin* have in common? Though drawn respectively from the purviews of the natural sciences, social sciences, and humanities, they can be fruitfully understood as behaviors of complex systems, and they all require interdisciplinary study. Thinking of each of them as behavior of a particular complex system can help interdisciplinarians better understand such phenomena; collectively, they can help us better understand the nature and conduct of interdisciplinarity. . . .

In order to justify the interdisciplinary approach, its object of study must be multifaceted, yet its facets must cohere. If it is not multifaceted, then a single disciplinary approach will do (since it can be studied adequately from one reductionist perspective). If it is multifaceted but not coherent, then a multidisciplinary approach will do (since there is no need for integration). To justify both elements of interdisciplinary study—namely that it draws insights from disciplines and that it integrates their insights—its object of study must be represented by a system [that] must be complex. (Newell, 2001, pp. 1–25)

This raises the question of why complexity should be a criterion for interdisciplinary studies. The answer involves revisiting the definition of interdisciplinary studies provided earlier, noting two of its key elements: Interdisciplinary studies *"draws on disciplinary perspectives and integrate[s] their insights."* The progression of thought, then, is as follows:

- Interdisciplinary studies draws on two or more disciplinary perspectives.

- Complex events or processes and behaviors have facets or parts that cohere.

- Each facet is typically the focus of a particular discipline.

- When the same facet is studied by more than one discipline, there are often conflicting insights generated.

- Understanding each facet involves drawing on the insights of the corresponding discipline(s).

- Understanding the complex phenomenon or behavior as a *whole* involves integrating insights from the relevant disciplines.

Interdisciplinary complexity theory also addresses the special case of the humanities and the arts. These disciplines are more concerned with behavior that is idiosyncratic, unique, and personal. The common practice in these disciplines is to practice **contextualization**. This is the practice of placing "a text, or author, or work of art into context, to understand it in part through an examination of its historical, geographical, intellectual, or artistic location" (Newell, 2001, p. 4). *Since complexity theory is concerned with the behavior of complex phenomena, and since contexts are themselves complex, the theory also provides a rationale for the interdisciplinary study of texts, artistic creations, and individuals that are unique and complex.*

Perspective Taking

Perspective taking is viewing a particular issue, problem, object, behavior, or phenomenon from a particular standpoint other than your own. As applied to interdisciplinary studies, **perspective taking** *involves analyzing the problem from the standpoint or perspective of each interested discipline and identifying their commonalities and differences.*

As developed by cognitive psychologists, perspective taking theory makes five important claims that are critical to your ability to engage in interdisciplinary work and function successfully in the contemporary world:

1. *Perspective taking reduces the human tendency to negatively stereotype individuals and groups* (Galinsky & Moskowitz, 2000). Assuming the position of the stereotyped individual, either virtually or actually (as John Howard Griffin did in *Black Like Me*), reverses your perspective. Holding a negative stereotype of an individual or group that is the object of study will certainly skew the interdisciplinary study and fatally compromise the resulting understanding. Stereotyping is inconsistent with good interdisciplinary practice.

2. *Perspective taking facilitates our ability to assemble new sets of potential solutions to a given problem* (Galinsky & Moskowitz, 2000; Halpern, 1996, pp. 1, 21). Here the old adage "there is wisdom in a multitude of counselors" applies: Examining the insights from the perspective of each interested discipline, even though they conflict, enriches your understanding of the problem and enables you to make creative connections (see Figure 1.1).

3. *Perspective taking heightens our awareness that we are biased in the direction of our own knowledge, whether it comes from our life experience or prior academic training.* In psychology, false-consensus bias is a cognitive bias whereby individuals tend to overestimate the extent to which their beliefs or opinions are typical of those of others (Fussell & Kraus, 1991; 1992). For example, after seeing a film, viewers who believe the film was excellent will tend to overestimate the percentage of people who thought that the film was excellent. The implication for interdisciplinary work is that we need to be aware of our biases, including disciplinary biases (which may have developed after majoring in a particular discipline), so that these do not prejudice (consciously or unconsciously) our analysis of the problem under study (Repko et al., 2020).

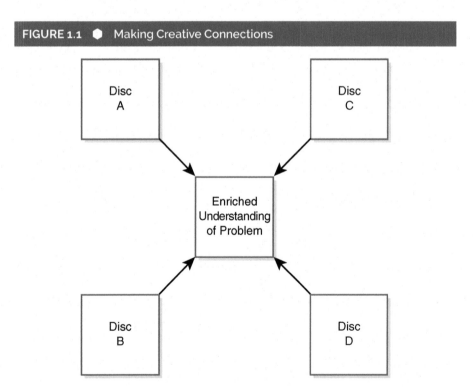

FIGURE 1.1 ● Making Creative Connections

Source: Allen F. Repko.

4. *Perspective taking invites us to engage in role taking* (Martin, Thomas, Charles, Epitropaki, & McNamara, 2005, p. 141). There are three role-taking aspects of perspective taking, each of which is pertinent to interdisciplinary work:

   - *Accurately perceive how others see and understand the world.* This involves seeing ourselves as role takers much as those in the theater arts do as they assume the role of a character in a play. To engage in the interdisciplinary research process, we must consciously assume the role, if only briefly, of a disciplinary expert and view the problem through the expert's eyes. This role-taking ability is particularly important for those engaged in non-Western cultural studies, race and ethnic studies, urban studies, women's studies, sexuality studies, and other programs that emphasize difference.

   - *View a situation broadly from multiple perspectives* (Martin et al., 2005, p. 141). The implications for interdisciplinary process are obvious: We must not limit our inquiries to only those disciplines with which we are familiar or to those expert views with which we agree.

   - "*Perceive the other's perspective in depth and have a full understanding of the other's perspective*" (p. 141, italics in original). In interdisciplinary work, *depth* and *full understanding* refer to disciplinary depth. We will see in later chapters that interdisciplinary scholars can achieve the necessary level of understanding of disciplinary insights if they appreciate disciplinary perspective. This holds special significance for those in the humanities and fine and performing arts, where the ability to understand and even assume or appropriate the identity of another is a critical skill.

5. *Perspective taking involves holistic thinking.* **Holistic thinking** is the ability to understand how ideas and information from relevant disciplines relate to each other and to the problem (Bailis, 2002, pp. 4–5). Holistic thinking differs from perspective taking in this important respect: Perspective taking is the ability to understand how each discipline would typically view the problem, whereas holistic thinking is the ability to see the whole problem in terms of its constituent disciplinary parts. In holistic thinking, the focus is on the relationships of parts to the whole and on the differences between and similarities to other parts. The object of holistic thinking is to view the problem inclusively in a larger context rather than under controlled or restrictive conditions favored by disciplinary specialists. But "larger context" does not mean the most encompassing context possible. One actually wants the narrowest context possible that still encompasses everything needed to address the problem as a whole. Holistic thinking allows for seeing characteristics of a problem that are not apparent when studying the problem in disciplinary isolation. For example,

an interdisciplinary study of community art, usually seen as separate from urban economic development, may show how the community benefits socially, culturally, and economically (i.e., holistically) from various kinds of art. The goal or the product of holistic thinking is a more comprehensive understanding of the problem (discussed below). Overcoming monodisciplinarity, which focuses on a single academic discipline, involves deciding that other disciplines—their perspectives, epistemologies, assumptions, theories, and methods—are worth considering when studying a particular problem. Indeed, interdisciplinarians eventually come to value and seek other perspectives.

Common Ground

Although *common ground* does not appear in the definition of interdisciplinary studies presented earlier, it is implicit in the concept of integration. The interdisciplinary concept of common ground comes from cognitive psychology's theories of common ground and the emerging field of cognitive interdisciplinarity. These theories are introduced here but discussed more fully in Chapters 8 and 11.

Noted cognitive psychologist Herbert H. Clark (1996) defines common ground in social terms as the knowledge, beliefs, and suppositions that each person has to establish with another person to interact with that person (pp. 12, 116).

Cognitive psychologist Rainer Bromme (2000) applies Clark's theory of common ground to communication between disciplines. Whether developing a collaborative language for interdisciplinary research teams or integrating conflicting insights, the theory of cognitive interdisciplinarity calls for discovering or creating the "common ground integrator" by which conflicting assumptions, theories, concepts, values, or principles can be integrated.

Working independently of Clark and Bromme, William H. Newell (2001) was the first inter-disciplinarian to define common ground in interdisciplinary terms. Common ground, he says, involves using various *techniques* to modify or reinterpret disciplinary elements (p. 20).

Newell's definition contains three ideas that are consistent with those of Clark and Bromme:

1. Common ground is something that the interdisciplinarian must create or discover.

2. Creating or discovering common ground involves modifying or reinterpreting disciplinary elements (i.e., concepts, assumptions, or theories) that conflict.

3. Modifying these elements to reduce the conflict between them involves using various techniques. (*Note:* These techniques are the subject of later chapters.)

Newell's particular contribution to understanding common ground is that it is what makes integration of disciplinary insights possible. In effect, Newell has illuminated the mysterious "black box" of interdisciplinary integration so that we can readily perceive how to create common ground and thus achieve integration.

A definition of common ground that integrates Newell's definition with the formulations of Clark and Bromme is as follows: **Common ground** is the shared basis that exists between conflicting disciplinary insights or theories and makes integration possible (Repko, 2012, pp. 56–57).

Integration

Integration is a process by which concepts, assumptions, or theories are modified to reconcile insights regarding the same problem from two or more disciplines. The purpose of interdisciplinary studies is not to choose one disciplinary concept, assumption, or theory over another, but to produce an even better understanding of the problem by integrating the best elements of competing concepts, assumptions, or theories. A primary focus of the debate over the meaning of interdisciplinary studies or interdisciplinarity concerns integration, which literally means "to make whole."

Practitioners are divided concerning the role of integration. **Generalist interdisciplinarians** understand interdisciplinarity loosely to mean "any form of dialog or interaction between two or more disciplines," while minimizing, obscuring, or rejecting altogether the role of integration (Moran, 2010, p. 14).[3]

Integrationist interdisciplinarians, on the other hand, believe that integration should be the *goal* of interdisciplinary work because integration addresses the challenge of complexity. Integrationists, pointing to a growing body of literature that connects integration with interdisciplinary education and research, are concerned with developing a distinctively interdisciplinary research process and describing how it operates (Newell, 2007a, p. 245; Vess & Linkon, 2002, p. 89). They advocate reducing the confusion about the meaning of *interdisciplinarity* and point to research in cognitive psychology that shows that the human brain is designed to process information integratively. *This book is aligned with the integrationist understanding of interdisciplinarity.*

The core of the **integrationist position** is that integration is achievable and that researchers should strive for the greatest degree of integration possible given the problem under study and the disciplinary insights at their disposal. Importantly, integrationists point to recent theories supportive of integration advanced by cognitive psychologists, curriculum specialists, teacher educators, and researchers. Moreover, they point to the increasing amount of interdisciplinary work characterized by integration.

The idea for interdisciplinary integration is grounded in Bloom's classic taxonomy of levels of intellectual behavior that are involved in learning. Drawing on theories on learning and cognitive development, an interdisciplinary team of researchers and educators updated Bloom's taxonomy in 2000. The team identified six levels within the cognitive domain, with simple recognition or recall of facts at the lowest level through increasingly more complex and abstract mental levels, leading ultimately to the highest order ability, creating, as shown in Figure 1.2.

The significance of this taxonomy for interdisciplinary studies is that it elevates the cognitive abilities of creating and integrating to the highest level of knowledge. **Creating** involves putting elements together—integrating them—to produce something that is new and useful. As noted earlier, integration is the distinguishing feature of interdisciplinary studies and is at the core of the interdisciplinary research process. We will find at many points in this book that the literatures on creativity and on the interdisciplinary research process intersect, students learning how to do interdisciplinary research will expand their creative capabilities more generally.

Interdisciplinary integration finds additional support in the work of linguists George Lakoff and Gilles Fauconnier, and cultural anthropologist Mark Turner. Lakoff

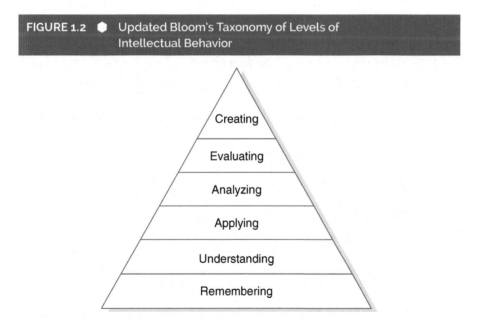

FIGURE 1.2 ● Updated Bloom's Taxonomy of Levels of Intellectual Behavior

Creating

Evaluating

Analyzing

Applying

Understanding

Remembering

Source: Anderson, L. W., Krathwohl, D. R., Airasian, P. W., Cruikshank, K. A., Mayer, R. E., Pintrich, P. R., Raths, J., & Wittrock, M. C. (2001). *A taxonomy for learning, teaching, and assessing: A revision of Bloom's Taxonomy of Educational Objectives* (p. 28). New York: Longman. Reprinted by permission of Pearson Education, Inc. New York, NY.

(1987) introduced the **theory of conceptual integration** to explain the innate human ability to create new meaning by blending concepts and creating new ones (p. 335). Fauconnier (1994) deepened our understanding of integration by explaining how our brain takes parts of two separate concepts and integrates them into a third concept that contains some properties (but not all) of both original concepts. For example, the nickname "Iron Lady," referring to former British prime minister Margaret Thatcher, represents a conceptual integration of the concept iron, a metal used in construction because of its strength, with the concept lady, a woman who holds political rank. The implicit claim of the metaphor is that Margaret Thatcher acted *as if* she were made of iron (p. xxiii). Conceptual blending is possible because certain commonalities exist in the two original concepts that provide the basis for the new integrated concept. This third concept is different from either of the two original concepts. Figure 1.3 depicts this process.

Turner (2001) extends the theory of conceptual integration still further by arguing that we cannot fully appreciate a concept without understanding its cultural or historical context (p. 17). Accordingly, concepts (discussed in depth in Chapter 10) should be analyzed in the context and theoretical framework of the disciplines from which they come.

FIGURE 1.3 ● Integrating Two Separate Concepts to Create a Third Concept

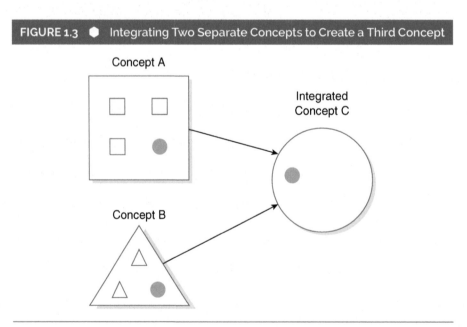

Source: Allen F. Repko.

From the discussion above, it is possible to construct a definition of integration as follows:

Integration is the cognitive process of critically evaluating disciplinary insights and creating common ground among them to construct a more comprehensive understanding. The new understanding is the product or result of the integrative process.

Epistemology of Interdisciplinary Studies

Epistemology involves questions such as "What can we know?" and "How can we know it?" Of the many ways that disciplinarity contrasts with interdisciplinarity, none is greater than their starkly different approaches regarding epistemology. Each disciplinary perspective involves a set of epistemological attitudes. Interdisciplinarity necessarily involves respecting these various epistemologies.

Some disciplines, especially in the natural sciences (but also economics to a considerable extent), believe that scholars can employ quantitative methods (notably experiments, statistical analysis, and mathematical modelling) to achieve very precise understandings of the phenomena that they investigate. Other disciplines, especially in the humanities, believe that scholarly understandings are always subjective to some degree and that the qualitative methods they employ (interviews, close reading of texts, surveys) cannot generate very precise understandings. Some scholars in these disciplines have come to doubt that any sort of objective understanding is possible: They see scholarship as only a game in which we argue for positions that we find congenial. (Note that all of these disciplines choose an epistemological outlook that reflects the nature of their favored methods. As noted above, disciplinary perspectives are internally consistent.)

Interdisciplinarity steers a path between two extremes. On the one hand, it rejects a "positivist" belief that scholarship advances by proving or disproving hypotheses. Philosophers of science now appreciate that it is always possible to interpret any research finding in multiple ways. On the other hand, interdisciplinary research must reject an alternative "nihilistic" belief that we are not able at all to advance human understanding through research. The middle-ground position, recommended by most but not all philosophers of science, is that scholarly understanding advances through careful amassing of evidence and argument. In the case of interdisciplinarity, we evaluate disciplinary insights, with a general expectation that these will be imperfect but contain some kernel of truth. We then seek a more comprehensive understanding that best fits our collective perception of the world (Szostak, 2007a). The interdisciplinary view that disciplinary insights are partial accords with contemporary philosophical understanding of epistemology (Welch, 2011).

Interdisciplinarians can practice epistemological pluralism, in which they respect the different epistemologies pursued in different disciplines (Welch, 2011). They can recognize that some disciplines may be too confident in their insights, and others perhaps too hesitant to reach firm conclusions. They can employ the interdisciplinary techniques of evaluation that we will outline in later chapters to critique insights from any discipline.

Note that epistemological pluralism supports a blend of instrumental and critical interdisciplinarity, as was advocated above: We are free both to draw upon and critique disciplinary insights and perspectives. Our interdisciplinary epistemological outlook is in turn grounded in an interdisciplinary ontology: our philosophical understanding of how the world works (as forcefully advocated by Bhaskar, Danermark, & Price, 2016). It is because the phenomena studied in one discipline interact in complex ways with the phenomena studied in other disciplines that we need interdisciplinary analysis to integrate across insights that can only be partial (see Henry, 2018).

DISTINGUISHING INTERDISCIPLINARITY FROM MULTIDISCIPLINARITY, TRANSDISCIPLINARITY, AND INTEGRATIVE STUDIES

Through articulating the nature of the interdisciplinary research process in later chapters, we can encourage rigor in interdisciplinary analysis. We have carefully defined and described interdisciplinary studies above to set the stage for discussion of that process. We can prevent unnecessary confusion with other terminology you may come across by carefully distinguishing here interdisciplinarity from multidisciplinarity, transdisciplinarity, and integrative studies.

Interdisciplinary Studies Is Not Multidisciplinary Studies

Some who are uninformed and outside the field mistakenly believe that *interdisciplinarity* and *multidisciplinarity* are synonymous. They are not. **Multidisciplinarity** refers to the placing side by side of insights from two or more disciplines. For example, this approach may be used in a course that invites instructors from different disciplines to present their perspectives on the course topic in serial fashion but makes no attempt to integrate the insights produced by these perspectives. "Here the relationship between the disciplines is merely one of proximity," explains Joe Moran (2010); "there is no real integration between them" (p. 14). Merely bringing insights from different disciplines together in some way but failing to engage in the additional work of integration is

multidisciplinary studies, not interdisciplinary studies. **Multidisciplinary research** "involves more than a single discipline in which each discipline makes a *separate* contribution [italics added]" (National Academies, 2005, p. 27).

Lawrence Wheeler's instructive fable of building a house for an elephant (Wheeler & Miller, 1970) illustrates a typical multidisciplinary approach to solving a complex problem:

> Once upon a time a planning group was formed to design a house for an elephant. On the committee were an architect, an interior designer, an engineer, a sociologist, and a psychologist. The elephant was highly educated too . . . but he was not on the committee.
>
> The five professionals met and elected the architect as their chairman. His firm was paying the engineer's salary, and the consulting fees of the other experts, which, of course, made him the natural leader of the group.
>
> At their *fourth* meeting they agreed it was time to get at the essentials of their problem. The architect asked just two things: "How much money can the elephant spend?" and "What does the site look like?"
>
> The engineer said that precast concrete was the ideal material for elephant houses, especially as his firm had a new computer just begging for a stress problem to run.
>
> The psychologist and the sociologist whispered together and then one of them said, "How many elephants are going to live in this house? . . . It turned out that *one* elephant was a psychological problem but *two* or more were a sociological matter. The group finally agreed that though *one* elephant was buying the house, he might eventually marry and raise a family. Each consultant could, therefore, take a legitimate interest in the problem.
>
> The interior designer asked, "What do elephants do when they're at home?"
>
> "They lean against things," said the engineer. "We'll need strong walls."
>
> "They eat a lot," said the psychologist. "You'll want a big dining room . . . and they like the color green."
>
> "As a sociological matter," said the sociologist, "I can tell you that they mate standing up. You'll need high ceilings."
>
> So they built the elephant a house. It had precast concrete walls, high ceilings, and a large dining area. It was painted green to remind him of the jungle. And it was completed for only 15% over the original estimate.

The elephant moved in. He always ate outdoors, so he used the dining room for a library . . . but it wasn't very cozy.

He never leaned against anything, because he had lived in circus tents for years, and knew that walls fall down when you lean on them.

The girl he married *hated* green, and so did he. They were *very* urban elephants.

And the sociologist was wrong too. . . . They didn't stand up. So the high ceilings merely produced echoes that greatly annoyed the elephants. They moved out in less than six months! (Wheeler & Miller, 1970, n.p.)

This fable shows how disciplinary experts usually approach a complex task: They perceive it from the narrow perspective of their specialty and fail to take into account the perspectives of other relevant disciplines, professions, or interested parties (in this case, the elephant).

This story also illustrates how a multidisciplinary approach to understanding a problem merely juxtaposes disciplinary perspectives. The disciplines speak with separate voices on a problem of mutual interest. However, the disciplinary status quo is not questioned, and the distinctive elements of each discipline retain their original identity. In contrast, interdisciplinarity consciously integrates disciplinary insights to produce a more comprehensive understanding of a complex problem or intellectual question.

Multidisciplinarity and interdisciplinarity have this in common: They seek to overcome the narrowness of disciplines. However, they do this in different ways. Multidisciplinarity means limiting activity to merely appreciating different disciplinary perspectives. But interdisciplinarity means being more inclusive of what disciplinary theories, concepts, and

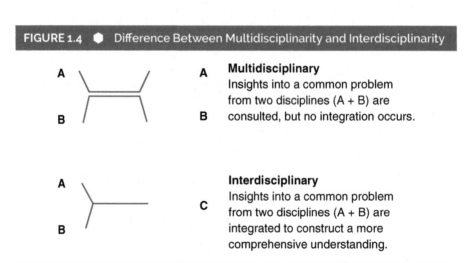

FIGURE 1.4 ● Difference Between Multidisciplinarity and Interdisciplinarity

A   A **Multidisciplinary**
Insights into a common problem from two disciplines (A + B) are
B   B consulted, but no integration occurs.

A   **Interdisciplinary**
Insights into a common problem
C from two disciplines (A + B) are
B integrated to construct a more comprehensive understanding.

Source: National Academy of Sciences, National Academy of Engineering, & Institute of Medicine. (2005). *Facilitating interdisciplinary research* (p. 29). Washington, DC: National Academies Press.

methods are appropriate to a problem. It also means being open to alternative methods of inquiry, using different disciplinary tools, and carefully estimating the degree of usefulness of one tool versus another to shed light on the problem (Nikitina, 2005, pp. 413–414).

Research is truly interdisciplinary, states the National Academies (2005), "when it is not just pasting two disciplines together to create one product but rather is an integration and synthesis of ideas and methods" (p. 27). Figure 1.4 shows the difference between multidisciplinarity and interdisciplinarity.

Interdisciplinary Studies Is Not Transdisciplinary Studies

Complementary to interdisciplinarity, **transdisciplinarity** involves the integration also of *insights generated outside the academy*, a team approach to research, the active involvement of nonacademic participants in research design, and a "case study" approach. For example, if investigating environmental degradation in a particular area, transdisciplinarians would seek insights from local people on both the causes and potential solutions of environmental challenges (Bergmann et al., 2012). Whereas an interdisciplinary scholar might tackle the general problem of economic development, a transdisciplinary scholar would more likely focus on development challenges in a particular locality.

None of these elements contradict the practice of interdisciplinarity, which can also involve insights, case studies, team research, and drawing on life experience and expertise outside the academy. We might think of transdisciplinarity as "interdisciplinarity plus," where additional constraints (noted above) are placed on the transdisciplinary researcher.

NOTE TO READERS

In this book, we focus on interdisciplinarity. However, we will have occasion to discuss team research. Some of the numerous examples of interdisciplinary analysis provided in what follows qualify as case studies. And some of these examples do indeed draw on insights generated beyond the academy.

Interdisciplinary Studies and Integrative Studies

Integrative studies is often used in the contemporary academy to indicate something more than just integrating insights from different disciplines. **Integrative studies** seeks to integrate various elements of student experience such as coursework and residential life (Hughes, Muñoz, & Tanner, 2015).

While integrative studies and interdisciplinarity do not share the same boundaries, they do share important points of overlap. They both stress integration. And integrative studies makes a point that should be stressed here: Integration, perhaps the most important step in the interdisciplinary research process, is called for in all aspects of our lives. The integrative skills that interdisciplinary students will master are useful more generally in life. These students will be well suited to the needs of employers for workers that can integrate diverse bits of information into a coherent strategy. They will be better prepared not only for the world of work, but also to participate as members of their community in facing today's complex challenges.

The Differences Between Multidisciplinarity, Interdisciplinarity, Transdisciplinarity, and Integrative Studies Summarized

- Multidisciplinarity studies a topic from the perspective of several disciplines at one time but makes no attempt to integrate their insights.

- Interdisciplinarity studies a complex problem by drawing on disciplinary insights (and sometimes stakeholder views) and integrating them. By employing a research process that subsumes the methods of the relevant disciplines, interdisciplinary work does not privilege any particular disciplinary method or theory.

- Transdisciplinarity is best understood as a type of interdisciplinarity that stresses team research, a case study approach, and especially integrating not just across disciplines but also beyond the academy.

- Integrative studies seeks to integrate various elements of student experience such as coursework and residential life.

Chapter Summary

Interdisciplinary studies and interdisciplinarity are evolving and dynamic concepts that are now mainstream in the academy. The chapter focuses on the meaning of each term, unpacking the field's DNA in terms of its assumptions, theories, and epistemology. It examines various conceptions of interdisciplinarity including generalist, integrationist, critical, and instrumental. And it discusses how interdisciplinarity differs from multidisciplinarity, transdisciplinarity, and integrative studies.

Chapter 2 introduces the disciplines and their perspectives, describes how knowledge is typically reflected in the organization of the academy, and presents an in-depth discussion of disciplinary perspective.

Notes

1. For the limited purposes of this book, references to *disciplines* are limited to the traditional lists of major disciplines rather than the much fuller contemporary taxonomies unless otherwise noted. References to specific interdisciplines and schools of thought (e.g., feminism, Marxism) are appropriately identified.

2. In the humanities, students are required to choose a definition of meaning: artist intent, audience reaction, and so on. However, Rick Szostak (2004) argues that the interdisciplinary conception of "meaning" should urge students to embrace all possible definitions and the causal links they imply. Students "could still choose to specialize with respect to one of these (or not) without needing to assume the others away" (p. 44).

3. Some generalists such as Moran see the terms *interdisciplinarity* and *integration* as synonymous with *teamwork* as in team teaching and cross-disciplinary communication on research projects (Davis, 1995, p. 44; Klein, 2005b, p. 23; Lattuca, 2001, p. 12). Other generalists such as Lisa Lattuca (2001) prefer to distinguish between types of interdisciplinarity by focusing primarily on the kinds of questions asked rather than on integration (p. 80). Still other generalists such as Donald G. Richards (1996) go so far as to reject any definition of interdisciplinary studies that "necessarily places priority emphasis on the realization of synthesis [or integration] in the literal sense" (p. 114).

Exercises

Defining for Clarity

1.1 You saw in this chapter the importance of defining the controversial and misunderstood term *interdisciplinary studies* to reveal its true meaning. Can you think of another controversial or misunderstood term whose true meaning could be clarified by studying its definition in a similar manner?

What and How

1.2 Definitions of some terms contain both a *what* and a *how* component. This is true of the integrated definition of interdisciplinary studies that appears in this chapter. Identify which part of the definition is the *what*, and which is the *how*.

Dominant Forms

1.3 Which form of interdisciplinarity, instrumental or critical, would most likely yield a more comprehensive understanding of why newly arrived immigrants typically resist (at least initially) assimilating into the majority culture?

(Continued)

(Continued)

Assumptions

1.4 Is the assumption that the complex reality beyond the university makes interdisciplinarity necessary justified? If so, why? If not, why not?

1.5 This chapter has argued that interdisciplinarity should be viewed as complementary to the disciplines rather than as a threat to them. In your view, what is the most compelling argument that can be made for a "both/and" rather than an "either/or" position?

1.6 Why should a person's life experience be considered less or more valid than a disciplinary insight published in a scholarly journal?

1.7 Identify a health issue or a health service challenge that could benefit from an interdisciplinary approach (see quote in Box 1.2).

Complexity

1.8 In interdisciplinary work, why must the object of study be complex?

Perspective Taking

1.9 Explain the relationship between perspective taking and holistic thinking.

Integration

1.10 Explain why creating is so closely associated with interdisciplinary studies.

Epistemology

1.11 Explain why epistemological pluralism is considered a key component of interdisciplinarity.

Inventory

1.12 Examine your university's undergraduate and/or graduate curriculum to determine how much interdisciplinary activity exists on campus.

1.13 How might your institution's general education curriculum be made more interdisciplinary?

Building Houses for Elephants

1.14 The fable of the elephant house is instructive to those who are engaging in a complex enterprise such as building a house. Think of another complex enterprise that is planned or already under way in your community and apply the lessons of the elephant house to it.

1.15 Is there a transdisciplinary aspect to the elephant house project? If so, what is it, or if not, what should it be?

Image by Mohamed Hassan from Pixabay.

©iStockphoto.com/tintin75

2

INTRODUCING THE DISCIPLINES AND THEIR PERSPECTIVES

LEARNING OUTCOMES

By the end of this chapter, you will be able to

- Explain the concept of disciplinary perspective
- Describe how disciplinary knowledge is typically reflected in the organization of the academy
- Explain how to use disciplinary perspective
- Identify the defining elements of disciplinary perspective

GUIDING QUESTIONS

What is a discipline?

What is disciplinary perspective? Why is it important?

How has the academy been organized around disciplines?

CHAPTER OBJECTIVES

Before you can engage in interdisciplinary work by identifying disciplines that are relevant to the problem (Chapter 4), you must have a firm grasp of the disciplines and their perspectives on reality.

We explain the concept of disciplinary perspective and describe how disciplinary knowledge is typically reflected in the organization of the academy. We next explain how to use disciplinary perspective and introduce the defining elements of disciplines (i.e., their phenomena, epistemology, assumptions, concepts, theory, and methods). This information, presented in easily accessible tables, is foundational to interdisciplinary research and critical to developing adequacy in relevant disciplines as called for in STEP 5 of the research process (see Chapter 6).

The definition of interdisciplinary studies presented in Chapter 1 implies that interdisciplinarity has a high degree of dependence upon and interaction with the disciplines. Therefore, understanding the role of the disciplines and their perspectives on reality is essential to fully understand interdisciplinarity and successfully engage in interdisciplinary inquiry.

DEFINING DISCIPLINARY PERSPECTIVE

In an overall sense, disciplinary perspective is each discipline's unique view of reality.[1] Raymond C. Miller (1982), the first to assert that disciplines have distinct perspectives or worldviews that are pertinent to interdisciplinary understanding, states that perspective should be "the primary means of distinguishing one discipline from another" (p. 7). We agree. A discipline's "perspective" is the lens through which it views reality. Each discipline filters out certain phenomena so that it can focus exclusively on phenomena that interest it. Disciplines such as history and biology are not collections of certified facts; rather, they are *lenses* through which we look at the world and interpret it (Boix Mansilla, Miller, & Gardner, 2000, p. 18). In the sciences, disciplines are most easily distinguished by the phenomena they study. A conventional physicist, for example, would not be interested in studying the declining salmon populations in the Columbia and Snake Rivers, but a biologist would. A conventional sociologist would not be interested in theological representation in a fifteenth-century oil painting, but an art historian would. Similarly, a conventional historian would likely not be interested in the regulatory hurdles involved in the building of a new oil refinery, but a political scientist would.

Some have a narrow conception of the term *perspective*, viewing it as but one of several elements that define a discipline, the others being its phenomena, epistemology, assumptions, concepts, theories, and methods.[2] This book shares the broader conception of perspective, viewing it as the *source* of all other disciplinary elements.[3] Rick Szostak (2004), for example, explains how disciplinary perspective both reflects and influences a discipline's choice of phenomena, theory, and method. These are the **defining elements of a discipline's perspective:**

- The phenomena it studies

- Its epistemology or rules about what constitutes evidence

- The assumptions it makes about the natural and human world

- Its basic concepts or vocabulary

- Its theories about the causes and behaviors of certain phenomena

- Its methods (the way it gathers, applies, and produces new knowledge)

Together, these elements comprise a discipline's cognitive map (Klein, 2005a, p. 68). From it, the discipline frames the "big" questions or "perennial issues and problems" that give the discipline its definition and signature characteristics (Becher & Trowler, 2001, pp. 26, 31). Each discipline's community of scholars substantially agrees on what constitutes an interesting and appropriate question to study, what constitutes legitimate evidence, and what a satisfactory answer to the question should look like (Boon & Van Baalen, 2019; Choi & Richards, 2017).

A clarified definition of disciplinary perspective is presented here:

Disciplinary perspective is a discipline's view of reality in a general sense that embraces and in turn reflects the ensemble of its defining elements that include phenomena, epistemology, assumptions, concepts, theories, and methods.

This definition of disciplinary perspective is consistent with the definition of interdisciplinary studies that emphasizes drawing on the disciplines and integrating their insights and theories to construct a more comprehensive understanding. We shall see that interdisciplinary scholars must evaluate disciplinary insights in the context of disciplinary perspective. The definition thus captures the messy reality of what occurs in actual interdisciplinary work—drawing not just on disciplinary perspectives in a general sense, but more particularly on those defining elements of disciplines (assumptions, concepts, and theories) that relate most directly to the problem being investigated.

HOW KNOWLEDGE IS TYPICALLY REFLECTED IN THE ORGANIZATION OF THE ACADEMY

Before discussing how you will use disciplinary perspectives, it is useful to understand how knowledge is typically reflected in the organization of the academy.

About Disciplines and Disciplinarity

Disciplines are intellectual communities deeply devoted to the study of a particular subject, say biology. Disciplines also involve an institutional structure of graduate (MA and PhD) programs, departmental hiring, and disciplinary journals. Disciplinary fields and interdisciplines are not truly disciplines until they have their own PhDs and hiring communities. Most academic departments typically represent a particular discipline. Clusters of related disciplines form larger administrative units called colleges, schools, or faculties such as the college of science, the school of social sciences, or the faculty of arts. In most university settings, academic departments are foundational to the institution's structure.

Disciplinary knowledge is produced in the form of books, journals, and conference presentations all of which are vetted by the disciplines. Departments and programs pass on that knowledge to the next generation through their majors, create new knowledge, and guide the careers of faculty members who do the teaching and conduct the research in the discipline. Disciplinary departments determine the curriculum or the courses that are taught, and influence research (i.e., subject matter and method) and mode of teaching. The institutional structure of most universities thus reinforces disciplinary perspective. Those who do not reflect the perspective of their discipline will face difficulty in completing their degrees, getting hired, publishing, and gaining tenure or promotion.

The widely used term **disciplinarity** refers to the system of knowledge specialties called **disciplines**, which is little more than a century old. *Discipline* is used throughout this book as an umbrella term that also includes subdisciplines and interdisciplines, which are defined as follows:

- A **subdiscipline** is a subdivision of an existing discipline. The discipline of anthropology, for example, has developed several subdisciplines, including cultural anthropology, physical anthropology, anthropology of religion, urban anthropology, and economic anthropology. Subdisciplines have many of the characteristics of disciplines—a shared subject matter, theories, methods— but lack complete control over PhDs and hiring. They also can have quite different sets of questions, theories, and methods from the broader discipline. (Choi and Richards [2017] note that some disciplines are also characterized by an important divide between practitioners and theorists.)

- An interdiscipline literally means the space "between disciplines"—that is, between the intellectual content of two or more disciplines (Karlqvist, 1999, p. 379). An interdiscipline may begin as an interdisciplinary field, but over time, it may become like a discipline, developing its own curriculum, journals, professional associations, and most important for interdisciplinary studies, perspective. The interdisciplines of biochemistry and neuroscience, for example, emerged as interdisciplinary fields that eventually grew to become their own mainstream disciplines.

The Changing Character of Disciplines

Today's disciplines exhibit three characteristics about which you should be mindful as you study their defining elements described below.

First, disciplines are constantly evolving and taking on new elements: studying new phenomena or applying new theories or methods. This necessarily means that there is some diversity within the discipline at any point in time. But the institutional structure of disciplines ensures that there is still coherence.

Second, some disciplines are characterized by **cognitive discord**, meaning disagreement among a discipline's practitioners over the defining elements of the discipline. The American Sociological Association (ASA, n.d.), for example, states on its website, "Sociology provides many distinctive perspectives on the world." These "distinctive perspectives" within sociology, openly acknowledged by the ASA, are reflective of sociologists having aligned themselves with various theories and schools of thought that currently inform the discipline. But in general, they apply these theories using the same methods used with old theories.

Cognitive discord also characterizes art history, a discipline experiencing divisive theoretical conflicts. Consequently, art historian Donald Preziosi (1989) says that there is no such thing as "an Olympian perspective" in the discipline, despite what might be inferred from numerous textbooks (p. xi). Indeed, some scholars go so far as to claim that a dominant perspective, as defined in interdisciplinary literature, is lacking in almost every discipline in the social sciences and humanities (Dogan & Pahre, 1989).

This raises the question of whether some disciplines, such as art history and sociology in their fragmented states, even have a general perspective on reality. The answer is yes because the very idea of a discipline as something entirely coherent, in terms of strict adherence to its defining elements (assumptions, concepts, theories, methods, etc.), is an idealization. The reality of disciplinarity, past and present, is ferment and fragmentation.[4] *Counterbalancing these centripetal forces to a large degree is an **intellectual center of gravity** that enables each discipline to maintain its identity and have a distinctive overall perspective. As long as disciplines bestow PhDs and make hiring decisions, there will be strong pressure to decide what a suitable sociologist or art historian is.*

A third characteristic of the modern disciplines is the growing practice of crossing disciplinary boundaries by disciplinarians themselves. Disciplines borrowing concepts, theories, and methods from one another, says Klein (1999), skew the picture of knowledge depicted in conventional maps of the academy. She observes, for example, how textuality, narrative, and interpretation were once thought to belong within the domain of literary studies. Now, she says, they appear across the humanities and the social sciences, including science studies, and the professions of law and psychiatry. Similarly, research on the body and on disease occurs in disciplines as varied as art history, gerontology, and biomedicine. The movement of methods and analytical approaches across disciplinary boundaries, she contends, has become an important feature of knowledge production today (p. 3). (Note that interdisciplinary scholarship encourages such borrowing and thus potentially enhances the ability of disciplines to answer disciplinary research questions.) However, these new developments do not mean the end of disciplines.

Implications for Interdisciplinary Work

Interdisciplinarians at all levels should approach disciplines not as self-contained repositories of information, but as being open to a wider range of concepts, theories, and methods that transcend their traditional boundaries. That is, researchers should not only examine the characteristic elements of relevant disciplines for insights into the problem, but also search for information from sources that transcend disciplines such as the categories of phenomena (appearing in Table 2.3) and schools of thought (referenced in Table 2.9).

CATEGORIES OF DISCIPLINES

Table 2.1 presents a conventional classification of the disciplines that includes traditional disciplines (by no means all of them) but that excludes the applied fields and professions.[5] A discipline may be considered part of one category at one university but belong to a different category at another. History, for example, is considered a discipline within the social sciences in some institutions but part of the humanities at others. Though history has elements of both social science and humanities, this book follows the traditional taxonomy of including history in the humanities.

DISCIPLINARY PERSPECTIVES

Given how important disciplinary perspective is in the interdisciplinary research process, it is useful to sketch the perspectives of at least the most important disciplines. Students can then construct perspectives for other disciplines that they may encounter.

The disciplinary perspectives in Table 2.2 are separated into the three categories of traditional disciplines and are stated in the most general terms. These are not comprehensive

TABLE 2.1 ● Categories of Disciplines	
Category	**Discipline**
The Natural Sciences	Biology Chemistry Earth Science Mathematics Physics
The Social Sciences	Anthropology Economics Political Science Psychology Sociology
The Humanities	Art and Art History History Literature (English) Music and Music Education Philosophy Religious Studies

TABLE 2.2 ● Overall Perspectives of Natural Sciences, Social Sciences, and Humanities Disciplines Stated in General Terms	
Discipline	**Overall Perspective**
Natural Sciences	
Biology	Biology views the living physical world, including that of humans, as a highly complex and interactive whole governed by deterministic principles that explain behavior (such as genes and evolution).
Chemistry	Chemistry views the physical world as a complex interplay of distinctive properties of the elements, individually and in compounds, and their interactions. Chemistry sees larger-scale objects, organic as well as inorganic, in terms of their constituent elements and compounds.
Earth Science	Earth science views Planet Earth as a large-scale physical system that includes the four subsystems and their interactions: the lithosphere (the Earth's hard, outermost shell), the atmosphere (the mixture of gases that envelop the Earth), the hydrosphere (the subsystems that contain the Earth's water), and the biosphere (the realm of all living things, including humans).
Mathematics	Mathematics views the world through abstract quantitative creations with postulates, assumptions, axioms, and premises and explores these by proving theorems.
Physics	Physics see the world as consisting of basic physical laws that connect objects (atoms and subatomic particles, quanta) and forces (gravity, electromagnetic, strong nuclear, and weak nuclear) that often cannot be directly observed. These laws and forces establish the underlying structure of observable reality and cosmology (the form, content, organization, and evolution of the universe).
Social Sciences	
Anthropology	Cultural anthropology sees individual cultures as organic integrated wholes with their own internal logic, and culture as the set of symbols, rituals, and beliefs through which a society gives meaning to daily life. Physical anthropology views former cultures through the artifacts it uncovers.
Economics	Economics views the world as a complex of market interactions with the individual functioning as a separate, autonomous, rational entity, and perceives groups (even societies) as the sum of individuals within them.
Political Science	Political science views the world as a political arena in which individuals and groups make decisions based on the search for or exercise of power. Politics at all levels and in all cultures is viewed as a perpetual struggle over whose values and interests will prevail in setting priorities and making collective choices.
Psychology	Psychology sees human behavior as reflecting the cognitive constructs individuals develop to organize their mental activity. Psychologists also study inherent mental mechanisms, both genetic predisposition and individual differences.
Sociology	Sociology views the world as a social reality that includes the range and nature of the relationships that exist between people in any given society. Sociology is particularly interested in voices of various subcultures, analysis of institutions, and how bureaucracies and vested interests shape life.

(Continued)

TABLE 2.2 ● (Continued)	
Discipline	**Overall Perspective**
	Humanities
Art History	Art history views art in all of its forms as reflecting the culture in which it was formed and therefore providing a window into a culture. Art history can also investigate whether there are universal aesthetic tastes.
History	Historians view any historical period as a complex interplay of trends and developments leading up to it, and past events as the result of both societal forces and individual decisions.
Literature (English)	Literature believes that cultures, past and present, cannot be adequately understood without understanding and appreciating the literature produced by the culture.
Music Education	Music educators believe that a critical component of culture past and present cannot be adequately understood without understanding the music produced by the culture.
Philosophy	Philosophy relies on careful argumentation—though only rarely formal proofs of theorems—to grapple with a set of "big questions" such as What is the nature of reality? How can we understand that reality? And What is the meaning of life?
Religious Studies	Religious studies views faith and faith traditions as human attempts to understand the significance of reality and cope with its vicissitudes through beliefs in a sacred realm beyond everyday life.

Note: This **taxonomy** or systematic and orderly classification of selected disciplines and their perspectives raises the question of how students can find perspectives of disciplines, subdisciplines, and interdisciplines not included in this book. Certainly, a good place to obtain leads is this chapter, which has tables that define elements of disciplines (their epistemologies, theories, methods, etc.). Also, the chapter references standard authoritative disciplinary sources. Researchers may consult content librarians who specialize in certain disciplines. Another strategy is to ask disciplinary experts to recommend sources. This combined approach should produce aids that are authoritative and useful. The issue of finding scholarly research aids is addressed more fully in Chapter 5.

generalizations about each discipline, but central tendencies that are a matter of consensus. In later sections, we will describe individual elements (epistemology, theory, method, etc.) of each discipline's perspective.

When Disciplinary Perspectives Are Used

Disciplinary perspectives are used in two circumstances. The first is near the beginning of the research process where the focus is on identifying disciplines that are *potentially* interested in the problem. (*Note:* How to identify these disciplines is the focus of STEP 3 and the subject of Chapter 4.) Once a discipline's overall perspective on reality is known, it is relatively easy to apply the perspective to the problem. It is common to work with disciplines within a particular cluster such as the humanities, although some problems require consulting disciplinary literatures from two or more clusters. *A rule of thumb is to let the problem dictate which categories and disciplines within each category are most relevant to it.* Identifying potentially interested disciplines early on helps to narrow the disciplinary literatures that need to be consulted when performing the full-scale literature search that STEP 4 calls for (Chapter 5).

The second is in performing STEP 5, developing adequacy in relevant disciplines (Chapter 6), and STEP 6, analyzing the problem (Chapter 7). Here it is important to note that *a discipline's perspective is not identical with the insights the discipline produces.* A discipline's experts produce insights and theories concerning a problem or class of problems. These insights and theories typically reflect the discipline's perspective. Interdisciplinarians draw on these insights and theories, analyze them (asking in particular whether the insights are biased by the disciplinary perspective), identify how they conflict, modify them by creating common ground among them, integrate them, and construct a more comprehensive understanding of the problem.

UNPACKING THE DEFINING ELEMENTS OF A DISCIPLINE'S PERSPECTIVE

Here we unpack the meaning of each element of a discipline's perspective and provide detailed tables of how these elements are associated with certain disciplines. *The tables are intended to illustrate each element and provide useful resources as you pursue your particular research topic or question. You should generally not have to acquaint yourself with each entry in each table.*

Phenomena

Phenomena are enduring aspects of human existence that are of interest to scholars and are susceptible to scholarly description and explanation. For example, individuals may differ in terms of personality, but a set of personality characteristics is always with us (Szostak, 2004, pp. 30–31).

The sorting out of distinctions between disciplines in this chapter does not imply that disciplines are static. Their character is ever changing, and their borders are elastic and porous. This reality and the absence of a logical classification of phenomena to guide the disciplines have produced two unfortunate effects. The first is that several disciplines may share a phenomenon, often unmindful of the efforts of other disciplines to comprehend it. For example, psychology and religious studies share an interest in the phenomenon of terrorism, but one rarely finds in their work references to the theories and research of the other discipline. The second effect is that the disciplines may ignore a particular phenomenon altogether. An example is the causes of economic growth, which has been a focus of economists but has not been studied by history or political science.

Interdisciplinary scholars, like their disciplinary counterparts, must identify the phenomena relevant to the research question. They can attempt this in one of two ways: approach the disciplines *serially* in hopes of locating a particular phenomenon in one or more of them, or *focus on the phenomenon itself*. Table 2.3 presents the traditional approach of first identifying relevant disciplines and searching their literatures in hopes of finding insights on a particular phenomenon. The success and speed of this search naturally depend on the researcher's familiarity with each discipline. Table 2.3 links the disciplines to illustrative phenomena of interest to them. These phenomena are linked to particular disciplines for the purpose of helping you identify which disciplines are relevant to the problem to decide which of their literatures to mine for insights.[6] The classifications provided in this table and elsewhere in this book should help advanced undergraduate and graduate students see how each discipline's perspective contributes to an overall understanding of a multifaceted problem.

TABLE 2.3 ⬢ Disciplines and Their Illustrative Phenomena		
Category	**Discipline**	**Phenomena**
The Natural Sciences	Biology	Cells, genes, tissues, organs, biological systems, classifications of flora and fauna
	Chemistry	Chemical elements, molecules, compounds, chemical bonds, molecular structure, crystal structures
	Earth Science	Rocks, soils, fossils, ecosystems, tectonic plates, climate
	Mathematics	Abstract entities—numbers, equations, sets, vectors, topological spaces, geometric shapes, curves
	Physics	Atoms, subatomic particles, waves, quanta; but also stars, star clusters, galaxies, etc.
The Social Sciences	Anthropology	The origins of humanity, the dynamics of cultures worldwide
	Economics	The economy: total output (price level, unemployment, individual goods and services), income distribution, economic ideology, economic institutions (ownership, production, exchange, trade, finance, labor relations, organizations), the impact of economic policies on individuals
	Political Science	The nature and practice of systems of government and of individuals and groups pursuing power within those systems
	Psychology	The nature of human behavior as well as the internal (psychosociological) and external (environmental) factors that affect this behavior
	Sociology	The social nature of societies and of human interactions within them

Category	Discipline	Phenomena
The Humanities	Art History	Nonreproducible art—painting, sculpture, architecture, prose, poetry—and reproducible art—theater, film, photography, music, dance
	History	The people, events, and movements of human civilizations past and present
	Literature	Development and examination (i.e., both traditional literary analysis and theory as well as more contemporary culture-based contextualism and critique) of creative works of the written word
	Music Education	Development, performance, and examination (i.e., both traditional musicological analysis and theory as well as more contemporary culture-based contextualism and critique) of creative works of sound
	Philosophy	The search for wisdom through contemplation and reason using abstract thought
	Religious Studies	The phenomena of humans as religious beings and the manifestations of religious belief such as symbols, institutions, doctrines, and practices

Source: Szostak, R. (2004). *Classifying science: Phenomena, data, theory, method, practice* (pp. 26–29, 45–50). Dordrecht: Springer. With kind permission from Springer Science+Business Media.

Phenomena Classified

Until recently, only the **perspectival approach** (i.e., relying on each discipline's unique perspective on reality as presented in Table 2.2) was available to interdisciplinarians because no system of classifying all human phenomena existed. Szostak (2004) meets this need in his pioneering work that classifies phenomena about the human world. His classification approach, shown as Table 2.4, moves left to right, from the most general phenomena to the most specific. A practical benefit of his approach is that all phenomena can be linked rather easily to particular disciplines, provided that one knows the discipline's general perspective and the phenomena it typically studies.

Using Table 2.4 should facilitate linking most topics readily to one or more of the particular phenomena in the table. For example, the phenomenon of freshwater scarcity concerns the nonhuman environment. Moving from left to right, one can see multiple links to a wide array of subphenomena (center column) that may pertain to the problem. These subphenomena, in turn, provide links to other phenomena identified in the right-hand column that may be of further interest. Reading the literature pertaining to the several subphenomena may lead the researcher to broaden the investigation to include the categories of economics and politics and their respective subphenomena. In short, using Szostak's classification

TABLE 2.4 ⬢ Szostak's Categories of Phenomena About the Human World[a]		
First Level	**Second Level**	**Third Level**
Genetic predisposition	Abilities	Consciousness, subconsciousness, vocalization, perception (five senses), decision making, tool making, learning, other physical attributes (movement, eating, etc.)
	Motivations	Food, clothing, shelter, safety, sex, betterment, aggression, altruism, fairness, identification with group
	Emotions	Love, anger, fear, jealousy, guilt, empathy, anxiety, fatigue, humor, joy, grief, disgust, aesthetic sense, emotional display
	Time preference	
Individual differences	Abilities • Physical abilities • Physical appearance • Energy level • Intelligences	• Speed, strength, endurance • Height, weight, symmetry • Physical, mental • Musical, spatial, mathematical, verbal, kinesthetic, interpersonal
	Personality • Emotionality (stable/moody) • Conscientiousness • Affection (selfish/agreeable) • Intellectual orientation • Other dimensions • Disorders • Sexual orientation • Interpersonal relationships	• Contentment, composure vs. anxiety, self-pity • Thoroughness, precision, foresight, organization, perseverance vs. carelessness, disorderly, frivolous • Sympathetic, appreciative, kind, generous vs. cruel, quarrelsome, fault finding • Openness, imagination, curiosity, sensitivity vs. closed-mindedness • Dominant/submissive, strong/weak, in/dependent, humor, aggression, future/present oriented, happiness • Schizophrenia, psychoticism . . . ? • View of self, others, causal relationships • Parent/child, sibling, employee/r, romance, friendship, casual acquaintance
Economy	Total output	Price level, unemployment, individual goods and services
	Income distribution	
	Economic ideology	
	Economic institutions	Ownership, production, exchange, trade, finance, labor relations, organizations
Art	Nonreproducible	Painting, sculpture, architecture, prose, poetry
	Reproducible	Theater, film, photography, music, dance

First Level	Second Level	Third Level
Politics	Political institutions	Decision-making systems, rules, organizations
	Political ideology	
	Nationalism	
	Public opinion	Issues (various)[b]
	Crime	Versus persons/property
Culture	Languages	By descent
	Religions	Providence, revelation, salvation, miracles, doctrine
	Stories	Myths, fairy tales, legends, family sagas, fables, jokes, and riddles
	Expressions of cultural values • Goals • Means • Community • Everyday norms	Rituals, dance, song, cuisine, attire, ornamentation of buildings, games • Ambition, optimism, attitudes to wealth, power, prestige, beauty, honor, recognition, love, friendship, sex, marriage, time preference, physical and psychological well-being • Honesty, ethics, righteousness, fate, work, violence, vengeance, curiosity, innovation, nature, healing • Identity, family vs. community, openness to outsiders, trust, egalitarianism, attitude to young and old, responsibility, authoritarianism, respect for individuals • Courtesy, manners, proxemics, tidiness, cleanliness, punctuality, conversational rules, locomotion rules, tipping
Social structure	Gender	
	Family types/Kinship	Nuclear, extended, single parent
	Classes (various)	Occupations (various)
	Ethnic/Racial divisions	
	Social ideology	
Technology and science	Fields (various)	Innovations (various)
	Recognizing the problem	
	Setting the stage	

(Continued)

TABLE 2.4 ⬡ (Continued)

First Level	Second Level	Third Level
	Act of insight	
	Critical revision	
	Diffusion/Transmission	Communication, adoption
Health	Nutrition	Diverse nutritional needs
	Disease	Viral, bacterial, environmental
Population	Fertility	Fecundity, deviation from, maximum
	Mortality	Causes of death (various)
	Migration	Distance, international, temporary
	Age distribution	
Nonhuman environment	Soil	Soil types (various)
	Topography	Land forms (various)
	Climate	Climate patterns (various)
	Flora	Species (various)
	Fauna	Species (various)
	Resource availability	Various resources
	Water availability	
	Natural disasters	Flood, tornado, hurricane, earthquake, volcano
	Day and night	
	Transport infrastructure	Mode (various)
	Built environments	Offices, houses, fences, etc.
	Population density	

Source: Szostak, R. (2004). *Classifying science: Phenomena, data, theory, method, practice* (pp. 27–29). Dordrecht: Springer. With kind permission from Springer Science+Business Media.

a. Close examination of the table shows that there are only 11 categories of phenomena and relatively small sets of second-level phenomena. The third-level phenomena in the table can sometimes be further unpacked into subsidiary phenomena. Szostak says that the table was developed using a mix of deduction and induction and that thus it can be extended if/when new phenomena are discovered. Students wanting more detail can visit Szostak's Basic Concepts Classification at https://sites.google.com/a/ualberta.ca/rick-szostak/research/basic-concepts-classification-web-version-2013/the-classification-of-things-phenomena.

b. *Various* here and elsewhere in this table means that there are many subsidiary phenomena. Identifying these will require the student to consult more specialized disciplinary literatures.

approach should facilitate making connections to neighboring phenomena that may touch on the research question. Making these connections quickly not only aids the research process, as will be demonstrated in later chapters, but it also enables researchers to confirm their selection of potentially relevant disciplines. This table may appear daunting at first glance, but you need only understand its basic structure (that is, you need not and should not memorize every element) to be able to utilize it once you have a research topic in mind.

Epistemology

Epistemology is the branch of philosophy that studies how one knows what is true and how one validates truth (Sturgeon, Martin, & Grayling, 1995, p. 9). An epistemological position reflects one's views of *what* can be known about the world and *how* it can be known. Literally, an epistemology is a theory of knowledge (Marsh & Furlong, 2002, pp. 18–19). Each discipline's epistemology is its way of knowing that part of reality that it considers within its research domain (Elliott, 2002, p. 85). As we shall see, a discipline's epistemology influences (and is influenced by) the assumptions it makes and the theories and especially the methods it employs.

The **epistemic norms of a discipline** are agreements about how researchers should select their evidence or data, evaluate their experiments, and judge their theories. Philosopher of science Jane Maienschein (2000) states, "It is epistemic convictions that dictate what will count as acceptable practice and how theory and practice should work together to yield legitimate scientific knowledge" (p. 123). For example, the experimental approach (favored by the natural sciences) is based on the epistemological assumption that stresses the value of experimental control and replicability, whereas the field approach (favored by some social sciences) is based on the value of studying the "messy, muddled life-in-its-context" (p. 134).

We noted in Chapter 1 that interdisciplinary scholarship pursues a middle ground between "positivist" and "nihilist" epistemological extremes. We appreciated at that time that interdisciplinary scholarship is thus consistent with most but not quite all contemporary thinking in the philosophy of science. It is useful here to recognize that many scholars, especially in the natural sciences and economics, are still very positivist in outlook, aspiring to achieve very precise understandings that can be established beyond reasonable doubt. We can describe an attitude that falls short of stressing the possibility of absolute proof/disproof as modernist. Many scholars in the humanities tend toward a nihilistic view that objective knowledge is impossible. (*Postmodernism* is a term used to describe both nihilists and scholars who hold skeptical views of the possibilities of scholarship but stop short of nihilism.) These two epistemological perspectives have spread widely in the academy. The social sciences, in particular, although largely pursuing an epistemological middle ground, possess scholars with both positivist and nihilist outlooks (Bell 1998, Creath & Maienschein 2000; Rosenau 1992, Szostak 2007a).

It is worthwhile to briefly recognize some key contrasts in these epistemological positions:

- Is there an external reality that we can perceive, or do we "construct" reality in our minds? The middle ground here is to accept that there is an external reality but that humans are limited in their perceptive and cognitive abilities to comprehend this.

- Can we objectively understand reality or not? A middle ground here is to recognize that many biases (including disciplinary biases) can affect scholarship, but that these can be confronted through careful analysis and attempts to integrate across conflicting insights.

- Can we prove or disprove hypotheses or is scholarship a matter of opinion? A middle ground here rejects the idea of proof/disproof (beyond the realm of mathematics and logic) but accepts that scholars can amass a body of argument and evidence such that certain hypotheses are accepted.

- Is language clear or hopelessly ambiguous? A middle ground here recognizes that language is inherently ambiguous, but that humans have recourse to various strategies (including classification and interdisciplinary practices) that limit ambiguity.

- Are there empirical regularities in the world, or is this ever-changing? A middle ground here recognizes that it is challenging to identify empirical regularities precisely because all phenomena influence each other. There may be a regularity in how A affects B, but this is hard to establish because C and D also influence each of them.

We will in what follows describe the most common epistemological outlook within particular disciplines, but you should recall that there is diversity within most if not all disciplines. Students should not just rely on the epistemological perspective of the discipline, but seek to identify, if possible, where authors stand with respect to the key contrasts identified above.

NOTE TO READERS

The statements on epistemologies in Tables 2.5, 2.6, and 2.7 are not definitive but central tendencies. Any way of classifying the epistemological positions of the disciplines can be contested.[7] These tables draw heavily from disciplinary experts, with the recognition that no two scholars may give precisely the same description of their disciplines.

TABLE 2.5 ⬢ Epistemologies of the Natural Sciences

Discipline	Epistemology
Biology	Biology stresses the value of classification, observation, and experimental control. The latter is the means of identifying true causes, and therefore privileges experimental methods (because they are replicable) over all other methods of obtaining information (Magnus, 2000, p. 115).
Chemistry	Chemists use both empirics and theory (especially thermodynamics). Even more than physics, chemistry relies on lab experiments and computer simulations. Chemistry involves less fieldwork than Earth science and biology do.
Earth Science	In much of Earth science, the theory of uniformitarianism (that all geologic phenomena may be explained as the result of natural laws and processes that have not changed over time) is prominent.
Mathematics	Mathematical truths are numerical abstractions that are discovered through logic and reasoning. These truths exist independently of our ability or lack of ability to find them, and they do not change. These truths or forms of "invariance" enable us to categorize, organize, and give structure to the world. These mathematical structures—"geometric images and spaces, or the linguistic/algebraic expressions"—are "grounded on key regularities of the world or what we 'see' in the world" (Longo, 2002, p. 434).
Physics	Like all the physical sciences, physics is empirical, rational, and experimental. It seeks to discover truths or laws about two related and observable concepts—matter and energy—by acquiring objective and measurable information about them (Taffel, 1992, pp. 1, 5). It stresses experiments far more than biology or Earth science.

TABLE 2.6 ⬢ Epistemologies of the Social Sciences

Discipline	Epistemology
Anthropology	Epistemological pluralism characterizes anthropology. Empiricists hold that people learn their values and that their values are therefore relative to their culture. Both physical and cultural anthropologists embrace constructivism, which holds that human knowledge is shaped by the social and cultural context in which it is formed and is not merely a reflection of reality (Bernard, 2002, pp. 3–4).

(Continued)

(Continued)

TABLE 2.6 ⬢ (Continued)	
Discipline	**Epistemology**
Economics	The epistemological dominance of modernism is being challenged by postmodernism that generates a pluralistic understanding of reality. Postmodernists see reality, and the self, as fragmented. Therefore, human understanding of reality is also fragmented. Nevertheless, the beliefs of economists are still largely determined by empirical evidence in direct relation to the mathematical theories and models they use. Empiricists stress fixed definitions of words, use a deductive method, and examine a small set of variables (Dow, 2001, p. 63).
Political Science	Political science embraces a modernist epistemology. However, positivists in the discipline are trying to cast the "science" of politics in terms of finding some set of "covering laws" so strong that even a single counterexample would suffice to falsify them. But human beings, according to others in the discipline, while they are undeniably subject to certain external forces, are also in part intentional actors, capable of cognition and of acting on the basis of it. Consequently, these scholars study "belief," "purpose," "intention," and "meaning" as potentially crucial elements in explaining the political actions of humans (Goodin & Klingerman, 1996, pp. 9–10).
Psychology	The epistemology of psychology is that psychological constructs and their interrelationships can be inferred through discussion and observation and applied to treatment (clinical) or a series of experiments with slight variations (experimental). A critical ingredient of a good experiment is experimental control that seeks to eliminate extraneous factors that might affect the outcome of the study (Leary, 2004, p. 208).
Sociology	Modernist (i.e., positivist) sociology shares a modernist epistemology with the other social sciences, but this epistemology is opposed by critical social theory, a theory cluster that includes Marxism, critical theory, feminist theory, postmodernism, multiculturalism, and cultural studies. What unites these approaches in the most general sense is their assumption that knowledge is socially constructed and that knowledge exists in history that can change the course of history if properly applied (Agger, 1998, pp. 1–13).

TABLE 2.7 ⬢ Epistemologies of the Humanities	
Discipline	**Epistemology**
Art History	Modernists determine the value of works of art by comparing them with standards of aesthetics and expertise. But practitioners of the new art history that emerged in the 1960s determine the value of works of art in relation to contestation between values of competing groups; that is, it understands them in social and cultural contexts (Harris, 2001, pp. 65, 96–97, 130–131, 162–165, 194–196, 228–232, 262–288). Postmodern critics (active from about 1970 to the present) "argue that the supposedly dispassionate old-style art historians are, consciously or not, committed to the false elitist ideas that universal aesthetic criteria exist and that only certain superior things qualify as 'art'" (Barnet, 2008, p. 260).

Discipline	Epistemology
History	Modernists focus on the authenticity and appropriateness of how an event, a person, or a period is interpreted by evaluating the work in terms of its faithfulness to appropriate primary and secondary sources. "Truth," they believe, "is one, not perspectival" (Novick, 1998, p. 2). Believing that "structure" is fundamental to understanding the past, social historians focus on structure and infrastructure—on material structure, on the economy, on social and political systems—but do not eliminate the individual. More recently, some social historians have begun to employ "micro history" or the new cultural history (a blend of social history and intellectual history) as a way of studying ideological structures, mental structures (such as notions of family and community), isolated events, individuals, or actions, borrowing from anthropology the ethnographic method of "thick description," which emphasizes close observation of small details, carefully listening to every voice and every nuance of phrase (Howell & Prevenier, 2001, p. 115).
Literature	In general, modernists focus on the text and employ text-based research techniques. Newer approaches see meaning making as a relational process. The close reading of texts is being informed by background research into the context of the text, such as the circumstances surrounding its production, content, and consumption. Other newer approaches abound. For example, notions of auto/biographic writing have shifted from an idea of presenting "the truth" about someone to presenting "a truth." Oral history is viewed as a means of understanding the workings of "literary and cultural phenomena in and on people's imagination." Critical discourse analysis examines patterns in language use to uncover the workings of an ideology to see how it exerted control or how it was resisted. Quantitative researchers are using computers to calculate the frequency with which certain words appear in a text so that they can better interpret its meaning (Griffin, 2005, pp. 5–14). Deconstructionists find ambiguities in all texts.
Music Education	For modernist music educators, knowledge is often primarily technical knowledge. They assume, therefore, that empirical investigation produces verifiable and objective "knowledge" and "truth" irrespective of context. Postmodernist music educators embrace a much more pluralistic view of knowledge, viewing it as elusive, fragile, temporary, and conjectural. They assert that there are an infinite number of potentially "true" statements that can be developed about any phenomenon and that no single form of research can possibly account for the complete "truth" or reality of anything. "The goal of research, then, is continuously to seek relevant descriptions and explanations of a phenomenon based on the best and most complete knowledge we can garner about that phenomenon" (Elliott, 2002, p. 91).
Philosophy	Recently, philosophical questions about perception have become more important. For both empiricist and rationalist positions, one of the major concerns is to ascertain whether the means of getting knowledge are trustworthy. The chief concerns of epistemology in this regard are memory, judgment, introspection, reasoning, "a priori–a posteriori" distinction, and the scientific method (Sturgeon, Martin, & Grayling, 1995, pp. 9–10).

(Continued)

(Continued)

TABLE 2.7 ● (Continued)	
Discipline	**Epistemology**
Religious Studies	Religious studies is concerned about the "assumptions and preconceptions that influence the analysis and interpretation of data, that is, the theoretical and analytical framework, even personal feelings, one brings to the task of organizing and analyzing facts" (Stone, 1998, p. 6). Though all humanities disciplines are concerned about the problem of subjectivity, few are as self-critical as religious studies (p. 7).

Epistemologies of the Natural Sciences, the Social Sciences, and the Humanities

Epistemologies of the Natural Sciences. Empiricism dominates the natural sciences.[8] Empiricism assures us that observation and experimentation make scientific explanations credible, and the predictive power of its theories is ever-increasing (Rosenberg, 2000, p. 146). However, the epistemologies of the sciences make scientific approaches inadequate for addressing value issues (Kelly, 1996, p. 95).

Epistemologies of the Social Sciences. The disciplines in the social or human sciences, more so than in the natural sciences, tend to embrace more than one epistemology, as shown in Table 2.6. For example, reflecting the growing postmodernist criticism of positivism's empiricism and value neutrality, most social scientists now agree that knowledge in their disciplines is generated by the "continual interplay of personal experience, values, theories, hypotheses, and logical models, as well as empirical evidence generated by a variety of methodological approaches" (Calhoun, 2002, p. 373).

Epistemologies of the Humanities. The humanities, even more so than the social sciences, embrace epistemological pluralism as shown in Table 2.7. This development is explained by the rise of the "new generalism" or "critical humanities" (feminism, critical theory, postcolonial studies, cultural studies, gender studies, postmodernism, poststructuralism, deconstructionism, etc.), which is discussed in greater detail below. The humanities prize diversity of perspective, values, and ways of knowing.

The Interdisciplinary Position on These Approaches

Interdisciplinarity seeks to avoid both the extremes of modernist optimism and postmodernist pessimism: If we doubt that enhanced understanding is possible, then there is no use in doing interdisciplinary research. But if we doubt the importance of perspective,

then interdisciplinarity is unnecessary. Interdisciplinarians should respect diverse epistemologies, but should not think that "anything goes" (Szostak, 2007a).

Good interdisciplinary work requires a strong degree of **epistemological self-reflexivity**
(Klein, 1996, p. 214). This is awareness of how epistemological choices tend to influence one's selection of research methods that, in turn, influence research outcomes (Bell,
1998, p. 101). Accordingly, interdisciplinarians should take care that their embrace of
certain assumptions, epistemologies, theories, methods, and political views do not bias
the research process and thus skew the resulting understanding.

As noted above, interdisciplinary researchers should be wary of certain epistemological attitudes but should otherwise be respectful of all epistemologies. As elsewhere in this book where
a both/and approach (that is, seeing value in alternative approaches rather than choosing one
over the other) is recommended, interdisciplinary scholars are guided to integrate the best of
both epistemologies rather than limit themselves to one. Interdisciplinary analysis is possible as
long as we back away from the most extreme postmodern arguments (as most postmodernists
themselves do) and desirable as long as we back away from extreme modernist assumptions.

Assumptions

From each discipline's epistemology (and ethics, etc.) flows a set of assumptions that tend
to characterize research in that discipline. An assumption is something taken for granted,
a supposition. **Assumptions** are the principles that underlie the discipline as a whole and
its overall perspective on reality. As the term implies, these principles are accepted as
the truths upon which the discipline's theories, concepts, methods, and curriculum are
based. Stated another way, it is the interplay of assumptions and empirical evidence that
shapes a discipline's theories, concepts, and insights.

Grasping the underlying assumptions of a discipline as a whole provides important clues
to the assumptions underlying its particular insights and theories. Assumptions underlying specific insights are important to the integrative part of the interdisciplinary process
that calls for identifying *possible* sources of conflict between insights. If conflicts between
sets of insights or theories exist, one can then work to modify the conflict(s) by creating
common ground (STEP 8). There are two kinds of assumptions: "basic" assumptions
that scientists across disciplinary clusters typically make, and more focused or "hallmark"
assumptions that are made by scientists working in a particular cluster of disciplines.

Basic Assumptions

The particular *combination* of assumptions is unique to each discipline, but disciplines can
share assumptions. The assumptions in Tables 2.8, 2.9, and 2.10 are not comprehensive
generalizations, but central tendencies and, thus, can be challenged by disciplinarians who
might prefer different representational selections. The purpose in presenting these tables
is twofold: (1) to help researchers decide which disciplines are relevant to the problem so

that their literatures can be mined for insights and (2) to identify assumptions that will be useful in performing later STEPS, particularly STEP 8.

Hallmark Assumptions of the Natural Sciences. The *hallmark assumptions* made by those working in the natural sciences are two. The first is that scientists can transcend their cultural experience and make definitive measurements of phenomena (things). The second is that "there are no supernatural or other a priori properties of nature that cannot potentially be measured" (Maurer, 2004, pp. 19–20).[9] This assumption is reflected to varying degrees in the characterizations of disciplinary assumptions underlying the natural sciences noted in Table 2.8. (*Note:* The sources cited in this and in following tables are good starting points for further reading.)

TABLE 2.8 ● Assumptions of Disciplines in the Natural Sciences	
Discipline	**Assumptions**
Biology	Biologists assume that the hypothetico-deductive approach (i.e., deductive reasoning used to derive explanations or predictions from laws or theories) is superior to description of pattern and inductive reasoning (Quinn & Keough, 2002, p. 2).
Chemistry	The function of the whole is reducible to the properties of its constituent elements and compounds and their interactions. "All living organisms share certain chemical, molecular, and structural features, interact according to well-defined principles, and follow the same rules with regard to inheritance and evolution" (Donald, 2002, p. 111).
Earth Science	The principle of uniformitarianism leads geologists to assume that the present is the key to understanding the past. Earth processes have not significantly changed during the several billion years that Earth has been a dynamic planet similar in many ways to the other planets constituting the solar system.
Mathematics	Assumptions (or axioms) in mathematics form the starting point for logical proofs of its theorems. They constitute the "if" part of a statement: "If A, then B." The consequences of the assumptions are found through logical reasoning, which leads the mathematician to discover the conclusion, "then B" (B. Shipman, personal communication, April 2005).
Physics	Logical empiricism assumes the existence of a finite set of laws that governs the behavior of the universe and that there is an objective method for discovering these truths. Natural realism, by contrast, assumes (1) that the universe works in a law-like manner, though the nature of the universe may be extremely complex and much of it may even be unfathomable; and (2) "scientists can build models that approximate nature sufficiently to allow further progress in understanding particular phenomena" (Maurer, 2004, p. 21). This atomistic approach to knowledge further assumes that separate parts together constitute physical reality, that these separate parts are lawfully and precisely related, and that physics events can be predicted.

Source: Adapted from Gerring, J. (2001). *Social science methodology: A critical framework*. Boston: Cambridge University Press. AND Stoker, G., & Marsh, D. (2002). Introduction. In D. Marsh & G. Stoker (Eds.), *Theory and methods in political science* (2nd ed., pp. 1 -16). New York: Palgrave Macmillan. AND Elliott, D. J. (2002). Philosophical perspectives on research. In R. Colwell & C. Richardson (Eds.), *The new handbook of research on music teaching and learning* (pp. 85–102). Oxford: Oxford University Press.

Hallmark Assumptions of the Social Sciences. The social sciences are grounded in essentially the same set of basic assumptions that characterize the natural sciences (Frankfort-Nachmias & Nachmias, 2008, p. 5). Assumptions in the social sciences are closely related to the research methods, theories, and schools of thought embraced by members of each discipline's community of scholars. For example, a popular textbook on behavioral research methods (psychology, communication, human development, education, marketing, social work, and the like) states the assumption underlying the scientific approach and systematic empiricism as these methods are applied to the behavioral sciences: "Data obtained through systematic empiricism allow researchers to draw more confident conclusions than they can draw from casual observation alone" (Leary, 2004, p. 9). Modernists share a "grizzled confidence" in such ideas as "progress" and "knowledge" grounded in empirical and replicable data (Cullenberg, Amariglio, & Ruccio, 2001, p. 3). This modernist assumption is present, to varying degrees, in many of the social science disciplines but is being challenged by postmodern notions. Both sets of assumptions—modern and postmodern—are noted in Table 2.9.

TABLE 2.9 ● Assumptions of Disciplines in the Social Sciences	
Discipline	**Assumptions**
Anthropology	Cultural relativism (the notion that people's ideas about what is good and beautiful are shaped by their culture) assumes that systems of knowledge possessed by different cultures are "incommensurable" (i.e., not comparable and not transferable) (Whitaker, 1996, p. 480). Cultural relativism has been the driving ethic of anthropology for generations, but it is being challenged by feminists, postcolonialists, and advocates for other marginalized groups on the grounds that relativism supports the repressive status quo in other cultures (Bernard, 2002, p. 73).[a]
Economics	Modernist approaches predominate. Modernist economists assume that the same dominant human motivation (rational self-interest) transcends national and cultural boundaries, in the past as in the present. Also, they assume that both usefulness and value are implicit in rational choices (on which they prefer to focus) under conditions of scarcity. Postmodernists assume that all things, including economic motivation and behavior, are intimately bound up with the situatedness (i.e., the cultural, political, and technological context) of those engaged in these activities and thus are not generalizable (Cullenberg et al., 2001, p. 19).
Political Science	Political science has been influenced primarily by history, but more recently, it is being influenced by theories from sociology, economics, and psychology. Consequently, its assumptions reflect whichever discipline and theory it is drawing from at the moment. Modernists assume rationality: "Human beings, while they are undeniably subject to certain causal forces, are . . . in part intentional actors, capable of cognition and acting on the basis of it" (Goodin & Klingerman, 1996, pp. 9–10). Behavioralists (who are also modernists) assume that political science can become a science capable of prediction and explanation (Somit & Tanenhaus, 1967, pp. 177–178). Proponents of the scientific method of research assume that empirical and quantitative, rather than normative and qualitative, analysis is the most effective way of knowing political reality (Manheim, Rich, Willnat, & Brians, 2006, pp. 2–3).

(Continued)

TABLE 2.9 ● (Continued)	
Discipline	Assumptions
Psychology	Psychologists assume that "data obtained through systematic empiricism allow researchers to draw more confident conclusions than they can draw from casual observation alone" (Leary, 2004, p. 9). Generalizations about larger populations may be inferred from representative sample populations. Psychologists also assume that group behavior can be reduced to individuals and their interactions and that humans organize their mental life through psychological constructs.
Sociology	Assumptions vary widely in this discipline. Empiricists assume an independent social reality exists that can be perceived and measured through gathering of data. Critics of modernism assume that our perceptions of social reality are filtered through a web of assumptions, cultural influences, and value-laden vocabularies, that individual human behavior is socially constructed, with rationality and autonomy playing modest roles at best; groups, institutions, and especially society have an existence independent of the individuals in them. People, they assume, are motivated primarily by the desire for social status (Alvesson, 2002, pp. 2–3).

a. Cultural relativism does not equate to ethical relativism (that all ethical systems are equally good since they are all cultural products), as Merrilee Salmon (1997) makes clear. Note that incommensurability, if true, would render interdisciplinarity infeasible.

Hallmark Assumptions of the Humanities. The humanities are grounded in a set of assumptions that differ greatly from those of the sciences. Over the course of the twentieth century and especially in recent decades, the older assumption of unified knowledge and culture has given way to a pluralistic and even conflicted set of assumptions. Klein (2010) lumps these new assumptions under the heading "the new generalism." This is not a unified paradigm, she explains, but "a cross-fertilizing synergism in the form of shared methods, concepts, and theories about language, culture, and history" (p. 30). The keywords of this new paradigm are *plurality* and *heterogeneity* (replacing *unity* and *universality*), and *interrogation* and *intervention* (supplanting *synthesis* and *holism*). The **new humanities**, she says, "interrogates the dominant structure of knowledge and education with the aim of transforming them" with the "explicit intent of deconstructing disciplinary knowledge and boundaries" (p. 30). This trend, Klein asserts, is especially apparent in cultural studies, women's and ethnic studies, and literary studies, where "the epistemological and political are inseparable" (p. 30).

> As humanities disciplines moved away from older paradigms of historical empiricism and positivist philology [the study of literature and the disciplines associated with literature], increasing attention was paid to contexts of aesthetic works and responses of readers, viewers, and listeners. The concept of culture was also expanded from a narrow focus on elite forms to a broader anthropological notion, and once discrete objects were reimagined as forces that circulate in a network of forms and actions. (Klein, 2010, p. 30)

TABLE 2.10 ● Assumptions of Disciplines in the Humanities	
Discipline	**Assumptions**
Art History	Modernists assume that the intrinsic value of the object is primary. Radical art historians—for example, Marxist, feminist, psychoanalytical, and poststructuralist—"share a broad historical materialism" of outlook: that all social institutions, such as education, politics, and the media, are exploitative and that "exploitation extends to social relations, based, for instance, on factors of gender, race, and sexual preference" (Harris, 2001, p. 264). In general, these critics assume that intrinsic values remain primary, but understanding the social context completes one's grasp of the work (p. 264).[a]
History	Modernist (positivist and historicist) historical scholarship rests on the idea that objectivity in historical research is possible and preferred (Iggers, 1997, p. 9). In general, social history (e.g., Marxian socioeconomic history, the Braudelian method, women's history, African-American history, and ethnic history) assumes that those whom traditional history writing had ignored (the poor, the working class, women, homosexuals, minorities, the sick) played an important but unappreciated role in historical change (Howell & Prevenier, 2001, p. 113).
Literature	Literature (broadly defined) or "texts" are assumed to be a lens for understanding life in a culture and an instrument that can be used to understand human experience in all of its complexity. Texts "encompass the continuous substance of all human signifying activities" (Marshall, 1992, p. 162). Another assumption is that these texts are "alien" to the reader, meaning that "something in the text or in our distance from it in time and place makes it obscure." The interpreter's task is to make the text "speak" by "reading" the text using extremely complex skills so as to give the text "meaning." *Meaning* is "an intricate and historically situated social process" that occurs between the interpreter and the audience (i.e., reader) that neither fully controls (Marshall, 1992, pp. 159, 165–166).
Music Education	Modernists assume that empirical investigation produces verifiable and objective "knowledge" (i.e., in the sense of infallible theories, laws, or general statements) and "truth" that is context free. Postpositivists (interpretivists, critical theory advocates, gender studies scholars, and postmodernists) deny the possibility of objectivity because human values are always present in human minds (Elliott, 2002, p. 99).
Philosophy	There are two schools of thought about how to get knowledge. Rationalists assume that the chief route to knowledge is the exercise of systematic reasoning and "looking at the scaffolding of our thought and doing conceptual engineering" (Blackburn, 1999, p. 4). The model for rationalists is mathematics and logic. Empiricists assume that the chief route to knowledge is perception (i.e., using the five senses of sight, smell, hearing, taste, and touch and the extension of these using technologies such as the microscope and telescope). The model for empiricists is any of the natural sciences where observation and experiment are the principal means of inquiry (Sturgeon, Martin, & Grayling, 1995, p. 9).
Religious Studies	Religious studies often queries faith, and the history of religions focuses on understanding humans as religious beings. One key assumption of the discipline is that there is something inherently unique about religion, and those who study it must do so without reducing its essence to something other than itself, as sociologists and psychologists tend to do. A related assumption is that even though religion is freighted with human emotion, objectivity is possible (Stone, 1998, p. 5).

a. Marxists assume that class struggle is the primary engine of historical development in capitalist society and that other forms of exploitation are either a product of the basic antagonism of class or peripheral to it. Feminists assume connections and causal links between patriarchal dominance within the society as a whole and its art. Psychoanalytic art historians assume that a full understanding of "the subject" requires inquiry into the complex nature of the embodied human psyche and its conscious and unconscious outworkings (Harris, 2001, pp. 262, 264, 195).

These keywords and trends provide important clues to the assumptions of humanities disciplines. Table 2.10 identifies the assumptions of modernism alongside the assumptions of the "new generalism."

NOTE TO READERS

Assumptions often play an important role in the process of creating common ground among conflicting disciplinary concepts and theories. Chapters 10 and 11 explain how to modify the assumptions underlying concepts and theories to prepare them for integration. For example, in examining theories explaining the causes of suicide terrorism, students in a class whose topic was terrorism found that an important assumption underlying scholarly insights from psychology and political science is that the behavior of the terrorists is rational (as defined by both disciplines), not irrational as many in the class had initially supposed.

Concepts

A **concept** is a symbol expressed in language that represents a phenomenon or an abstract idea generalized from particular instances (Novak, 1998, p. 21; Wallace & Wolf, 2006, pp. 4–5). For example, chairs come in various shapes and sizes, but once a child acquires the concept *chair,* that child will always refer to anything that has legs and a seat as a chair (Novak, 1998, p. 21).

Although *concept* is a key term used throughout this book, we do not provide examples of concepts favored by each discipline here for two reasons. First, the term lacks clarity as it relates to other terms such as *phenomena, causal link, theory,* and *method.* Szostak (2004) finds that many concepts can be defined in terms of phenomena, causal links (that is, relationships between phenomena), theory, or method. Some concepts, such as culture, are clearly phenomena. Others, such as oppression, are results of phenomena—in this instance, political decision making. Still others, such as revolution, globalization, and immigration, describe processes of change within or between phenomena (p. 41). But most of these can best be understood as illustrating features of causal links—for example, the link between art and human appreciation (pp. 42–43). It is thus best to deal with concepts when discussing phenomena, theories, and methods. Moreover, the same phenomenon (or concept more generally) might be called by different names in different disciplines.

In addition to the difficulty of differentiating concepts from phenomena and causal links is the more formidable challenge of dealing with the huge number of concepts that each discipline has generated. Perhaps this is why so few scholars have attempted exhaustive surveys of scholarly concepts in particular disciplines, let alone across entire disciplinary categories.

For this reason, this book makes no attempt to associate particular concepts with particular disciplines. Researchers will certainly encounter what are purported to be concepts and should consult Szostak's classification of phenomena presented earlier in this chapter to see whether the concept is, in fact, a phenomenon. If not, the interdisciplinarian should investigate whether the concept is or could be carefully defined in terms of causal links, theory, or method.

We might note here that some disciplines strive for very precise definitions of key concepts (e.g., mass and energy in physics), while other disciplines allow diffuse interpretations (there are over a thousand distinct definitions of culture, for example). Humanities scholar Mieke Bal (2002) agrees that concepts "need to be explicit, clear, and defined." She notes, however, that in interdisciplinary humanities, concepts "are neither fixed nor unambiguous" (pp. 5, 22, 23).

Theory

The root meaning of the word *theory* is "looking at or viewing, contemplating or speculating." There are two kinds of theory: scientific theory (about the world), which corresponds to the root meaning of theory just noted, and various types of philosophical theory (epistemological, ethical, etc.), which were dealt with in the section on epistemology. Confusion sometimes arises by the fact that some theories such as feminism or Marxism or literary theory operate as both scientific theory and philosophical theory: They make not only epistemological arguments, but also arguments about the world.

The Importance of Theory to Interdisciplinary Work

Interdisciplinarians need a basic understanding of theory, both scientific and philosophical, for four practical reasons. First, as Janet Donald (2002) emphasizes, for students to work in a discipline, they "must have the vocabulary and the *theory* of the field" because "each discipline requires a different mindset [italics added]" (p. 2).

Second, more than ever before, theory dominates the scholarly discourse within the disciplines and often drives the questions asked, the phenomena investigated, and the insights produced. Klein (1999) notes the increasingly common practice of disciplines borrowing theories and methods from other disciplines and, in some cases, making the borrowed theory or method their own (p. 3).

Third, since these theories explain particular or local phenomena, they provide many of the disciplinary "insights" into a particular problem, and it is these insights that students need to integrate to produce an interdisciplinary understanding of the problem.

Fourth, students need to develop a basic understanding of theory because of the interrelationship between theory and disciplinary research methods. In his discussion of how to do interdisciplinarity, Szostak (2002) emphasizes the importance of ascertaining "what theories and methods are particularly relevant to the research question. In the conduct of interdisciplinary work," he says, "there are complementarities such that borrowing a theory from one discipline will encourage use of its methods, study of its phenomena, and engagement with its worldview" (p. 106). As with phenomena, he cautions researchers to not ignore theories and methods that may shed some lesser light on the question.[10] He also cautions not to blindly accept the evidence for a theory from the methods preferred by that discipline. Disciplines choose methods that make their theories look good. It is that sort of synergy that makes disciplinary perspective so powerful (See Box 2.1).

NOTE TO READERS

Understanding each theory, even in general terms, will enable researchers to approach many topics with greater sophistication and depth of insight. Explanation of precisely how theory may actually be used in interdisciplinary work is reserved for later chapters, where we will discuss working with theories.

BOX 2.1

Szostak (2017b) noted that most disciplines posit some sort of stability among the phenomena that they investigate. This stability can be challenged by interactions with the phenomena studied by other disciplines, as when changes in consumer tastes or weather patterns shock the market prices studied by economists. Though disciplinary scholars may know that the systems they study are not always stable, they may nevertheless focus on theorizing stability and thus be hostile to interdisciplinary explanations of instability. Economists, for example, have rejected arguments that technological shocks may have been important in causing the Great Depression because they prefer to focus on interactions among economic variables that generally produce greater stability in economic outcomes. Scholarship as a whole needs to understand both stability and instability (why the Great Depression happened, but also why such calamities are rare) and may thus benefit from a symbiotic relationship between disciplinary and interdisciplinary research.

Method

Method concerns how one conducts research, analyzes data or evidence, tests theories, and creates new knowledge (Rosenau, 1992, p. 116).[11] Methods are ways to obtain evidence of how some aspect of the natural or human world functions (Szostak, 2004, pp. 99–100).

Each discipline tends to devote considerable attention to discussing the method(s) it uses, and it does this by requiring students majoring in the discipline to take a research methods course. The reason is simple: The methods a discipline favors reflect its epistemology and are well suited to investigating its favored theories. Interdisciplinarians should be particularly aware of this linkage between a discipline's methods and theories: There may be other methods that would shed less favorable light on the discipline's theories than the method(s) favored by that discipline.

The Importance of Disciplinary Methods to Interdisciplinary Work

The interest of interdisciplinarians in disciplinary methods and the kind of knowledge required of them varies considerably depending on how they work with methods. Those interdisciplinarians conducting basic research have to decide when and whether to use quantitative or qualitative methods, or both. Though the furor over this difference is dying down, disciplinary researchers remain divided about which approach is preferable. Interdisciplinarians engaged in basic research should be open to both approaches. The **quantitative approach**, such as the number of molecules and the size of the ozone layer, emphasizes that evidence can be expressed numerically over a specified time frame. The **qualitative approach** focuses on evidence that cannot easily be quantified, such as cultural mannerisms and personal impressions of a musical composition. In reality, the quantitative or qualitative distinction is becoming increasingly blurred. For example, theories in natural science that focus on nonintentional agents—such as the germ theory of disease or cell theory—are inherently qualitative. Scholars employing qualitative methods often quantify by using words such as *most* rather than percentages (Szostak, 2004, p. 111).

There has for some decades been a large and growing literature on "mixed methods research" or "multimethod research." This literature has stressed for the most part the value of mixing quantitative and qualitative methods. This sometimes means utilizing both simultaneously: One might both statistically analyze and do close readings of interview transcripts. It sometimes means using different methods in sequence: One might use the results of statistical analysis to suggest questions for a focus group. There is a large overlap between the mixed methods literature and the literature on interdisciplinarity (Szostak, 2015a, pp. 128–143).

Just as researchers must have at least a general knowledge of the theories informing the disciplines relevant to the problem, so too must they have a working knowledge of the methods used by these same disciplines. Interdisciplinary programs whose courses cross only a few disciplinary boundaries naturally emphasize only a few methods. Interdisciplinary programs or courses that take interdisciplinarity itself as a focus tend toward a much broader coverage of methods, though this coverage is far from exhaustive. The latter kind of program clearly demands that students read widely in the disciplinary literature

TABLE 2.11 ● Key Quantitative and Qualitative Methods	
Approach	**Methods**
Quantitative	Experiments
	Surveys
	Statistical analysis
	Mathematical modeling
	Classification
	Mapmaking
	Examination of physical traces
	Careful examination of physical objects (as when geologists study rocks)
Qualitative	Participant observation
	Interview
	Textual analysis
	Hermeneutics
	Intuition/experience
	Textual analysis

Source: Adapted from Szostak, R. (2004). *Classifying science: Phenomena, data, theory, method, practice.* Dordrecht: Springer.

to develop at least a general understanding of all the standard methods. Fortunately, the number of these is relatively small. Table 2.11 lists commonly used quantitative and qualitative methods. Analysis of the strengths and weaknesses of each method is reserved for Chapter 6.

Research Methods Associated With Disciplines

Tables 2.11, 2.12, and 2.13 associate particular disciplinary categories with particular methods. The methods associated with each category are not definitive and are stated in the most general terms. Any statement of disciplinary practices can be contested on the ground that it disguises the pluralistic and even conflicted nature of disciplinary practice. The following descriptions are written in awareness of the possible criticisms. The purpose of these tables is to help researchers decide which research method(s) are appropriate to the problem, or topic.

TABLE 2.12 ● Research Methods Associated With the Natural Sciences	
Discipline	**Methods**
Biology	The epistemological debate between the naturalist or field position and the experimental or laboratory position is also about which methods (i.e., lab or field) produce "good science" (Creath & Maienschein, 2000, p. 134). Laboratory (i.e., experimental design and data analysis) methods extract life from its natural ecological setting and examine specimens under controlled conditions using electron microscopy and positron-emission tomography (PET) to produce visual images of the structure of systems (Bechtel, 2000, p. 139). Systems ecologists and developmental biologists insist on studying life in its living, functioning, active form using "philosophical, sociological, anthropological, and cognitive explanatory schemata" (Holmes, 2000, p. 169). Biologists increasingly appreciate that the scientific method must take into consideration ethical limits to experimentation.
Chemistry	Chemistry differs from the other sciences by attempting to develop new materials using the foundational principles discovered and developed by chemistry via experiments. "Understanding the properties of a substance and the changes it undergoes leads to the central theme in chemistry: macroscopic properties and behavior, those we can see, are the results of submicroscopic properties and behavior that we cannot see" (Silberberg, 2006, p. 5). Experiments are the dominant method in chemistry.
Earth Science	Like physics and chemistry, Earth science relies on a variety of quantitative methods of displaying and analyzing data, including statistics, geographic information systems (GIS), computer modeling, X-ray diffraction and florescence, mass spectronomy, emission and absorption spectronomy, gravity and magnetic resonance, acoustic (seismic) wave propagation (reflection and refraction), remote sensing using the electromagnetic spectrum, and well logging techniques that include sonic, electrical resistivity, and neutron absorption. Increasingly, however, geologists are relying more on fieldwork because processes taking place in geologic time cannot be replicated (J. Wickham, personal communication, August 2006).
Mathematics	Mathematics is totally abstracted from the empirical world, though other disciplines that are empirical apply mathematics. The worlds mathematicians create are rational simply because rationality is a requirement mathematicians impose on themselves. Mathematics uses proven theorems about the properties (e.g., consistency, transitivity, completeness) of the abstract realities they create.
Physics	Like chemistry, physics takes objects apart to study their constituent parts (atoms and subatomic particles, quanta) to see how they are related; but unlike chemistry, it also studies overall characteristics such as mass, velocity, conductivity, and heat of evaporation. The methods of physics are split into theoretical and experimental. "Theoretical" physicists solve problems using mathematical modeling rather than experimentation. Experimental physicists use experiments and computers to measure and quantify objects and phenomena and to test and verify or falsify the theories produced by the theoretical physicists (Donald, 2002, pp. 32–33). In physics, the hypothesis often takes the form of a causal mechanism or a mathematical relation. Cosmology, the branch of physics that studies the origins and development of the universe, must generally rely on astronomical observation as its method.

TABLE 2.13 ● Research Methods Associated With the Social Sciences	
Discipline	**Methods**
Anthropology	Anthropology uses a wide variety of scientific and interpretive techniques to reconstruct the past including experiments, sampling, cultural immersion, fieldwork, interviewing (unstructured and semistructured), structured interviewing (questionnaires and cultural domain analysis), scales and scaling, participant observation, field notes, direct and indirect observation, thick description, analysis of human interaction, language, archaeology, and biology (Bernard, 2002). The most common method in cultural anthropology has long been detailed field observation, though this has changed in recent decades. Physical anthropology relies on the examination of the results of archaeological excavation.
Economics	Modernist methods include mathematical modeling and statistical analysis. What is distinctive about most economic datasets is that they are generated for other purposes (e.g., governmental policy) and often do not directly measure the variables of interest to economists, so economists end up working with inferential indicators more than direct measurements. Mainstream economics, however, is experiencing some degree of methodological fragmentation by postmodernists who oppose the reduction of human behavior and motives to a single purpose: individual gain. Concluding that "an overarching methodology is rendered impossible by the fragmented nature of discourse-based knowledge," postmodernism denies the role of methodology altogether. Recently, a corrective "synthetic" approach has adopted a pluralistic approach to methodology, holding that the methodology of each economic school of thought should be analyzed critically on its own terms (Dow, 2001, pp. 66–67).
Political Science	Political science does not have a single big methodological device all its own, the way that many disciplines do. Rather, "political science as a discipline is defined by its substantive concerns, by its fixation on 'politics' in all its myriad forms" (Goodin & Klingerman, 1996, p. 7). More specifically, practitioners describe governments and examine ideas, normative doctrines, and proposals for social action (Hyneman, 1959, p. 28). Political scientists rely heavily on mathematical modeling and statistical testing. A method distinctive to political science is polling data on voter behavior. Like other social sciences, political science believes that "research should be theory oriented and theory directed," and that "findings [should be] based upon quantifiable data" (Somit & Tanenhaus, 1967, p. 178).
Psychology	There are two primary types of research: basic research to understand psychological processes, the primary goal of which is to increase knowledge, and applied research to find solutions for certain problems such as employee morale. Experiments are commonly employed, especially in basic research. Other applied researchers conduct evaluation research to assess the effects of social or institutional programs on behavior (Leary, 2004, p. 4).

Discipline	Methods
Sociology	The intellectual labor of sociology, not unlike other disciplines, is divided among theorists, methodologists, and researchers who use surveys, interviews, and observation. The effect of this cognitive separation in sociology is "that theorists do not deal with the relationship of theory to evidence" and, thus, to method (Alford, 1998, pp. 11–12). Methodologists are usually divided into quantitative and qualitative specialties. "Quants" are further divided between applied and theoretical statisticians. "Quals" are divided into ethno methodologists, symbolic interactionists, grounded theorists, historical methodologists, and ethnographers, each having its own specialized terminology and research techniques. Researchers analyze the substantive problems defined as part of the discipline's subfields of criminology, demography, social stratification, political sociology, the family, education, and the sociology of organizations (Alford, 1998, pp. 1, 11). Though sociology has been long dominated by modernist approaches to research, this is being seriously challenged by methodologies inspired by the humanities that are qualitative (i.e., meaning-based), constructionist, interpretative, narrative, and contextualized (situated in power, race, and gender). Qualitative research methods do not rely heavily on mathematical and statistical analysis but "study people in their natural setting and attempt to make sense of phenomena in terms of the meanings that people bring to them" (Dorsten & Hotchkiss, 2005, p. 147).

Source: Adapted from Szostak, R. (2004). *Classifying science: Phenomena, data, theory, method, practice.* Dordrecht: Springer.

Methods of the Natural Sciences. All the natural sciences use what is often called the "scientific method."[12] Interdisciplinarians should be aware that there are more than a dozen scientific methods used in the scholarly enterprise (see "Methods" in Table 2.11). The phrase *scientific method* can loosely be understood to mean careful, quantitative, hypothesis-driven research, but often is interpreted to recognize only experimental research.

The **scientific method**, defined narrowly, has four steps: (1) observation and description of phenomena and processes; (2) formulation of a hypothesis to explain the phenomena; (3) use of the hypothesis to predict the existence of other phenomena, or to predict quantitatively the result of new observations; and (4) execution of properly performed experiments to test those hypotheses or predictions. The scientific method is based on beliefs in empiricism (whether the observation is direct or indirect), quantifiability (including precision in measurement),[13] replicability or reproducibility, and free exchange of information (so that others can test or attempt to replicate or reproduce).

The scientific method assumes that there is a single explanation of how phenomena that appear to be separate entities are intrinsically unified (Donald, 2002, p. 32). Similarly, the assumption underlying interdisciplinarity is that conflicting disciplinary insights into a complex problem can be intrinsically unified by modifying or creating an underlying common ground concept, assumption, or theory. This assumption is unlike the assumption underlying the "scientific method," however, in that the resulting disciplinary general "law" is applicable to all similar phenomena, whereas the resulting interdisciplinary understanding is "local" and limited to the problem at hand.

Not all the sciences use the scientific method in the same way. The physical sciences, such as physics and chemistry, use experiments to gather numerical data from which relationships are identified and conclusions are drawn. Yet geologists and cosmologists can generally not employ experiments and thus rely instead on careful observation of physical objects. Among the differences that Table 2.12 addresses are what each discipline considers to be data and how each gathers and processes data. For example, chemistry's approach to research is quite similar to that of the other physical sciences, such as physics and Earth science, in that it seeks to measure and describe observed phenomena.

Methods of the Social Sciences. The social sciences use modernist scientific techniques, such as mathematical models and statistical analysis of empirical data, in conducting much of their research. The more descriptive social sciences, such as anthropology, may use qualitative methods that involve gathering information by making visual observations or interviewing and using "thick (that is, detailed) description" to record this information.

But modernist and quantitative approaches have lost force in recent decades largely because of developments in the philosophy of science and the rise of postmodernism. It is in methodology that postmodernism is having its greatest impact on the social sciences and the humanities by "deflating the confidence previously held in the capacity to identify best practice" (Dow, 2001, p. 66). Today, says H. Russell Bernard (2002), "the differences *within* anthropology and sociology with regard to methods are more important than the differences *between* those disciplines" (p. 3). Consequently, the description of methods in Table 2.13 reflects both modernist and postmodernist approaches.

Methods of the Humanities. Researchers working in the humanities draw on fields of scholarship in which different beliefs hold. Table 2.14 shows that the humanities rarely insist on quantifying observations. Part of the challenge of interdisciplinary integration (introduced in Chapter 8) is reducing conflict between insights by modifying their concepts and/or assumptions. Once these insights are prepared for integration, constructing a more comprehensive understanding is possible. Whereas the natural and social sciences leave the integration of knowledge out of the scientific method altogether and the humanities leave it up to the reader, viewer, or listener to integrate knowledge, interdisciplinary studies strives to achieve integration. The scholarly enterprise needs both specialized and integrative research.

Discipline	Methods
TABLE 2.14 ● Research Methods Associated With the Humanities	
Art History	Modernist art historians examine art objects in terms of the artists' mastery of appropriate technique, their structure and meaning within particular historical, political, psychological, or cultural contexts. Formalist analysis of a work of art, for example, considers primarily the aesthetic effects created by the component parts of the design, while iconography studies focuses on content rather than form. Two methodological reactions against formalism are Marxism, which studies the economic and social context of art, and feminism, which is predicated on the idea that gender is an essential component to understanding art. Biographical and autobiographical methods rely on texts (if they exist) and approach works of art in relation to the artist's life and personality. Semiotics, a recent methodological approach derived from linguistics, philosophy of language, and literary criticism, assumes that cultures and cultural expressions such as language, art, music, and film are composed of "signs" and that each sign has a meaning beyond, and only beyond, its literal self (Bal & Bryson, 1991, p. 174). Other approaches include deconstruction, which assigns meaning according to contexts that themselves are continually in flux, and the complex psychoanalytic method that deals primarily with unconscious significance of works of art (Adams, 1996). Postmodern critics, who see the artist as deeply implicated in society, "reject formal analysis and tend to discuss artworks not as beautiful objects produced by unique sensibilities but as works that exemplify society's culture, especially its politics" (Barnet, 2008, p. 260).
History	Historians engage in research that involves identification of primary source material from the past in the form of documents, records, letters, interviews, oral history, archaeology, and so forth, or secondary sources. They also practice critical analysis involving interpretation of historical documents and forming these into a picture of past events or the quality of human life within a particular time and place. To write good history, historians need a combination of well-reasoned arguments based on solid evidence combined with objectivity and interpretive scrutiny. In the twentieth century, the narrative, event-oriented history characteristic of nineteenth-century professional historiography gave way to "various kinds of social science–oriented history spanning the methodological and ideological spectrum from quantitative sociological and economic approaches and the structuralism of the Annales School to Marxist class analysis" (Iggers, 1997, p. 3). As applied to history, postmodernists question whether there are objects of historical research accessible to clearly defined methods of inquiry, asserting that every historical work is a literary work because historical narratives are verbal fictions, the contents of which are more invented than found (pp. 8–10).
Literature (English)	Research methods emphasize the centrality of texts and include auto/biographical, oral history, critical discourse analysis (i.e., analyzing patterns in language) for exploring visual signs (e.g., illuminations of manuscripts, graphic novels, photographs), computer-aided discourse analysis, ethnography (concerns cultural and social practices), quantitative analysis (i.e., how numbers are used as interpretive tools and as a means of calculating the frequency with which certain words occur and the contexts in which they are set), textual analysis that sees meaning making as a relational process (which relies on other research methods such as feminist and deconstructionist ones), interviewing living authors, and creative writing (which must be accompanied by a theoretical piece of writing) (Griffin, 2005, pp. 1–14). Literary theories are also approaches to literature and include New Criticism (that insists on the preeminence of the text itself and its literary properties), psychoanalytic criticism, reader-response criticism, structuralism, deconstructionist criticism, Marxist criticism, feminist criticism, Bakhtinian criticism, Foucaultian criticism, and multicultural criticism that takes seriously the cultural perspectives of minorities (Bressler, 2003).

(Continued)

TABLE 2.14 ⬡ (Continued)	
Discipline	**Methods**
Music Education	Music education research is multimethod. Positivist scholars use expertise (i.e., mastery of techniques involved in the production of works of art) and criticism (i.e., interpretation of compositions in terms of their aesthetic qualities, techniques employed, and their meaning within specific historical, political, psychological, or cultural contexts). However, the basic trend in music education research methods is on interpretivist forms of inquiry (i.e., phenomenology, action research, ethnography, narrative inquiry that focuses on human actions, beliefs, values, motivations, and attitudes), critical theory (that stresses that teaching and learning are deeply related to social practices and injustices), feminist or gender studies (that argues that gender issues are implicit in all research methods and interpretations, although most in the field reject the idea of a distinctly feminist research methodology), and postmodernism (that rejects the idea of "methods" altogether and holds that there are no rules of procedure that must be followed and no "right" procedures of investigation, though it does embrace introspection, individualized interpretation, and deconstruction) (Elliott, 2002, pp. 85–96).
Philosophy	The method of philosophy is the making and the questioning of distinctions (a distinction is a difference displayed). Philosophy explains by distinguishing between concepts, as for example, how responsible action is to be distinguished from the irresponsible (Sokolowski, 1998, p. 1). Philosophers use a variety of techniques to examine a written composition including dialectic, syllogism and logic, contemplation, linguistic/symbolic analysis, argument and debate, and also thought experiments.
Religious Studies	Scholars of religion employ a variety of research and analytical methods that cut across disciplinary lines when examining religious phenomena, religious actions, religious groups, and religious ideas. The methods used by researchers are largely dictated by the questions they ask and the issues they seek to explore. The common ground among scholars of religion is their efforts to describe and explain religious phenomena as an aspect of human culture and experience and do this by engaging in self-reflection, self-criticism, self-censorship, and self-control (Stone, 1998, pp. 6–8).

NOTE TO READERS

Undergraduate interdisciplinarians are highly unlikely to apply disciplinary methods themselves. Their challenge is to critically analyze, interpret, and apply insights produced by disciplinarians wielding those methods. Graduate students and even more senior scholars acting as solo interdisciplinarians may still not apply disciplinary methods themselves other than to identify and examine linkages among the insights of contributing disciplines. However, they would apply disciplinary methods if, as part of their integrative work, they choose to conduct their own basic research. Interdisciplinary teams may well employ such methods as they conduct basic research.

Chapter Summary

This chapter provides information that is foundational to interdisciplinary practice and critical to developing adequacy in contributing disciplines (STEP 5; see Chapter 6) and evaluating their insights (STEP 6; see Chapter 7). It explains the role of the disciplines and defines disciplinary perspective to mean a discipline's worldview as well as its defining elements (i.e., phenomena, epistemology, assumptions, concepts, theory, and method). How perspective is used depends on what STEPS are being performed. The chapter also provides two ways of beginning interdisciplinary inquiry. One is Szostak's classification approach that involves linking the topic to the appropriate phenomena. The virtue of this approach is that it enables researchers to identify more readily neighboring phenomena that may otherwise be overlooked but may well be relevant to the problem. Researchers, then, can broaden their investigation without focusing, at least initially, on particular disciplines. The other, the traditional perspectival approach, involves linking the problem to those disciplines whose perspectives embrace it. Researchers can profitably use both approaches to identify disciplines relevant to the problem and then delve deeply into their scholarship, thus countering the occasional criticism that interdisciplinary studies is shallow and lacks rigor. Using both approaches shows that interdisciplinary analysis can be systematic and cumulative.

Armed with this basic knowledge of the disciplines, their perspectives, and their defining elements, students are now able to identify the disciplines relevant to the problem. Making this decision is STEP 3, the subject of Chapter 4.

Notes

1. "Most disciplines tend to think that what they study is the most important part of reality (that is, their worldview isn't just about what they study but [is about] its role in the larger whole)" (Rick Szostak, personal communication, January 11, 2011).

2. Early on, interdisciplinarians such as Newell and Green (1982) opted for a narrow definition: "Disciplines are distinguished from one another by the questions they ask about the world, by their perspectives or world view, by the set of assumptions they employ, and by the methods which they use to build up a body of knowledge (facts, concepts, theories) around a certain subject matter" (p. 24). According to this definition, "perspective" is but one of four primary disciplinary elements and is co-equal with the questions that the disciplines ask about the world, the set of assumptions they employ, and the methods they use to build up a body of knowledge (facts, concepts, theories).

3. Newell (1992) argues that "perspective" should be defined in broader terms, even suggesting that it is the source of all other disciplinary elements. He refers to "perspective" as that "from which those concepts, theories, methods, and facts emerge" (p. 213). He adds, "The interdisciplinary researcher must understand how the relevant concepts, theories, and methods underlying each-discipline's perspective are operationalized" (1998, p. 545). Janet Donald (2002) apparently agrees, emphasizing that "to understand a field of study [i.e., a discipline], students must learn its perspectives and processes of inquiry" (pp. xii–xiii). By "perspective," she means a discipline's epistemology,

(Continued)

(Continued)

vocabulary, theory, and methods or processes of inquiry (pp. 2, 8). Jill Vickers (1998) states interdisciplinarians "must accept that the different disciplines have different cognitive maps" (p. 17). For Hugh Petrie (1976), reliable borrowing from the disciplines requires that the interdisciplinarian know quite a lot about the "cognitive and perceptual apparatus utilized" (p. 35).

4. Rogers, Scaife, and Rizzo (2005) explain that much of this discord within disciplines "may owe more to internal political agendas than we would like [to admit]" (p. 268). The reason why scholarly disciplines do not become inert and settled, explains Marjorie Garber (2001), is "the disciplinary libido," meaning the ways in which disciplines seek to differentiate themselves from each other while at the same time desiring to become "its nearest neighbor, whether at the edges of the Academy (the professional wants to become an amateur and vice versa), among the disciplines (each covets its neighbor's insights), or within the disciplines (each one attempts to create a new language specific to its objects, but longs for a universal language understood by all)" (p. ix).

5. Among the applied fields and professions, Geertz (1983) includes education, communications, criminal justice, management, law, and engineering (p. 7). Elsewhere, Geertz (2000) characterizes these broad categories as "rather baggy" because of their indeterminacy (p. 156). Mary Taylor Huber and Sherwyn P. Morreale (2002) use the term *disciplinary domains*, referring to the humanities, the social sciences, and the sciences (p. 8). The use of the term *core disciplines* implies a hierarchy of knowledge that many would contest (Salter & Hearn, 1996, p. 6).

6. Members of various disciplines will likely find the descriptions of their respective disciplines not comprehensive enough. But experience using these descriptions in interdisciplinary classrooms validates their intended purpose: to point the student to those disciplines that are potentially relevant to the problem. Once these are identified, the student should then consult each discipline's research aids, many of which are cited in tables in this chapter. These include handbooks, companions, journals, and bibliographies to validate the relevance of each discipline to the problem.

7. For example, Alan Bryman (2004) states, "There is no agreement on the epistemological basis of the natural sciences" (p. 439). Competing epistemological values in biology, for example, are fueling the debate over how much can be learned in the laboratory versus how much can be learned in the field—in other words, what constitutes "good science." Admittedly, there is some overlap between assumptions, epistemology, and preferred method in the tables.

8. Empiricism has come under fire from postmodernists, particularly from feminist philosophers of science who identify a role for value judgments in science and advocate tolerance and willingness to encourage a variety of approaches and multiple judgments of significance to the same scientific problem (Rosenberg, 2000, p. 183).

9. R. N. Giere (1999) calls this philosophical approach "naturalistic realism" and states that it is closest to the actual mindset that most scientists take.

10. Polkinghorne (1996) says that philosophers of science, if not practicing scientists, now accept that scientific methods can neither prove nor disprove any theory (or even any narrow hypothesis). Nevertheless, the application of scientific methods to theories provides scientists with invaluable, if imperfect, evidence with which they can judge whether a theory is in accord with reality (pp. 18–19).

11. Szostak (2004), in *Classifying Science: Phenomena, Data, Theory, Method, Practice*, is careful to distinguish methods "from techniques or tools, such as experimental design or instrumentation, or particular statistical packages" (p. 100). Tools and techniques and so on are a subset of methods.

12. Alexander Taffel (1992) states, "The combination of activities in which scientists engage to achieve the understanding they seek is sometimes called the scientific method. There is however no single method of science, but rather a variety of activities in which scientists use different combinations of these to solve difficult problems. Scientific activities include recognizing and defining problems, observing, measuring, experimenting, making hypotheses and theories, and communicating with other scientists" (p. 5).

13. Modern science relies heavily on statistical methods in the testing of hypotheses (Rosenberg, 2000, p. 112). Taper and Lele (2004) discuss the two schools of statistical thought, frequentist and Bayesian, and how these approaches impact quantitative statements. "There cannot be such a thing as quantification of support for a single hypothesis," they argue. Scientific evidence "is necessarily comparative," meaning that "one needs to specify two hypotheses to compare, and data may support one hypothesis more than the other" (p. 527).

Exercises

About Disciplinary Perspective

2.1   This chapter has said that a discipline's perspective is like a lens through which it views reality.

Identify three relevant disciplinary perspectives and describe how they might view each of the following:

- Offshore drilling for oil and gas
- Urban sprawl (e.g., building subdivisions and shopping centers on farmland)
- Income inequality
- Border security

How Knowledge Is Typically Reflected in the Organization of the Academy

2.2   How is knowledge reflected in the organization of your university? Where do the applied fields such as hospitality, architecture, etc., fit into the organization?

Disciplinary Perspective

2.3   How does juxtaposing different or even conflicting perspectives aid one's understanding of a complex problem, event, or behavior?

Phenomena

2.4   Disciplines, we have said, often share interest in the same phenomenon. Which disciplines would likely share an interest in these phenomena?

(Continued)

(Continued)

- Extreme drought in sub-Saharan Africa
- The Israeli–Palestinian conflict
- A performance of Shakespeare's *Hamlet*

2.5 Using Szostak's "Categories of Phenomena About the Human World" (see Table 2.4), identify subphenomena that are likely connected to these problems, topics, or issues:

- Gang violence
- The disintegration of the Ross Ice Shelf in Antarctica
- Student debt

Epistemological Approaches

2.6 What are the logical limits of postmodernist epistemology?

2.7 Describe the strengths and weaknesses of *modernist* and *postmodernist* epistemology when each tries to explain the rise of religious fundamentalism in the Middle East?

Assumptions

2.8 What might be the assumptions of these researchers:

- Those working in the natural sciences about the cause of increasing volcanic activity in Indonesia?
- Those working in the social sciences about the cause of population decline in the developed nations of the world?
- Those working in the humanities and influenced by "the new generalism" about the meaning (in a general sense) of violent lyrics in some genres of music?

Quantitative and/or Qualitative

2.9 Here are research topics that might be addressed by either quantitative or qualitative methods. For each one, describe how you would conduct either a quantitative study or a qualitative study, and explain which approach would most likely lead to a more comprehensive understanding of the topic:

- Policing in urban high-crime neighborhoods
- High unemployment among 18- to 24-year-olds

DRAWING ON DISCIPLINARY INSIGHTS

PART II

Image by Pexels from Pixabay.

BEGINNING THE RESEARCH PROCESS

LEARNING OUTCOMES

By the end of this chapter, you will be able to

- Describe the integrated model
- Describe STEP 1
- Describe STEP 2

GUIDING QUESTIONS

What is the interdisciplinary research process?

How do you perform the first two STEPS of this process?

(*Note:* This material is best learned while being applied.)

CHAPTER OBJECTIVES

Today, there are many types of interdisciplinarity being practiced in the United States, Canada, Europe, Australia, and elsewhere. We might thus ask whether it makes sense to speak of one interdisciplinary research process. The answer is yes. Different types of interdisciplinarity essentially make different decisions *within an overarching research process.*

The chapter presents the integrated model of the interdisciplinary research process (IRP) and explains its defining characteristics. We shall see that interdisciplinary research can be performed by individuals or teams. The chapter also introduces the first two "STEPS" or decision points that the model calls for: Define the problem or state the research questions (STEP 1), and justify using an interdisciplinary approach (STEP 2).

THE INTEGRATED MODEL OF THE INTERDISCIPLINARY RESEARCH PROCESS

When driving to an unfamiliar place away from home, travelers rely on Global Positioning System (GPS) to avoid unproductive, time-consuming detours. Similarly, when proceeding from a problem to an understanding of the problem, interdisciplinarians need a map

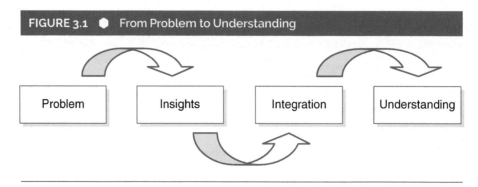

FIGURE 3.1 ● From Problem to Understanding

to guide them through the interdisciplinary research process or IRP. The IRP in its most simplified form is shown in Figure 3.1.

Although helpful, Figure 3.1 lacks the detail necessary to proceed from the problem to the understanding. This book presents a detailed model of the IRP (shown in Table 3.1) to serve as a GPS. The model presented here integrates the prominent models of the IRP.[1] Using 10 STEPS, it provides a proven approach to conducting interdisciplinary research, finding new meaning, and creating new knowledge. Unlike a GPS that tells you when to turn and which way, the IRP can only tell you when to make a decision.

The 10 STEPS clarify the "points of decision" or "operations" that are taken in almost any interdisciplinary research project. While those working in the "softer" social sciences and in the humanities may stress the elements of intuition, creativity, and art in the research process over STEPS, the IRP, especially the integrative part of it, involves intuition *and* method, creativity *and* process, art *and* strategic decision making.

Each STEP will be carefully explained in subsequent chapters. The first couple of STEPS focus on developing a good research question. The next four STEPS involve identifying and evaluating disciplinary insights. STEPS 7 through 9 focus on integrating across disciplinary insights. The final STEP urges reflection, testing, and communication of findings. (*Note:* Descriptions of creative processes often speak of four steps—preparation, incubation, illumination, and verification—which bear a strong similarity to these four groups of steps. The IRP is thus a creative process [Szostak 2017a].)

Dividing what is essentially a fluid process into distinct STEPS may give the misleading impression that these STEPS do not overlap. They often do. For example, the cursory literature search begins during STEP 1 and continues over the next STEPS until the full-scale search is completed in STEP 4 (see Chapter 5). Some researchers begin conducting the full-scale literature search (shown as STEP 4) as soon as STEP 1, and some continue

TABLE 3.1 ● The Integrated Model of the Interdisciplinary Research Process

A. Drawing on disciplinary insights[a]
1. Define the problem or state the research question.
2. Justify using an interdisciplinary approach.
3. Identify relevant disciplines.
4. Conduct the literature search.
5. Develop adequacy in each relevant discipline.
6. Analyze the problem and evaluate each insight or theory.
B. Integrating disciplinary insights
7. Identify conflicts between insights and their sources.
8. Create common ground between insights.
9. Construct a more comprehensive understanding.
10. Reflect on, test, and communicate the understanding.

Source: Repko, A. F. (2006). Disciplining interdisciplinarity: The case for textbooks. *Issues in Integrative Studies, 24*, 112–142.

a. The term *disciplinary insights* includes insights from disciplines, subdisciplines, interdisciplines, and schools of thought.

the search while performing later STEPS. *It is good to consider STEP 4 as a fluid process within the overall research process, especially in its early phases.*

For each STEP, we will provide a set of strategies or guidelines that have proven useful to researchers in the past, and we will provide examples of their use by scholars, practitioners, and students across the natural sciences, the social sciences, and the humanities (see Mokari Yamchi et al. [2018] for one recent recommendation of this approach in the field of food security). There is a set of common challenges—such as grappling with differences in disciplinary perspective (Chapter 2)—that are faced by interdisciplinary researchers addressing any topic, and thus a common set of strategies for confronting these. These STEPS and strategies reflect considerable consensus within the community of scholars of interdisciplinarity, although of course additional useful strategies may be discovered in the future. Some scholars are nevertheless hesitant to embrace an entire interdisciplinary research process for fear that this will somehow constrain the freedom of interdisciplinary research (Box 3.1). Significantly, the STEPS and strategies outlined in this book are shown to be inherently flexible.

BOX 3.1

Interdisciplinarians generally agree on the need to specify, at least to some extent, how to draw on disciplinary expertise and, especially, how to integrate disciplinary insights and theories. Those who oppose any greater specificity in research methodology do so reasoning that it might constrain freedom of activity, stifle creativity, or prevent interdisciplinarity from functioning as the antidote to restrictive disciplinary perspectives (Szostak, 2012, p. 4). Neither freedom nor creativity is compromised by providing some structure and direction to the research process. What these critics overlook is that all research, including interdisciplinary research, uses some method or strategy to approach a problem. While disciplinary methodologies generally involve a preference for certain theories, methods, and phenomena, the IRP encourages researchers to cast their gaze across *all* relevant theories, methods, phenomena, and insights. The IRP does not constrain research in the way that disciplines do.

KEY CHARACTERISTICS OF INTERDISCIPLINARY RESEARCH

Interdisciplinary research is a decision-making process that is heuristic, iterative, and reflexive. Each of these terms—*decision making, process, heuristic, iterative,* and *reflexive*—requires explanation.

It Involves Decision Making

Decision making, a uniquely human activity, is the cognitive ability to choose among alternatives. Decision making is complicated by the prevalence of complex problems in our personal lives, in business, in society as a whole, and in the international realm. Interdisciplinarity focuses on complex problems or questions. A characteristic of these is that there are many variables involved, each of which may be studied by a different discipline, subdiscipline, interdiscipline, or school of thought. The **interdisciplinary research process (IRP)** is a practical and demonstrated way to make decisions about how to approach these problems, decide which ones are appropriate for interdisciplinary inquiry, and construct comprehensive understandings of them (Newell, 2007a, p. 247).

It Is a Process

Doing interdisciplinary research, whether performed individually or collaboratively, is a process (Newell, 2007a, p. 246). **Process** means following a procedure or strategy.

Interdisciplinary research has in common with all disciplinary research an overall plan or approach. Reduced to its simplest terms, all applied research has these three steps in common:

- The problem is recognized as needing research.

- The problem is approached using a research strategy.

- The problem is solved or at least a tentative solution is devised.

Each discipline has developed its own methods and preferred research strategy, as noted in Chapter 2. Likewise, interdisciplinary studies has developed a research process that differs in important respects from disciplinary methods *and subsumes them,* as shown in Figure 3.2. *The IRP is an overarching research process (noted by the arching line) that draws on disciplinary perspectives and their insights that are relevant to the problem.* The process of interdisciplinary research is necessarily distinct from the processes employed in disciplinary research because integration is at the very core of interdisciplinary activity, whereas it is not at the core of disciplinary activity.

It Is Heuristic

A **heuristic** is an aid to understanding or discovery or learning. A heuristic does not provide you with an answer but guides you to seek solutions in an effective way. The IRP is heuristic in that it places you, the student, in the role of the discoverer of knowledge. You learn how to approach the problem either by yourself or as part of a group. The IRP aids discovery by asking you to make decisions at each STEP. You as the researcher will

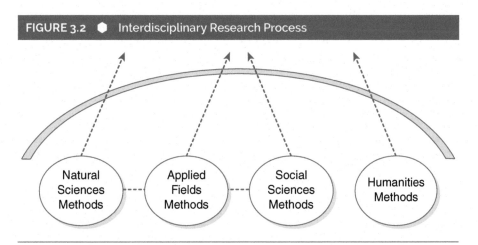

FIGURE 3.2 ● Interdisciplinary Research Process

Note: The dotted lines connecting the applied fields to the natural sciences and the social sciences show that the applied fields (such as education, criminal justice, communication, law, and business) use research methods drawn from these other disciplinary categories.

still have much scope for reason, experimentation, trial and error, and creativity in the exercise of each STEP. If some find interdisciplinary research "a tall order," it is probably because so much academic learning is "rote learning." This is learning that occurs when the learner memorizes new information without relating it to prior knowledge, or understanding the theory underlying it. Rote learning involves no attempt to integrate new knowledge with existing concepts, experience, or objects (Novak, 1998, pp. 19–20). The standard Western approach to education lacks guides to integration and holism, a deficiency that interdisciplinarity addresses.

The IRP is student friendly. Practitioners in several nations are successfully teaching the material in this book. Students who apply themselves will master this important and new way of approaching complex problems and framing new and creative solutions to them that otherwise would be impossible using a disciplinary or multidisciplinary approach.

It Is Iterative

The IRP is **iterative** or procedurally repetitive. Although the research process features decision making and STEP taking, the process is by no means linear. That is, the process is not a simple matter of moving from point *A* to point *B* to point *C* and on to the end. Rather, when you get to point *B,* you may discover that you need to revisit and revise the decision you made at point *A.* In fact, revising work performed at earlier STEPS is likely to happen at any given point in the process. For example, the process of selecting the most relevant disciplines (STEP 3; see Chapter 4) may lead to restating the problem identified in STEP 1. And you may revisit the literature search as you perform later STEPS. *Throughout the research process, you should expect to revisit earlier work.*

Interdisciplinary researchers rely on "systems thinking" to approach a problem creatively, thinking about it "outside the box" without being influenced by solutions attempted in the past, and viewing it from different perspectives. **Systems thinking** is a method of visualizing interrelationships within a complex problem or system by (1) breaking it down into its constituent parts, (2) identifying which parts different disciplines address, (3) evaluating the relative importance of different causal linkages, and (4) recognizing that a system of linkages is much more than the sum of its parts. (We discuss this approach further in Chapter 4.) **Feedback loops**, depicted in Figure 4.1, are central elements of systems thinking. They describe the process that requires the researcher to periodically revisit earlier activity. **Feedback** is corrective information about a decision, an operation, an event, or a problem that compels the researcher to revisit an earlier STEP. This corrective information typically comes from previously overlooked scholarship. As you proceed, periodically ask questions such as

- Have I defined the problem or the question too broadly or too narrowly?

- Have I correctly identified the parts of the problem?

- Have I identified the disciplines most relevant to the problem?

- Have I gathered the most important insights concerning the problem?

- Am I privileging one discipline's literature or terminology over another's simply because I am more comfortable working in the discipline?

- Have I allowed my personal bias to shape the direction of the study?

It Is Reflexive

The IRP is also **reflexive**, meaning that we become self-conscious or self-aware of our disciplinary or personal biases that may influence our inquiry and possibly skew our evaluation of insights and thus bias the end product. As you make decisions about which insights to use and which to discard, avoid the temptation to eliminate insights or theoretical approaches that are unfamiliar to you or that challenge your beliefs. In fact, you should expect that your biases will be challenged throughout the research process.

Two Cautions Concerning the STEPS

First, it is tempting to avoid difficult STEPS and leap ahead to later STEPS. By keeping in mind the STEPS of the model, researchers are more likely to realize that they have skipped over an important STEP and need to return to complete it. Since each STEP typically requires at least preliminary completion of previous STEPS, it is important to regularly reexamine the work done in earlier STEPS. For example, you might be tempted not to spend much time developing adequacy in the disciplines relevant to the problem (STEP 5; see Chapter 6) and proceed with analyzing the problem and evaluating disciplinary insights into it (STEP 6; see Chapter 7). This impatience to "get on with the project" can prove costly, however. Unless you know what specific information to look for when developing adequacy—the discipline's relevant concepts, assumptions, theories, and methods—the time and effort invested may fail to yield the quality information that you need to perform later STEPS. Ultimately, you have to develop adequacy in each relevant discipline before reading and comprehending the discipline's insights profitably. Avoiding difficult STEPS and decisions will make the task of modifying insights and then integrating them problematic.

Second, describing the IRP in terms of STEPS may give the impression that each relevant discipline is "mined separately for nuggets of insights before any integration takes place, and that when integration occurs, it happens all at once" (Newell, 2007a, pp. 248–249). Nothing could be further from actual practice. You should partially integrate as you go, meaning that you should incorporate disciplinary insights or theories into a broader understanding of the problem as you proceed (p. 249).

In the end, each interdisciplinary research project presents a unique combination of challenges and opportunities. The many examples of professional work and exemplary student projects threaded throughout this book clearly show the variety of ways to creatively do interdisciplinary research. The critical differences between interdisciplinary and disciplinary approaches to research are noted along the way.

NOTE ON TEAM RESEARCH

A guiding premise of this book is that individual students and scholars can perform interdisciplinary research. They need not have the same level of expertise in a discipline as a researcher who specializes in that discipline to draw intelligently on that discipline for the purposes of interdisciplinary analysis. It is nevertheless true that large-scale interdisciplinary research is often performed in teams. Team members may each bring expertise with respect to different disciplines or theories or methods or phenomena.

Although teams have the potential advantage of bringing diverse perspectives and expertise to bear on a single problem, they inevitably face all of the challenges inherent in interdisciplinary communication. Team members may attach different meanings to words. These differences may not always be obvious and may lead them to think that they agree when they do not, or that they disagree when they do not. Team members will also bring different perspectives. This is another important source of miscommunication, for one team member may make assumptions that are not made by others from a different discipline.

The Toolbox project (long centered at the University of Idaho) has tackled the second challenge. The project managers give questionnaires—regarding epistemological and methodological issues—to members of interdisciplinary research teams. They then discuss with the research team how and why their answers differ. When the questionnaire is given again, there is usually some convergence in answers: Team members come to respect the views of others and move away from extreme attitudes. Most importantly, team members report that the exercise enhanced team communication: Each member had a better sense of where other team members were coming from (see Looney et al., 2014; it is in a volume by O'Rourke et al. that addresses communication challenges more generally). The lesson for interdisciplinary communication is that it is very useful to be explicit about the nature of disciplinary (and other) perspectives. Much of this book seeks to do precisely that. Students will thus be better prepared for teamwork later in life.

In addition to these cognitive challenges—dealing with differences in definitions and perspectives—there are also psychological challenges. Team members must get along and respect each other, and individual team members must do their share of the work. These challenges may be exacerbated in interdisciplinary research if some team members feel—perhaps subconsciously—that their discipline (or favorite theory or method) is somehow superior.

Team members must respect other perspectives and be curious to learn about them. They must more generally have considerable intellectual curiosity. They must have the courage to reflect on their own hidden assumptions as they interact with others. They must be willing to cope with complexities and uncertainties (see Misra, Hall, Feng, Stipelman, & Stokols, 2011). Team members should also be collaborative, responsible, and have good time-management and information-management skills.

There are various strategies for encouraging positive team outcomes. There must be many collaborative conversations, but also clear tasks, for each team member to perform between conversations. There should be opportunities for the exploration of differences in definition and perspective. Individual team members should feel that they will be rewarded for their effort and collaboration. If a team leader is chosen, this should be a person who is respectful of all team members and good at providing constructive encouragement (and constructive criticism if necessary).[2] It is best if the team is formed at the start of the interdisciplinary research process: If the team does not agree upon—and fully understand—the research question(s) then collaboration in later steps is unlikely.

Instructors in interdisciplinary courses may encourage students to work in teams. They may require a group project or group presentation in class. They might employ group exercises in class. For example, students in a group might each be asked to sketch an interdisciplinary research question on a piece of paper. As the papers are passed around, each subsequent student seeks to clarify the question. This exercise could be performed either before or after guidelines for a good interdisciplinary research question are addressed in class. Similar exercises can be pursued at each step in the interdisciplinary research process. Students will experience the advantages of having different minds work together on a single project. They will likely also experience some of the communication challenges inherent in interdisciplinary research. Even if explicit teamwork is not encouraged, class discussions regarding the challenges students are facing at each step in the interdisciplinary research process are an invaluable strategy for learning both about interdisciplinary research and about the value of bringing multiple perspectives to bear on a particular challenge. (More details on this sort of group exercise can be found at https://i2insights.org/2019/03/12/idea-tree-brainstorming-tool/.)

STEP 1: DEFINE THE PROBLEM OR STATE THE RESEARCH QUESTION

This graphic shows the STEPS of the research process. We highlight STEP 1 and bullet point the decisions that it involves.

A. DRAWING ON DISCIPLINARY INSIGHTS

1. **Define the problem or state the research question.**
   - o **Select a problem or pose a question that is complex and requires drawing on insights from more than one discipline.**
   - o **Define the scope of the problem or question.**
   - o **Avoid three tendencies that run counter to the IRP.**
   - o **Follow three guidelines for stating the problem or posing the question.**
2. Justify using an interdisciplinary approach.
3. Identify relevant disciplines.
4. Conduct the literature search.
5. Develop adequacy in each relevant discipline.
6. Analyze the problem and evaluate each insight or theory.

B. INTEGRATING DISCIPLINARY INSIGHTS

7. Identify conflicts between insights and their sources.
8. Create common ground between insights.
9. Construct a more comprehensive understanding.
10. Reflect on, test, and communicate the understanding.

Defining the problem or stating the research question is the first and most basic activity that one undertakes in conducting research or engaging in problem solving of any kind. It is also the STEP that often takes considerable time and effort because you do not yet know much about the problem or even if it is researchable in an interdisciplinary sense. For this reason, you should expect to revisit your definition of the problem or statement of the research question as you take additional steps.

Select a Problem or Pose a Question That Is Complex and Requires Drawing on Insights From More Than One Discipline

A problem is ripe for interdisciplinary study when

- it is complex (i.e., requires insights from more than one discipline), and

- it is **researchable in an interdisciplinary sense** (i.e., authors from at least two disciplines have written on the topic or at least on some aspect of it).

If you have trouble telling *in advance* whether a problem is complex, a useful *initial test* is to ask whether there is more than one legitimate way to look at the problem and, if so, which disciplines would likely be interested in it. Referring to Tables 2.2 and 2.3 in Chapter 2 will aid you in making this tentative determination. (*Note:* A more detailed discussion of complexity as a criterion for interdisciplinary inquiry follows below.)

To decide with confidence whether a problem is researchable requires conducting the literature search (the subject of Chapter 5). A problem may be complex but for some reason has failed to generate scholarly interest outside a particular discipline. Such is the case with the problem concerning the effects of physician shortages on society. The problem appears to be complex and is certainly important to society. But for whatever reason, it has failed to attract much scholarly attention outside the field of medicine (although it is a subject of discussion in multiple arenas that draw on economic, sociological, political, and demographic perspectives). (*Note:* The discovery of such gaps in research opens the door to potentially fruitful interdisciplinary inquiry, although undergraduates may want to avoid such questions, while graduate students and scholars may see an opportunity.)

Frequently, the research problem that we would *like* to investigate cannot be the problem that we *can* investigate because the cursory literature search has failed to reveal relevant insights concerning it from two or more disciplines. Consequently, we must revise the problem, question, or topic based on material that the search has revealed.

How, though, do we identify a research question in the first place? Some students may begin with some burning question that has long troubled them, and find that it is suitable for interdisciplinary inquiry. Some instructors may encourage certain research questions. But many students will find it difficult to formulate a suitable question. They can take heart from the fact that even seasoned scholars sometimes struggle to identify good research questions. Students, like scholars, can develop questions by reading the existing literature in some area that they are curious about. What questions do authors raise but not answer? Do you have doubts about the conclusions they reach? If you read works on the same topic from multiple disciplines, do there seem to be disagreements or gaps in understanding between these? Note that this exploratory reading by its nature is a bit unstructured: You cannot know in advance what sort of question you might come up with. But you should have confidence that if you do read widely, you will come up with something. You are more likely to develop a good question if you care about the topic in which you are reading. We have hinted above that interdisciplinary research is a creative process: You may find even at this earliest step that an idea will suddenly pop into your head. But this will only happen if you are relaxed and confident enough that subconscious thought processes can generate a good idea. And good ideas only come to the prepared mind: You need consciously to think and read about an area of interest before you can subconsciously develop a good research question. It can be useful to map the connections among phenomena identified by different authors: Do you see novel connections, or perhaps a novel system of connections that you might investigate? Yet you may well find that you develop a question that is tangential to your original reading: Some side issue or related topic may catch your interest. Last but not least, you may be inspired by conversation: Talk to friends and fellow students and your instructor (or perhaps collaborators in a community service learning initiative) about your interests and you may collaboratively achieve a good research question. The sort of group exercise mentioned in our note on team research above may be useful here.[3]

Define the Scope of the Problem or Question

Once you have identified the topic or problem, your next decision is to define its scope. **Scope** refers to the parameters of what you intend to include and exclude from consideration. In other words, you are telling your readers how much of this problem you plan to investigate. For example, if the problem is repeat spousal battery, how will you approach it? Will you focus on the *causes* of repeat spousal battery or on the *prevention* of repeat spousal battery? Will you research the *treatment* of the perpetrator and/or the victim of spousal battery? Or will you focus on the *effects* of repeat spousal battery on a particular demographic, say the children? Though all these options are clearly related to the overall problem of repeat spousal battery, narrowing the scope of the problem at the outset, to the extent possible, will facilitate the literature search and provide focus to subsequent STEPS in the research process. The extremes to be avoided are conceiving the problem too broadly so that it is unmanageable (such as investigating both the causes and effects of repeat spousal battery), and conceiving the problem too narrowly so that it is not interdisciplinary or researchable (such as focusing just on the psychological effects of spousal battery on the children).

Subsequent STEPS in the research process may require revisiting your initial statement of the problem or focus question and modifying it in some way. Here is an example (developed in a class) of how to transition from the *very broad topic* of "ways to prevent domestic violence" to a narrower and more focused *interdisciplinary statement* of the problem:

> The problem of domestic violence is broad, and developing strategies to prevent one of its most insidious manifestations—repeat spousal battery—is a pressing social need. Whereas single disciplinary approaches focus on only a single aspect of repeat spousal battery, an interdisciplinary approach that takes into account all aspects of the problem will hopefully lead to interventions that will mitigate this social scourge.

This transition from broad to narrow was possible after the class had read more widely about domestic violence and had begun to understand its complexity. This statement, which appeared in the introductory paragraph, was the product of several iterations, each of which was made after the class had taken additional STEPS in the IRP.

Avoid Three Tendencies That Run Counter to the IRP

In defining the problem or stating the research question, three tendencies should be avoided that may be acceptable in some academic settings but that run counter to the IRP: disciplinary bias, disciplinary jargon, and personal bias.

Disciplinary Bias

The first tendency is to engage in **disciplinary bias**, which means to state the problem using words and phrases that connect it to a particular discipline. For example, the problem statement "The Responsibility of Public Education for Sex Education" is biased in favor of education. Stating the problem in *discipline-neutral* terms makes it easier to justify using an interdisciplinary approach. By removing the disciplinary bias in the above example, the problem could be stated like this: "Sex Education in Public Education: An Interdisciplinary Analysis." Adding "An Interdisciplinary Analysis" alerts the reader that the problem is going to be approached from multiple disciplinary perspectives, not just the perspective of education.

Disciplinary Jargon

The second tendency is to use **disciplinary jargon**, which means using technical terms and concepts that are not generally understood outside a particular discipline. If a technical term or concept must be used in the statement introducing the problem, then it must be defined in the next sentence or two. *A rule of thumb is to assume that the reader is unfamiliar with a technical term or concept.* Here is an example of a statement that introduces a problem that is appropriate to interdisciplinary inquiry but contains disciplinary jargon: "The recidivism of domestic batterers is a significant problem in the United States because of the short-term and long-term psychological effects on the victim." This statement contains three technical terms that are probably unfamiliar to most readers and thus require definition: *recidivism, domestic batterers,* and *psychological effects.* (If the researcher wants to limit the investigation to "psychological effects on the victim," then a simple disciplinary approach will do. Otherwise, the statement should omit the term *psychological* to expand the search to other disciplines. Researchers must learn what terms mean and factor them into the interdisciplinary frameworks they construct.) Even disciplinary experts working on interdisciplinary research teams must first develop a common vocabulary before the work of research can begin.

The following are student-written examples of discipline-neutral statements introducing a problem that involves multiple disciplines:

- "Euthanasia is the intentional killing by act or omission of a dependent human being for his or her alleged benefit, be it voluntary or involuntary. The controversy over euthanasia was rekindled in the 1993 case of *Sue Rodriguez v. British Columbia (Attorney General)*, which involved a woman in her forties who was suffering from Lou Gehrig's disease and who wanted to choose the time and manner of her inevitable death." This student carefully defines what would otherwise be disciplinary jargon. From this wording, the reader can readily discern that the three disciplines deemed by the student as most relevant to the problem of euthanasia are ethics, medicine, and law.

- "Recent ACT scores show that a growing number of students are failing to grasp basic scientific knowledge. Science and technology play an integral role in modern society. Without scientifically and technically trained students, there will be a shortage of trained professionals in critical fields such as medicine, biology, engineering, and information technology. Even fields that are not normally thought of as scientific, including business, agriculture, journalism, and sociology, now rely heavily on science and technology." The disciplines that the student found to be most relevant to this topic are biology, psychology, and education.

In summary, the statement of an interdisciplinary problem should not privilege any one discipline. Using (perhaps unconsciously) disciplinary jargon or terminology tacitly favors one disciplinary perspective at the expense of another.

Personal Bias

A third tendency is to inject **personal bias** or one's own point of view when introducing the problem. While appropriate in many academic contexts, injecting your personal bias is not appropriate in most interdisciplinary contexts where the goal is quite different: to construct a more comprehensive understanding. Some fall into the trap of collecting evidence from various disciplines that supports their bias on the issue. This only adds their personal bias to the biased insights of disciplinary authors. *An interdisciplinary understanding cannot be "more comprehensive" if your personal bias dominates and if you exclude insights that differ from yours.*

We note personal bias in this introduction of the problem: "Taxpayer dollars should not be used to finance sports complexes for professional teams." The student obviously believes this and would evidently prefer to write a paper advancing this point of view. However, the interdisciplinarian is not to play the role of prosecuting attorney or defense counsel for the accused. *The role of the interdisciplinarian is to produce an understanding of the problem that is more comprehensive and more inclusive than the narrow and skewed understandings that the disciplines have produced.* This calls for approaching the problem with a frame of mind that is decidedly different from that of the disciplinarian. This frame of mind is one of neutrality (or at least suspended judgment) and objectivity until all the evidence is in. This means open-ness to different disciplinary insights and theories, even if these challenge your deeply held beliefs. *A defining characteristic of interdisciplinary work should be to mitigate conflict by finding common ground among conflicting perspectives, including your own.* A neutral question such as "Should taxpayer dollars subsidize sports complexes for professional teams?" would guide the researcher to evaluate competing arguments and seek common ground among them.

Follow Three Guidelines for Stating the Problem or Posing the Question

If a problem appears suitable for interdisciplinary inquiry, it should be phrased in conformity to these important guidelines:

- *The problem should be stated clearly and concisely.* This statement demonstrates lack of clarity: "The majority of complaints registered by the Childcare Licensing Agency (CLA) concern unsafe child care facilities." It is unclear what the focus of the investigation is: the complaints, whether or not they are valid, lack of enforcement of safety regulations by the CLA, lack of funding of the CLA by the federal government, or lack of legislation that establishes strict enforcement procedures. Sometimes greater clarity can be achieved by stating the problem as a question.

- *The problem or focus question should be sufficiently narrow to be manageable within the specified limits of the essay.* The problem of "Securing the southern border of the United States" was too broad for an essay requiring only three disciplinary perspectives. Upon discovering that the literature on border security was vast, the student narrowed the problem to the more manageable one of "Perspectives on securing the southern border of the United States against human smuggling: An interdisciplinary study."

- *The problem should appear in a context (preferably in the first paragraph of the introduction) that explains why it is important—that is, why the reader should care.* The following introduction (developed in class) places the problem of wife battery in a context that not only engages the reader but, more important, indicates why the problem warrants the reader's interest:

Wife battery is a widespread problem in the United States. It is urgent that a solution be found because of its devastating effects on the victim, including debilitating depression and redirected violence against her children. The wife's extended family and associates also feel the effects of her physical and emotional pain. Most tragically, studies show that children who grow up in abusive homes tend to be abusive to their own children, thus perpetuating a vicious cycle of violence.

Examples of Statements of an Interdisciplinary Problem or Question

The following are examples from published work and student projects of well-written statements introducing the problem that illustrate the above criteria. The student projects are identified by an asterisk (*). Students should note how each example accords with each of the guidelines outlined above.

From the Natural Sciences. Dietrich (1995), *Northwest Passage: The Great Columbia River.* William Dietrich introduces the problem of how dams on the Columbia River system in the Northwest are impacting the salmon populations and the people who depend on them for their livelihood:

> To a Pacific Northwest journalist such as myself, the river was inescapable as a subject. Its energy powered the region and its history dictated the region's history. . . . Many of the people I encountered, however, looked at the river from the narrow perspective of their own experience. One colleague said it was as if everyone was looking at the Columbia River through a pipe. . . . Each interest group looked at the Columbia and saw a different river.

> That experience dictated the approach of this book. One of the mistakes of the past . . . has been the tendency to focus narrowly on development of some part of a river without considering the consequences for the whole. "When we [whites] are confronted by a complex problem, we want to take a part of the complexity and deal with that," remarked Steve Parker, a fish biologist hired by the Yakima Indian tribe. The Henry Ford assembly line is an example of this kind of specialization, Parker said. Its economic success is why narrow focus and admiration of specialists became ingrained in American culture. (pp. 23–24)

From the Natural Sciences. Smolinski* (2005), *Freshwater Scarcity in Texas.* Joe Smolinski introduces the problem of freshwater scarcity in Texas in this clearly written introductory paragraph:

> There is little doubt among experts that freshwater is one of the most valuable natural resources in the state of Texas. Experts, in a variety of disciplines, have not yet been able to reach agreement as to the cause and effect of the widespread freshwater shortages currently experienced across the state. With population predictions calling for a dramatic increase in the number of residents over the next fifty years, the competition between these uses will only become more intense. How we address the use and allocation of water will have a dramatic impact on the environment and the quality of life for all Texans. (p. 1)

From the Social Sciences. Fischer (1988), "On the Need for Integrating Occupational Sex Discrimination Theory on the Basis of Causal Variables." Charles C. Fischer introduces the problem of occupational sex discrimination (OSD) in the workplace as follows:

> The majority of complaints filed with the Equal Employment Opportunity Commission under Title VII of the Civil Rights Act involve sex discrimination. Complaints of sex discrimination pertain mainly to pay discrimination, promotion (and transfer) discrimination, and occupation discrimination. Occupational sex discrimination (OSD) is particularly serious since other forms

of sex discrimination are, to a large degree, symptomatic of a lack of female access to "male" occupations—those occupations that pay good wages, that are connected to long job ladders (that provide opportunities for vertical mobility via job promotion), and that offer positions of responsibility. (p. 22)

From the Social Sciences. Delph* (2005), *An Integrative Approach to the Elimination of the "Perfect Crime."* Janet B. Delph introduces the growing problem of unsolved homicides, which she calls "perfect crimes," in these stark terms:

> Modern day criminal investigation techniques do not eliminate the possibility of the "perfect crime." . . . A "perfect crime" is one that will go unnoticed and/or for which the criminal will never be caught (Fanton, Tolhet, & Achache, 1998). The public is all too aware of these likely outcomes and consequently feels unsafe and vulnerable. Parents experience silent fear each time their child wanders beyond their reach. While "men are afraid women will laugh at them, women are afraid that men will kill them" (DeBecker, 1997, p. 77). Deviant minds should not be allowed to think that they can commit murder without suffering the gravest consequences. (p. 2)

From the Humanities. Bal (1999), "Introduction," *The Practice of Cultural Analysis: Exposing Interdisciplinary Interpretation.* Mieke Bal's introduction serves two purposes. The first is to decipher the complex meaning of the object she is subjecting to interdisciplinary scrutiny: an enigmatic love poem written in yellow paint on a red brick wall (e.g., a graffito) in post–World War II Amsterdam, the Netherlands. The second and closely related purpose is to introduce the reader to the interdisciplinary process of cultural analysis, of which she is a leading practitioner, and illustrate its ability to reveal new meaning in an object or a text like the graffito.

> Cultural analysis as a critical practice is different from what is commonly understood as "history." It is based on a keen awareness of the critic's situatedness in the present, the social and cultural present from which we look, and look back, at the objects that are already of the past, objects that we take to define our present culture. . . .
>
> This graffito, for example, has come to characterize the goals of the Amsterdam School for Cultural Analysis (ASCA). . . . In the most literal translation the text means:
>
> Note
>
> I hold you dear
>
> I have not
>
> thought you up
>
> This graffito fulfills that function because it makes a good case for the kind of objects at which cultural analysis would look, and—more importantly—how it can go about doing so. (pp. 1–2)

From the Humanities. Silver* (2005), *Composing Race and Gender: The Appropriation of Social Identity in Fiction.* Lisa Silver writes an informal personal narrative of how she became interested in her subject, the appropriation of social identity in fiction writing. Her story begins with her trip to Mexico during spring break of her junior year. When she returned, she had a story due in her creative writing class, so she tried writing about the people she met in the mountain villages in Oaxaca, Mexico.

> And that's when the interdisciplinarity kicked in. . . . In Mexico we learned it would be offensive for us, as outsiders, to assume we could fix their problems. What could we, carrying our Nalgene bottles, comprehend of the effects of water privatization and pollution? How could we listen to the plight of maquiladora workers while wearing Nikes and stonewashed jeans? How could I understand the lives of the indigenous Oaxacan villagers enough to write about them—especially from their own points of view? I couldn't separate the sociological and political lessons I'd learned in Mexico from my fiction. I ended up writing my story from the first-person peripheral perspective of a white college-aged female looking in at the village. I got good critiques in class, but was never personally satisfied with the story. It felt like I'd written a nonfiction piece. I wanted to create characters with backgrounds unlike my own, but suddenly didn't know how. (p. 2)

Summary of STEP 1

All of these examples conform to the above criteria: They are *appropriate* to interdisciplinary inquiry, they carefully *define* the scope of the problem, and they *avoid the three tendencies* that run counter to the interdisciplinary process: disciplinary bias, disciplinary jargon, and personal bias. They also *follow the three guidelines* for introducing the problem: The problems are stated clearly and concisely, are sufficiently narrow in scope to be manageable (depending on the scale of the writer's project), and appear in a context that explains why the problem should interest the reader.

As further STEPS are taken, you are likely to encounter new information, receive new insights (including flashes of intuition), or encounter unforeseen problems that will require revisiting the initial STEP and modifying the research question. This is a normal part of the interdisciplinary research process.

CREATIVITY AND STEP 1

We established in Chapter 1 that the IRP is a creative process. While we can identify useful strategies for each STEP, students still need to think creatively at various STEPS. With respect to the research question, students (or scholars) face a tradeoff. A narrow question may prove easier to answer, but the answer may not be particularly exciting. A broader question may prove harder, but it may yield an answer that is both novel and useful. We recommend above that students read a bit on a topic of interest to identify a good question. If students read in quite different disciplines (physics and literature, say), they may find it hard to make a useful connection. But they may make a very creative connection: one that has never been made before. If they instead read only in disciplines that are somewhat similar (sociology and anthropology, say), they may find it easier to develop a manageable research question but leave less scope for creativity. While undergraduate students should generally stress manageability, graduate students and scholars may want to lean toward creativity. If so, they may find it useful to ask, "What is the real problem?" Such a question may guide them to look for deeper causes of the problem they are addressing. It may also guide them to worry about the adoption of their solution: Creative solutions almost inevitably face resistance and so the creative researcher (especially) may want to reflect on the barriers to adopting solutions (why is inner city poverty so intractable?) even as they frame their research question (Szostak, 2017a).

STEP 2: JUSTIFY USING AN INTERDISCIPLINARY APPROACH

 A. DRAWING ON DISCIPLINARY INSIGHTS

1. Define the problem or state the research question.
2. **Justify using an interdisciplinary approach.**
   - **Determine that the problem is complex.**
   - **Determine that important insights concerning the problem are offered by two or more disciplines.**
   - **Determine that no single discipline has been able to explain the problem comprehensively or resolve it satisfactorily.**
   - **Determine that the problem is an unresolved societal need or issue.**
3. Identify relevant disciplines.
4. Conduct the literature search.
5. Develop adequacy in each relevant discipline.
6. Analyze the problem and evaluate each insight or theory.

STEP 2 is to justify using an interdisciplinary approach. Though typically absent from professional writings, this STEP is worthwhile for undergraduates (and even graduate students) to take because it provides an opportunity to see if their projects meet the four criteria (bullet pointed in the graphic) commonly used for justifying an interdisciplinary approach and supported by the National Academies (2005).

Determine That the Problem Is Complex

The **operational definition of complexity** used in this book is that the problem has multiple parts studied by different disciplines. The definition of interdisciplinary studies appearing in Chapter 1 states that *complexity requires interdisciplinarity.* We know of no way other than interdisciplinarity to study specific complex problems such as climate change, freshwater scarcity, and terrorism. That is, *interdisciplinarity is necessary for the study of complexity* (Newell, 2001, p. 2). The criterion of complexity also extends to problems that those in the humanities typically examine, such as the contextual meaning of an object or a text.[4]

Examples of complex questions include these: What is consciousness? What is freedom? What is a family? What does it mean to be human? Why does hunger persist? Admittedly, these problems are so fundamental and complex, requiring sophisticated analysis from so many disciplines, that they are beyond the capacity of most undergraduates to address comprehensively. Nevertheless, movement toward a more comprehensive understanding of these questions is possible even if students are limited to using only a few relevant disciplines.

Confirmation of complexity will be forthcoming as additional STEPS are taken, especially STEP 3 that involves mapping the problem to reveal its disciplinary parts (see Chapter 4), and STEP 4 that calls for conducting a full-scale literature search (see Chapter 5).

Determine That Important Insights Concerning the Problem Are Offered by Two or More Disciplines

A problem that is controversial, such as climate change, has likely generated interest from two or more disciplines, each offering its own insights or theories in the form of books and journal articles. This condition makes the problem researchable. Sometimes, however, scholars from the disciplines you plan to consult have not yet published on the problem because its occurrence is recent.

NOTE TO READER

Undergraduates should work on problems that have been studied by more than one discipline. Graduate students and especially senior scholars may be able to project how a hitherto silent discipline might address the problem and what insights it might offer into the problem. In this circumstance, they may choose to conduct basic research on the problem themselves and then integrate their insights or theory with existing disciplinary insights or theories.

Determine That No Single Discipline Has Been Able to Explain the Problem Comprehensively or Resolve It Satisfactorily

A problem is ripe for interdisciplinary inquiry if no single discipline has been able to explain it comprehensively or resolve it satisfactorily. For example, several disciplines consider terrorism within their respective domains, but no one discipline has been able to create a single comprehensive theory explaining terrorism in all of its complexity, let alone propose a holistic solution to it. For instance, political scientists typically use rational choice theory to explain terrorist behavior, but the theory fails to address religious and cultural variables. Other topics that no single discipline has been able to address comprehensively include undocumented immigration, human cloning, and genetically engineered food. The value of an interdisciplinary approach over a single disciplinary approach is that it can address complex problems in a more comprehensive way.

Determine That the Problem Is an Unresolved Societal Need or Issue

Societal/public policy problems necessitate what is widely referred to as **problem-based research**, which focuses on unresolved societal needs, practical problem solving, and intellectual problems that are the focus of the humanities, such as the meaning of some artifact. What distinguishes problem-based research from other applied research is its holistic focus that involves more than one discipline.

Examples of Statements That Justify Using an Interdisciplinary Approach

The rationale for using an interdisciplinary approach should be made explicit in the introduction to the research project. After all, this rationale distinguishes truly interdisciplinary research from multidisciplinary, not to mention disciplinary, research. Stating the rationale has the added benefit of alerting the researcher to possible problems with the topic. Spending extra time in carefully screening a potential topic according to these criteria will minimize the possibility of investing in an enterprise that later may prove unprofitable.

Satisfied that the proposed problem or topic meets one or more of the above criteria, it is then possible to present a clear rationale for using an interdisciplinary approach. Common practice is to include this statement of justification in the introduction to the study, as shown in these examples of professional work and student projects (marked with an asterisk) from the natural sciences, the social sciences, and the humanities.

From the Natural Sciences. Dietrich (1995), *Northwest Passage: The Great Columbia River.* Dietrich is struck by how narrowly people continue to look at the Columbia River. This narrowness of perspective and the lack of systems thinking provide his justification for taking an interdisciplinary approach, as follows:

My work as a writer on environmental issues, particularly the old-growth forests of the Pacific Northwest, had introduced me to the idea of ecosystems and the interrelationships of many parts to a greater whole. I wanted a comprehensive understanding of the river embracing history, Earth science, biology, hydrology, economics, and contemporary politics and management. (pp. 23–24)

From the Natural Sciences. Smolinski* (2005), *Freshwater Scarcity in Texas.* Smolinski is concerned that after years of study, disciplinary experts have not been able to reach agreement on the cause and effect of the worsening problem of freshwater scarcity. This failure provides ample justification for taking an interdisciplinary approach.

The causes and effects of freshwater scarcity across Texas are beyond the ability of any single discipline to explore. A review of the professional literature in political science, Earth science, and biology shows that these disciplines are most relevant to the problem. Each has produced its own well-defined theories about how the shortages impact the state of Texas and its communities. While each of these theories reflects the perspective of its particular discipline, none of these explanations comprehensively addresses the issues posed by the statewide shortage of freshwater. (p. 3)

From the Social Sciences. Fischer (1988), "On the Need for Integrating Occupational Sex Discrimination Theory on the Basis of Causal Variables." Fischer provides an example of professional work from the social sciences that presents a clear rationale for taking an interdisciplinary approach.

It appears that the problem of OSD is a good candidate for an IR [interdisciplinary] approach. OSD is a problem that a number of disciplines have separately analyzed, yet it is a problem of such complexity and breadth that its division among individual disciplines leads to incomplete and naïve views.

Another important advantage of IR is that it can . . . lead to [a] more complete understanding by providing a dynamic, holistic view of the problem. (p. 37)

From the Social Sciences. Delph* (2005), *An Integrative Approach to the Elimination of the "Perfect Crime."* Having introduced the topic and explained its importance, Delph justifies using an interdisciplinary approach.

To achieve the level of expertise necessary to solve more crimes, the criminal justice system must integrate a wide range of skills from multiple disciplines. This synthesis of skills and insights could serve as a strong deterrent to crime and result in safer communities. (p. 2)

From the Humanities. Bal (1999), "Introduction," *The Practice of Cultural Analysis: Exposing Interdisciplinary Interpretation.* The topic of the graffito is not a societal problem; it is an intellectual one that cries out for interdisciplinary understanding or meaning. Bal (1999) sees cultural analysis as an interdisciplinary practice and the field as a counterweight to critics who charge that interdisciplinarity makes objects of inquiry "vague and methodically muddled" (p. 2). Seeking to correct this mistaken view, she justifies using cultural analysis, an interdisciplinary approach, to find meaning in the graffito.

> As an object, it requires interdisciplinarity [and calls for] an analysis that draws upon cultural anthropology and theology [and] reflection on aesthetics, which makes philosophy an important partner. . . . [T]he humanistic disciplines . . . brutally confront scholars with the need to overcome disciplinary hang-ups. . . . Museum analysis requires the integrative collaboration of linguistics and literary, of visual and philosophical, and of anthropological and social studies. . . . Instead of speaking of an abstract and utopian interdisciplinarity, then, cultural analysis is truly an interdiscipline, with a specific object and a specific set of collaborating disciplines. (pp. 6–7)

From the Humanities. Silver* (2005), *Composing Race and Gender: The Appropriation of Social Identity in Fiction.* From her fiction class experience, Silver (2005) discovered that she did not know how to write authentically about the people in the Mexican village whose backgrounds were very different from her own. Frustrated and disappointed with the artificial characters she had created for her fiction piece, she decided to use the topic of character appropriation for her senior project. Character appropriation refers to a writer's attempt to write about, or an actor's attempt to assume, another person's identity. As Silver read, she developed "a sense of what different disciplines— sociology, psychology, cultural studies, and creative writing—[said] about the matter" (p. 2). Finding that each of these disciplines offered an important perspective on an important subject, she determined that an interdisciplinary approach was clearly called for (pp. 1–6).

Each of these examples conforms to one or more of the above criteria. In most cases, the writer also identifies the disciplines relevant to the problem that informs the reader which disciplinary insights the author will draw upon.

Chapter Summary

This chapter introduces the integrated model of the interdisciplinary research process (IRP) and its use of STEPS. It explains the importance of the research process to interdisciplinarity, describing it as a decision-making process that is overarching, heuristic, iterative, and reflexive and issues cautions concerning the use of the STEPS. STEP 1, "define the problem or state the research question," involves making four decisions on how to select a problem, define its scope, avoid three tendencies that run counter to good interdisciplinary practice, and follow three guidelines for stating the problem.

STEP 2, "justify the use of an interdisciplinary approach," involves meeting one or more criteria: (1) the problem is complex, (2) important insights concerning the problem are offered by at least two disciplines, (3) no single discipline has been able to address the problem comprehensively or resolve it, and (4) the problem is an unresolved societal need or issue.

Even after subjecting the proposed problem to these criteria, it is still too early in the research process to know with any certitude that the problem is researchable. This question can be resolved only by taking subsequent STEPS in the research process.

Once the problem is defined (STEP 1) and the justification for using an interdisciplinary approach is stated (STEP 2), you must decide which disciplines are relevant to the problem which is STEP 3 (see Chapter 4). Making this decision requires that you understand the disciplines and the concept of disciplinary perspective as explained in Chapter 2.

Notes

1. Scholarly consensus exists on the following STEPS: The problem or focus question should be defined; relevant disciplines and other resources must be identified; information from these disciplines (concepts, theories, methods, etc.) must be gathered; adequacy in each relevant discipline must be achieved; the problem must be studied, and insights into the problem must be located and evaluated; conflicts between insights must be identified, and their sources must be revealed; disciplinary insights must be integrated; and a new understanding must be constructed or new meaning achieved. The models disagree on the number, order, and identity of STEPS, leaving students and instructors alike without a clear road map of the overall interdisciplinary research process. Of special concern is the lack of consensus on how many STEPS are involved in the integrative part of the process. Welch (2003) notes that when the participants in a Delphi Study recommended that students be provided "basic integrational methods," the question arose as to which model and/or which particular STEPS within these models should be provided (p. 185).

2. Students interested in learning more about teamwork are urged to consult the websites of both Science of Team Science at www.scienceofteamscience.org/scits-a-team-science -resources and td-net at www.naturwissen schaften.ch/topics/co-producing_knowledge. Cooke and Hilton (2015) survey the literature. Dan Stokols has written extensively about team science. The About Interdisciplinarity section of the website of the Association for Interdisciplinary Studies (AIS) also surveys the literature on team research.

3. We thank Sharon Woodhill for making these points at the 2018 conference of the Association for Interdisciplinary Studies.

4. "What is contested is whether interdisciplinarity studies only complexity, or whether interdisciplinarity can appropriately study problems/issues/questions that are not complex as well. Some practitioners say that interdisciplinarity studies only complexity, but others remain unconvinced. Thus, the debate is not over whether interdisciplinarity is necessary for complexity, but whether complexity is necessary for interdisciplinarity" (William H. Newell, personal communication, January 7, 2011).

Exercises

The Best Approach

3.1 This chapter compared and contrasted the interdisciplinary research process to the disciplinary approach (in a general sense) and argued that both have utility, depending on the problem. Below are short descriptions of a problem, question, or topic. In each case, decide which approach is probably more appropriate and why:

- What is the cost of building a high-speed rail system to connect two large cities?
- Should the city build a new performing arts center in an area characterized by low-income housing and mom-and-pop stores?
- What is the cause of obesity among teens?
- What is the meaning of the science fiction movie *Avatar*?

The Integrated Model

3.2 The chapter introduced the integrated model of the interdisciplinary research process. What parts of the model are most similar to and different from disciplinary approaches to research?

Is It Researchable?

3.3 The chapter presented criteria for determining if a problem, topic, or question is researchable in an interdisciplinary sense. Which of the following meets one or more of these criteria?

- The psychological dimension of Alzheimer's disease
- The loss of manufacturing jobs to China
- The effects of closing fine arts programs in public schools

Stating the Problem

3.4 The following is an example of student work on the topic of the underachieving child. Based on the discussion of STEP 1, how could this introduction to the problem be stated differently so that it conforms to the criteria and guidelines set forth?

(Continued)

(Continued)

Many school-age children underachieve. Underachievement is when the performance of a child falls below what is expected and the ability of the child. Underachievement means to perform academically below the potential indicated by tests of one's ability or aptitude.

Justify Using an Interdisciplinary Approach

3.5   The chapter notes that it is common practice for practitioners to justify using an interdisciplinary approach. Compare the various examples and identify their commonalities. What would you change, if anything, in any of the statements?

3.6   In addition to justifying using an interdisciplinary approach, should you criticize the disciplines for taking narrow positions on the problem?

Image by Pexels from Pixabay.

Image by StockSnap from Pixabay.

4

IDENTIFYING RELEVANT DISCIPLINES

LEARNING OUTCOMES

By the end of this chapter, you will be able to

- Select *potentially* relevant disciplines
- Map the problem to reveal its disciplinary parts
- Reduce the number of potentially relevant disciplines to those that are *most* relevant

GUIDING QUESTIONS

How do you choose potentially relevant disciplines for your research question and then decide which are most relevant?

How and why do you map your research problem?

CHAPTER OBJECTIVES

The fable of building a house for an elephant (referenced in Chapter 1) shows the importance of taking into account *all* relevant disciplinary perspectives when trying to solve complex problems or understand complex systems. Interdisciplinary studies tends to focus on complex problems or systems such as global warming, gun violence in inner-city neighborhoods, or the meaning of a graffito. These problems are complex because they involve many variables that are typically studied by different disciplines.

After stating the research problem (STEP 1) and justifying an interdisciplinary approach (STEP 2), the next challenge is deciding which disciplines are *potentially* relevant to the problem and then which of these are *most* relevant (STEP 3). This chapter presents a way to quickly identify disciplines potentially relevant to the problem. It discusses the value of mapping the problem so that its constituent disciplinary parts are revealed. It then shows how to narrow the number of disciplines to those that are most relevant. Once these disciplines are identified, the full-scale literature search (STEP 4; see Chapter 5) can proceed.

A. DRAWING ON DISCIPLINARY INSIGHTS

1. Define the problem or state the research question.
2. Justify using an interdisciplinary approach.
3. **Identify relevant disciplines.**
   - **Select *potentially* relevant disciplines.**
   - **Map the problem to reveal its disciplinary parts.**
   - **Reduce the number of potentially relevant disciplines to those that are *most* relevant.**
4. Conduct the literature search.
5. Develop adequacy in each relevant discipline.
6. Analyze the problem and evaluate each insight or theory.

SELECT POTENTIALLY RELEVANT DISCIPLINES

In selecting disciplines from which to draw insights and theories, the challenge is to decide which disciplines contribute substantially to the problem or overall pattern of behavior you wish to study. A **potentially relevant discipline** is one whose research domain includes at least one *phenomenon* involved in the problem or research question, whether or not its community of scholars has recognized the problem and published its research. Determining which disciplines are potentially relevant to the problem is a relatively straightforward process beginning with a focus on phenomena and disciplinary perspectives.

Focus on the Phenomena

Refer to Table 2.4 (Chapter 2) "Szostak's Categories of Phenomena About the Human World" to see which disciplines focus on the phenomena relevant to the problem, topic, or question. This table should facilitate linking most topics to one or more of the particular phenomena that may touch on the topic. Recall (or reread in Chapter 2) the example of freshwater scarcity that concerns the nonhuman environment. Then refer to Table 2.3 "Disciplines and Their Illustrative Phenomena" to link the relevant phenomena to disciplines that study them. Scan the right column to locate the phenomena that touch on the problem and then note those disciplines in the center column that study them.

Draw on Disciplinary Perspectives (in a General Sense)

Consult Table 2.2 "Overall Perspectives of Natural Sciences, Social Sciences, and Humanities Disciplines Stated in General Terms" to see if the topic as a whole is included in the phenomena studied by two or more disciplines. Ask this of each disciplinary perspective: "Does it illuminate some aspect of the problem, topic, or question?" This questioning process should help you decide which disciplines are potentially relevant to the problem and explain how each illuminates some aspect of it. However, just because the problem falls within a discipline's perspective and research domain does not mean that the discipline's community of scholars has addressed the problem.

Once you have identified potentially relevant disciplines, skim their literatures to see if their community of scholars have written on the topic. Those new to interdisciplinary research often start out conceiving of the problem too narrowly because of their prior exposure to a particular discipline. Stating the problem more broadly at first will make your conception of the problem more inclusive of all relevant perspectives.

Note: As mentioned in Chapter 3, the literature review (STEP 4) is a fluid step that can begin with STEP 1 and continue through later STEPS. A student may search across their library's database for relevant works, taking note of which disciplines these represent. One challenge here, as we shall see in the next chapter, is that different disciplines often use different terminology for the same phenomenon or process.

An Example of How to Select Potentially Relevant Disciplines

Using phenomena *and* perspective approaches is illustrated in the example of a student who was interested in writing an interdisciplinary research paper on the topic of human cloning. After referencing Tables 2.4 and 2.3, the student consulted Table 2.2 on perspectives to see if the topic as a whole was included in the phenomena studied by two or more disciplines. Since it was, the student then skimmed these disciplinary literatures and found that the topic was ripe for interdisciplinary inquiry. But final confirmation had to wait until the student completed the full-scale literature search. If, however, the student had found that human cloning is embraced by the perspective of only one discipline, then the problem would likely not have merited interdisciplinary study at the undergraduate level, and the topic would have had to have been revised or abandoned.

In fact, the student discovered that no fewer than seven disciplinary perspectives embraced the topic of human cloning, as shown here in Table 4.1. The characterization of these disciplines as *potentially relevant* is based on the fact that their perspectives include some aspect of human cloning. The information in the right-hand column is

TABLE 4.1 ● Disciplines Potentially Relevant to the Problem of Human Cloning and How They Illuminate Some Aspect of It (Before the Full-Scale Literature Search)	
Discipline, Interdiscipline, and Applied Field	**How Each Illuminates Some Aspect of the Problem of Human Cloning**
Biology	The biological process of human cloning and rates of success or failure
Psychology	Possible psychological impact on the cloned person of a sense of personhood
Political Science	The role of the federal government
Philosophy	Ethical implications of cloning a human life and what it means to be human
Religious Studies	Sanction in sacred writings against the creation of a new form of human life
Law[a]	Legal rights and relationships of the cloned child and its "parents"
Bioethics[b]	Ethical implications of the technical procedures used to clone a human, particularly in the event of failure

a. Law is an applied field in many taxonomies.

b. Bioethics is an interdisciplinary field in many taxonomies.

derived from the more general information in Table 2.2 in Chapter 2. For example, it states in the overall perspective for biology that "when biologists venture into the world of humans, they look for physical, deterministic explanations of behavior (such as genes or evolution)." It is reasonable to conclude, then, that biology is likely to be interested in human cloning *as a biological process* and to be concerned with its rates of success or failure.

Identifying potentially relevant disciplines based on whether or not a discipline's perspective embraces the research problem is only a starting point. As noted earlier, broad statements of disciplinary perspectives are typically matters of scholarly contention. For this reason, the results of the perspectival approach should be verified, as this student did, by identifying phenomena that are typically of interest to disciplines. In this way, we might find different disciplines that investigate different parts of the complex problem.

NOTE TO READERS

There are several advantages to creating a table such as 4.1 that connects the topic to particular disciplinary perspectives. (1) It establishes whether each perspective includes the topic and how it illuminates some aspect of it. (2) It may reveal an instance of apparent overlap of perspectives. In the case of human cloning, there is obviously some overlap in perspective between bioethics and philosophy but not enough to warrant ignoring either perspective initially because each may lead to different insights. But one may very well choose to drop one perspective in the narrowing-down process if their insights turn out to overlap too much. (3) It may prompt shifting the focus of the problem or redefining it. Instead of focusing on the problem of human cloning in general, one may wish to narrow the topic and pose a research question such as "What should be the role of government in the development of human cloning technology?" Once again, we see the iterative nature of the interdisciplinary research process (IRP) where performing one STEP may prompt the revisiting of an earlier STEP. (4) The table is a convenient resource to consult later in the research process should it be needed.

MAP THE PROBLEM TO REVEAL ITS DISCIPLINARY PARTS

Mapping may well assist in identifying relevant disciplines. After selecting the disciplines potentially relevant to the research question, you need to identify the constituent parts of the problem, understand how these relate to each other and to the problem as a whole, and view the problem *as a system.* Using systems thinking to map the problem facilitates this understanding.[1]

The map may reveal a gap in your understanding of the problem or establish that you are placing too much emphasis on a few disciplinary components at the expense of other equally important components. Whereas the disciplinarian is often satisfied to focus on a single part or on a few "neighboring" parts of the problem, the interdisciplinarian is concerned with achieving an interdisciplinary understanding of the problem *as a whole.*

Maps profitable to interdisciplinary work include the system map, the research map, the concept or principle map, and the theory map. Mapping the problem may occur as early as STEP 1, but should occur before conducting the full-scale literature search (STEP 4).

Once you see the problem as a complex whole, you can then shorten the list of disciplines to those that are essential before conducting the full-scale literature search. The results of the search should confirm the completeness and accuracy of the map.

Systems Thinking and the System Map

Part II of the IRP is designed to help you deconstruct the problem and understand it in all of its complexity. An important and useful tool to accomplish this is to cultivate systems thinking. Systems thinking, defined in Chapter 3, helps us understand the relationship between a whole and its parts. Systems thinking involves not just visualizing interrelationships, but also thinking in terms of positive and negative feedback loops, spotting patterns in the behavior of the whole at a point in time and over time. Systems thinking adds a powerful dimension to interdisciplinary research, integration, and problem solving.[2]

Although systems thinking is typically associated with the quantitatively oriented fields of engineering, operations management, computer science, and environmental science, it increasingly is being applied to a widening range of qualitatively oriented problems such as the actions of emergency workers at the World Trade Center on September 11, 2001.

A primary analytical tool of systems thinking is the **system map**, a visual that shows all the parts of the system (complex problem) and illustrates the causal relationships among them (that is, which phenomena influence which others). Such a map helps the researcher to visualize the system as a complex whole. Generally, each part of the complex problem is studied by a different discipline. Constructing the system map will help reveal disciplinary parts of the problem that were not initially obvious (Newell, 2007a, p. 246). The mapping exercise is iterative: You should sketch the system near the outset but likely will add important elements or clarifications as the research progresses.

Systems maps have two key elements: the *phenomena* relevant to a complex problem, and *arrows* that indicate visually how these influence each other. (*Note:* Students should appreciate that these maps can be built up slowly by reflecting upon [and later, reading about] the elements of the complex problem [see Repko, Szostak, & Buchberger, 2020, pp. 173–178].) Two types of system maps of particular interest to researchers are causal loop diagrams and stock and flow diagrams. These diagrams facilitate our understanding of the inner workings of a system by visualizing the behavior that occurs in the system, as well as the likely outcome(s) that the system will produce.

For example, to understand why communities experience traffic congestion on a road system just after it has been expanded requires an understanding of the relationship between the key actors in the system. Figure 4.1 shows a causal loop diagram that illustrates how individuals select driving routes and the relationships between population, driving, air quality, and word-of-mouth communication.

FIGURE 4.1 ● Causal Loop Diagram

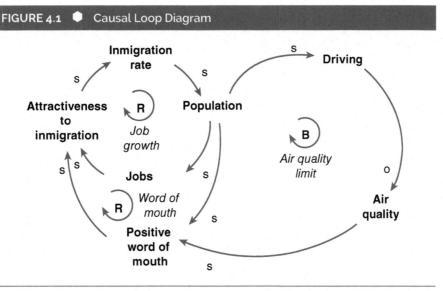

Source: Mathews, L. G., & Jones, A. (2008). Using systems thinking to improve interdisciplinary learning outcomes. *Issues in Integrative Studies, 26,* 73–104, 77.

Specifically, Figure 4.1 shows how population affects vehicle traffic in a region that in turn affects air quality negatively.

Leah Greden Mathews and Andrew Jones (2008) explain the subscripts:

> The subscripts *s* and *o* on the relationship arrows in a causal loop diagram show whether the two phenomena are moving in the same or opposite direction. Reduced air quality tends to negatively impact public perceptions of a region's livability, which, in turn, has the effect of reducing or even reversing population growth. This type of offsetting relationship is said to be a "balancing loop," and is labeled with a B (for balancing). There are other factors that serve to reinforce the population trend simultaneously which are labeled with an R (for reinforcing) in the causal loop diagram. As population grows, positive word of mouth can also grow as people express enthusiasm about the region to their friends and relatives. The positive "buzz" can lead to "in-migration" which reinforces population growth. (p. 77)

In addition to breaking down a problem into its constituent parts, the causal loop diagram helps identify which parts of the system are likely to be addressed by different disciplines, subdisciplines, and interdisciplines. For example, the disciplines interested in the relationship between a region's road system and its air quality include biology, economics (the economics of transportation), and political science (government agencies at the local,

FIGURE 4.2 ● Stock and Flow Diagram Showing the Population-Level Flows of People Through Stages of Diabetes, With the Public Health Interventions

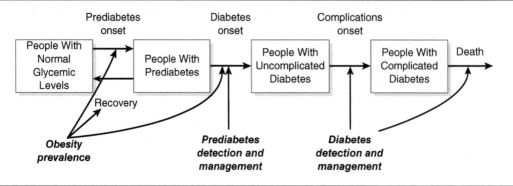

Source: Mathews, L. G., & Jones, A. (2008). Using systems thinking to improve interdisciplinary outcomes. *Issues in Integrative Studies, 26,* 73–104, 79.

state, and federal levels responsible for monitoring air quality), as well as the interdisciplinary field of environmental science. Knowing the key parts of a system and understanding how they relate to each other and to the problem as a whole enables the researcher to undertake the full-scale literature search with greater focus and efficiency.

Szostak (2017b) recognized that disciplines often posit "balancing loops" among the phenomena they study such that there is some tendency toward stability in the systems they examine. Phenomena studied in other disciplines often serve to destabilize these systems (as when changes in weather or consumer tastes destabilize a market studied by economists). Disciplinary scholars may wish to ignore these destabilizing influences, and may thus be resistant to interdisciplinary understandings. Students performing interdisciplinary research should be ready to question whether subsystems are as stable as disciplinary scholars may assert.

Figure 4.2 shows a stock and flow diagram that helps the researcher visualize the flows and accumulations (stocks) in the complex system of diabetes incidence. The usefulness of this diagram is that it makes distinctions between people's physiological condition relative to Type 2 diabetes within a given population.

Mathews and Jones (2008) explain the boxes:

> Each of the boxes represents stocks or accumulations of numbers of people with various diabetic conditions. The arrows represent flows or relationships between the stocks. The flow mapped in Figure 4.2 shows the flow of people through various stages of diabetes. For example, a fraction of *People with Normal Glycemic*

Levels will experience prediabetes onset and become part of the population of *People with Diabetes*. Some of these individuals will recover and return to the category of *People with Normal Glycemic Levels*. There is a reverse flow depicted between these two stocks. However, some *People with Prediabetes* will experience diabetes onset and become part of the accumulation of *People with Uncomplicated Diabetes*. (p. 78)

Benefits to Students of Using Systems Thinking and the System Map

There are five practical benefits to using systems thinking and drawing a system map:

- The map clarifies what part of the problem different authors or disciplines are addressing. Sometimes it will seem that authors or disciplines are disagreeing when in fact they are simply investigating different phenomena or relationships within the broader problem.

- Even if a project is limited to only two or three disciplines, and thus the entire system cannot be engaged, one still needs to understand which parts of the system each discipline illuminates. Researchers may discover that relationships at first thought to be of secondary importance are of primary importance.

- Drawing a system map and locating the disciplines on it that one believes are relevant enables the researcher to more easily identify other relevant disciplines that may have been overlooked earlier.

- Drawing a system map helps the researcher understand not only how the system operates the way it does but also *why* the system behaves the way it does. For example, simply knowing *how* coal contributes to acid rain does not explain *why* coal was chosen over other materials to fuel electric utilities to begin with. One should be as concerned with knowing the *why* of the problem as knowing the *how* (Motes, Bahr, Atha-Weldon, & Dansereau, 2003, pp. 240–242).

- A system map can reveal the existence of a system that did not initially seem to be a system. Systems include not only physical phenomena such as roads or climate, but also cultural phenomena such as faith traditions and gender roles. A map may illustrate, for example, how cultural values support (or not) particular political institutions or practices.

Note: The system map will prove to have further advantages in later STEPS. Researchers can reflect on what theories or methods might best illustrate different relationships among phenomena (useful when evaluating disciplinary research or when graduate students or scholars plan their own research). And a map like this can spark inspiration regarding how to integrate disciplinary insights. If the research is seeking a solution to a social problem, the map may indicate which phenomena or relationships it is best to act upon. It may be necessary to act simultaneously upon many elements of a complex problem.

The Similarity of Systems Thinking to Problem-Based and Inquiry-Based Learning

Systems thinking is similar to problem-based learning and inquiry-based learning in that they use a "scaffolding strategy" to help students move progressively toward stronger understanding while assuming increasing responsibility over the learning process. This strategy helps to break up the problem into its discrete parts (as the IRP does) and then move in step-like fashion toward understanding it as a whole (Mathews & Jones, 2008, p. 80). The systems methodology of learning and conducting research is summarized in Figure 4.3.

FIGURE 4.3 ● Systems Thinking Methodology

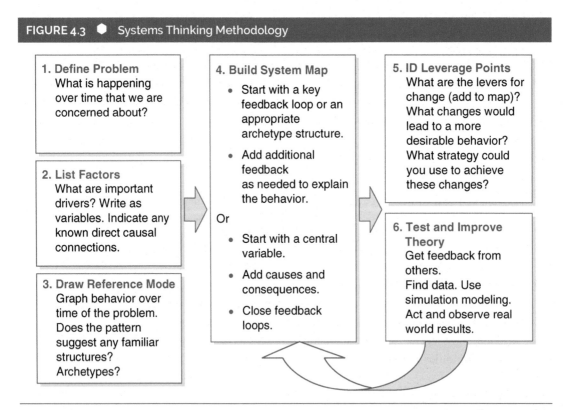

1. **Define Problem**
   What is happening over time that we are concerned about?

2. **List Factors**
   What are important drivers? Write as variables. Indicate any known direct causal connections.

3. **Draw Reference Mode**
   Graph behavior over time of the problem. Does the pattern suggest any familiar structures? Archetypes?

4. **Build System Map**
   - Start with a key feedback loop or an appropriate archetype structure.
   - Add additional feedback as needed to explain the behavior.

   Or

   - Start with a central variable.
   - Add causes and consequences.
   - Close feedback loops.

5. **ID Leverage Points**
   What are the levers for change (add to map)? What changes would lead to a more desirable behavior? What strategy could you use to achieve these changes?

6. **Test and Improve Theory**
   Get feedback from others. Find data. Use simulation modeling. Act and observe real world results.

Source: Mathews, L. G., & Jones, A. (2008). Using systems thinking to improve interdisciplinary outcomes. *Issues in Integrative Studies, 26,* 73–104, 81.

How Systems Thinking Promotes Interdisciplinary Learning and Facilitates the Research Process

Using systems thinking and system mapping promotes at least four skills appropriate to interdisciplinary learning and research: perspective taking, nonlinear thinking, holistic thinking, and critical thinking.

Perspective taking, as noted in Chapter 2, involves examining a problem from the standpoint of the interested disciplines and identifying the differences between them. Students have to employ perspective taking when asked to identify the potentially relevant disciplines to a problem. For example, in an interdisciplinary course on land use, students were asked to assume the role of an individual experiencing poor air quality who sets about collecting data to present to appropriate local, state, and federal agencies.

Systems thinking, as well as system mapping, fosters nonlinear thinking by looking for nonlinear relationships among the phenomena or actors in a system. Where they occur, a small change in one part of the system may lead to big changes elsewhere in the system—what Mathews and Jones (2008) refer to as "leverage points" (p. 80). These nonlinear relationships can be visualized as causal loops. Causal loops include balancing loops—also known as negative feedback loops (where phenomena are in an offsetting relationship)—and reinforcing loops—also known as positive feedback loops (where phenomena are moving in the same direction).[3]

Systems thinking and system mapping promote holistic or comprehensive thinking by asking students to view system parts in relationship to the system as a whole. Comprehensive thinking is required because students must (1) identify a problem, (2) break it down into its constituent parts (so that they connect parts to particular disciplines), (3) identify the causal factors that are important contributors to the problem, and (4) show how these factors relate to each other and to the problem as a whole. "Once students understand the dynamics of the system, they can identify and test hypotheses about where and when to intervene in the system. This will enable them to later propose solutions to the problem" (Mathews & Jones, 2008, p. 78).

Systems thinking also promotes critical thinking by requiring students to examine their assumptions and base their conclusions on evidence. That is, they must ask whether scholarly insights into each relationship on their map is supported by adequate argument and evidence. While all disciplines lay claim to critical thinking, such thinking is freighted with added meaning for interdisciplinary students who must examine assumptions (their own as well as those of the disciplines in which they are working) and evaluate and reconcile the conflicting claims of disciplinary experts (Mathews & Jones, 2008, p. 81).

The Research Map

The **research map** helps those new to interdisciplinary research visualize the research process from beginning to end. While creating a research map may at first appear as a diversion from the more important business of "getting on" with the project, experience shows that investing time in constructing the research map—formulating the problem and identifying its various components—increases efficiency in performing subsequent STEPS of the IRP. The sample research map, shown in Figure 4.4, reveals key components of the research map:

- It states the purpose of the research.

- It identifies what disciplines are *potentially* relevant.

- It states the perspective of each discipline on the problem.

- It identifies the assumptions of each discipline.

- It identifies nondisciplinary sources or interpretations.

Note: Students in some courses may be encouraged to consider nondisciplinary sources. For example, in a course with an environmental focus, it might be useful to learn what farmers or other local people think about a particular environmental challenge. This may prove more challenging than evaluating disciplinary insights because there may be no detailed consensus to draw upon. Nevertheless, locals may have insights quite distinct from those of any scholarly discipline.

The Concept or Principle Map

More advanced students working with more complex or larger-scale problems can benefit from using concept or principle maps. The **concept or principle map** organizes information about the problem showing meaningful relationships between the parts of the problem that requires thinking through all the parts of the problem and anticipating how these behave or function, as shown in Figure 4.5.

The Theory Map

The **theory map** describes a theory's supporting evidence and importance, and compares it to other theories. In the following chapters, we shall see that theories are important in most of the professional work and student projects used to illustrate aspects of the interdisciplinary research process. For example, a student investigating the causes of freshwater scarcity in Texas discovered that the insights from each of the relevant disciplines are couched in terms of theories that are well known within that discipline. If one or more disciplinary theories are involved in an inquiry, you must develop adequacy in each theory and in each discipline that produced it. We will discuss theory in more detail in STEP 5 (Chapter 6). Preliminary to developing adequacy, however, is mapping the problem. The theory map on Piaget's theory of cognitive development shown in Figure 4.6 can easily be modified to focus on additional aspects of any theory.

FIGURE 4.4 ● Research Map

Research Map

To fully understand a piece of research, the student must understand the purpose of the research, the particular methods used in the investigation, and the findings. The student should also understand the implications of the study: How do the findings fit with existing scientific knowledge, what impact did the study have on subsequent research on the particular topic, and what impact did the findings have on society? Finally, the student should also know about any alternative interpretations of the study.

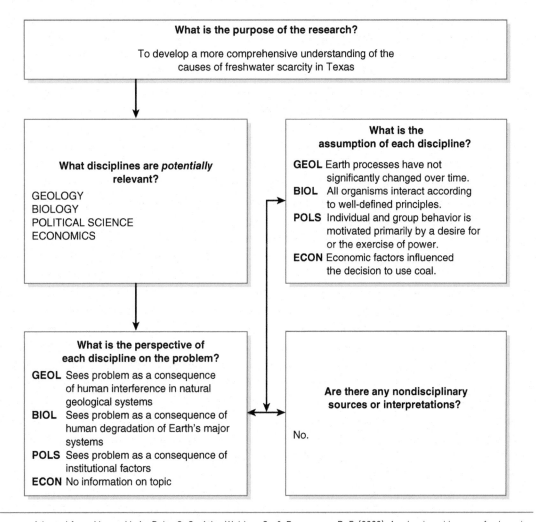

Source: Adapted from Motes, M. A., Bahr, G. S., Atha-Weldon, C., & Dansereau, D. F. (2003). Academic guide maps for learning psychology. *Teaching of Psychology, 30*(3), 240–242.

FIGURE 4.5 ● Concept or Principle Map

Concept or Principle Map

Concepts and principles are ubiquitous in science. To fully understand a concept or principle, the student must be able to describe it and must also know how the concept or principle fits with existing theories and what research has been conducted on the concept or principle. Additionally, the student should know how and why the concept or principle is important to science and society, as well as to any related concepts or principles.

What is the name of the concept or principle?
Implicit memory

What is a good description of the concept or principle?
Memory without awareness

Why is the concept or principle important?
Your thoughts and behavior can be influenced by an event WITHOUT your being aware of the influence!

Are there any related concepts or principles?
Implicit transfer
Covert attention

What theories are related to this concept or principle?
I don't really know:
Ask professor in next class!!

What research has been conducted?
Whole battery of implicit memory tests has been developed:
Verbal, e.g., anagram solving
Nonverbal, e.g., identification of fragmented pictures

Source: Adapted from Motes, M. A., Bahr, G. S., Atha-Weldon, C., & Dansereau, D. F. (2003). Academic guide maps for learning psychology. *Teaching of Psychology, 30*(3), 240–242.

FIGURE 4.6 ⬣ Theory Map

Theory Map

To fully appreciate a scientific theory, the student should be able to describe the theory, know evidence for and against the theory, know why the theory is important, and know whether there are any similar and competing theories.

What is the name of the theory?
Piaget's Theory of
Cognitive Development

What is a good description of the theory?

Our actions are based on our available schemas. Upon encountering new situations, we either assimilate the new information into old schemas (i.e., respond in old ways) or accommodate the new information by creating new schemas (i.e., respond in new ways).

Stages
Sensorimotor (0-2 yrs) reacts to sensory stimuli through reflexes; seems not to understand object permanence (according to Piaget)

Preoperational (2-7 yrs) develops language; can represent objects mentally by words and symbols; shows object permanence; lacks concept of conservation

Concrete Operations (7-11 yrs) understands conservation; can reason logically with regard to concrete objects

Formal Operations (11 yrs) can reason logically about abstract and hypothetical concepts

---

What evidence is there for and against this theory?

For Piaget's methods of testing each stage (object permanence and conservation tasks) reveal the child's ability or inability to accomplish certain tasks; Piaget found reasoning mistakes made on IQ tests

Against Stages are discontinuous; arguments about age estimates and about reasons for difficulty (e.g., object permanence appears much earlier, and conservation failures may be due to demand characteristics because rewording the questions changes the outcomes)

Conservation of volume is found much earlier if the researcher states a reason for changing from a short fat glass to a tall thin glass; studies of other cultures

---

Why is the theory important?

Major influence on and contributions to developmental and cognitive psychology

---

Are there any similar or analogous theories?

Neo-Piagetian theories concentrate on scientific or logical aspects, Fifth-Stage Theorists

---

Which theories compete with this one?

Bayley, Gessell, Vygotsky zone of proximal development, Info-Processing Theorists

---

Source: Adapted from Motes, M. A., Bahr, G. S., Atha-Weldon, C., & Dansereau, D. F. (2003). Academic guide maps for learning psychology. *Teaching of Psychology, 30*(3), 240–242.

REDUCE THE NUMBER OF POTENTIALLY RELEVANT DISCIPLINES TO THOSE THAT ARE MOST RELEVANT

Once you have selected the disciplines *potentially* relevant to the problem, you must decide which of these are *most* relevant. Conduct a cursory search of the literature of each potentially relevant discipline to identify the disciplines most relevant to the problem. The **most relevant disciplines** are those disciplines, often three or four, that are most directly connected to the problem, have produced the most important research on it, and have advanced the most compelling theories to explain it. These disciplines, or parts of them, will provide information about the problem that is essential to developing a comprehensive understanding of it.

Three Questions to Ask to Distinguish Between Potentially Relevant and Most Relevant Disciplines

To identify the most relevant disciplines, ask three questions of each discipline you selected as potentially relevant to the problem:

- Is the problem a natural focus for the discipline's perspective? Is it (or key components of it) a part of the subject matter of the discipline?

- Has it produced a body of research (i.e., insights and supporting evidence) on the problem of such significance that it cannot be ignored?

- Has it generated one or more theories to explain the problem?

Answer these questions as you conduct the cursory literature search.

Applying These Questions to the Disciplines Potentially Relevant to Various Topics

Question 1: Does the discipline have a well-defined perspective on the problem? At this early phase of the research process, you should be able to explain how each discipline's overall perspective illuminates the problem or some facet of it. One way to do this (and perhaps gain new insight into the problem) is to recast each perspective in terms of an overarching question about the problem as shown in Tables 4.2, 4.3, 4.4, and 4.5. Compile this table as you do cursory reading in each discipline's literature.

Question 2: Has the discipline produced a body of research (i.e., insights and supporting evidence) on the problem of such significance that it cannot be ignored? Here, the focus is on the

TABLE 4.2 ● Disciplines and How They Illuminate Some Aspect of the Problem of Human Cloning (Before the Full-Scale Literature Search)

Discipline, Interdiscipline, and Applied Field	Perspective Stated in Terms of an Overarching Question Asked About Human Cloning
Biology	What are the scientific consequences of human cloning?
Psychology	How will the discovery of being cloned affect the cloned person psychologically and the perceptions of others who know this about the person?
Political Science	What should be the role of government concerning this issue?
Philosophy	How will human cloning affect humanity, and what it means to be human?
Religious Studies	Does the science of human cloning conform to sacred writings and, more particularly, to the notion of what it means to be human?
Law	What are the legal implications of human cloning, and what are the rights of those who are participants in human cloning experiments?
Bioethics	What are the ethical implications of the biotechnology used in human cloning?

TABLE 4.3 ● Disciplines and How They Illuminate Some Aspect of the Columbia River Ecosystem

Discipline	Perspective Stated in Terms of an Overarching Question Asked About the Columbia River Ecosystem
Biology	What are the consequences of the dam system to native salmon populations?
Economics	What are the economic benefits and liabilities of the dam system on the people living in the region?
Earth Science	What are the implications of the dams on the region's hydrological system?
History	What does the damming of the Columbia River system tell us about the nation's confidence at this period of history?
Political Science (Politics)	What should be the role of government at all levels concerning the future of the dam system?

TABLE 4.4 ● Disciplines and How They Illuminate Some Aspect of Occupational Sex Discrimination (OSD)

Discipline and School of Thought	Perspective Stated in Terms of an Overarching Question Asked About Occupational Sex Discrimination
Economics	What is the economic motivation for OSD?
History	What is the historical context that would help explain OSD?

(Continued)

TABLE 4.4 ● (Continued)	
Discipline and School of Thought	**Perspective Stated in Terms of an Overarching Question Asked About Occupational Sex Discrimination**
Sociology	How is OSD a reflection of broader social relationships in society?
Psychology	How does the behavior of the perpetrators and victims of OSD reflect the psychological constructs individuals develop to make sense of their situations?
Marxism	How is OSD a necessary act of preserving capitalism?

TABLE 4.5 ● Disciplines and How They Illuminate Some Aspect of a Graffito (i.e., a Wall Writing)	
Discipline and Subdiscipline	**Perspective Stated in Terms of an Overarching Question Asked About the Graffito**
Anthropology (Cultural)	Is the graffito an expression of contemporary "popular" Dutch culture?
Art History	Is the graffito merely illustrative of the text about it?
Linguistics (Narratology)	What does the graffito represent?
Philosophy (Epistemology)	What does the graffito suggest is real and unreal?
Literature	What can the graffito be compared to in Dutch poetry?
Psychology	Is the graffito a text of psychic mourning for love lost?

Note: Cultural anthropology, narratology, and epistemology are subdisciplines.

significance of each discipline's published research. Since the goal of any interdisciplinary research effort is to achieve the most comprehensive understanding of the problem possible, it is advisable to include those disciplines producing important insights into the problem, whether the number of insights is one or several. In the example of human cloning, the disciplines include biology, psychology, political science, philosophy, religion, law, and bioethics.

Course requirements generally determine how many disciplines and how much reading in their literatures students can reasonably be expected to handle. Some students may need to limit the number of disciplines used to only three or four, based on the *comparative importance of their insights*. Ways to evaluate the importance of insights include

- seeing how often the insight is cited by other writers,

- consulting disciplinary experts, and

- noting the date of publication.

The last factor is particularly important when dealing with time-sensitive issues involving, for instance, rapidly evolving reproductive technologies. You should focus on research that is published in peer-reviewed journals, by university presses, and by academic presses. There is an abundance of material on the Internet, some of which is peer reviewed but much of which is not. Your instructor will provide guidelines on how to evaluate information that is not peer reviewed. (*Note:* These guidelines are discussed in Repko et al. [2020].)

When determining the relative importance of a discipline and its insights, you should not be influenced by the *quantity* of a discipline's research on the problem. If a discipline has just begun to address the problem, or if the problem is of recent origin, then it is not uncommon to find that its experts have published only one or a few insights. But those few may be extremely important because they are based on the latest research and may advance an important theory. A single treatment of the problem by a leading scholar in a discipline may impact the discussion in such a forceful way that one cannot ignore it. In this event, the next question may prove particularly important.

Question 3: Has the discipline generated one or more theories to explain the problem? Theories about the causes or consequences (real or possible) of a problem should be part of developing adequacy in each relevant discipline (STEP 5; see Chapter 6) and may be among the possible sources of conflict between disciplinary insights (STEP 7; see Chapter 9). Whether or not theories are involved can be answered only by conducting a full-scale literature search.

NOTE TO READERS

Advanced undergraduates and some graduate students who labor under time and other constraints must somehow reduce the number of potentially relevant disciplines to those that are the most relevant to the problem, and do so quickly and in a way that does not compromise the integrity of the end product. More senior scholars acting as solo interdisciplinarians conducting solo research have considerably more latitude in identifying relevant disciplines and their insights and theories. Reducing the number of disciplines is not as necessary for them as it is for graduate and undergraduate students who are expected to not overlook any important insight or theory. Interdisciplinary research is a process where the necessarily incomplete research of one scholar is built upon by others. No research is ever completely comprehensive. The process of narrowing

(Continued)

(Continued)

may occur also in collaborative research, where interdisciplinary teams conducting basic research are limited by their budget or by the availability of researchers from particular disciplines. Identifying the most relevant disciplines may involve revisiting the formulation of the research question (STEP 1) and undertaking the full-scale literature search (STEP 4).

Recall from Chapter 2 that there is a tradeoff between manageability and creativity at this STEP. Graduate students and scholars may wish to keep an eye on disciplines that seem only tangentially related to the problem *precisely because* these disciplines may have a novel

contribution to make. Any researcher that does not face really tight time constraints may find that they gain valuable and novel insights by looking further afield.

With respect especially to Questions 2 and 3, it is important to remember that particular disciplines may only have insights or theories relevant to just a subset of the relationships involved in your research problem as you have mapped this above. One of the purposes of mapping is indeed to guide you in identifying most relevant disciplines. You should make sure that you include a discipline that addresses each relationship that you want to address in your research.

Applying These Questions to the Problem of Human Cloning

Reading the literature on the problem of human cloning with these questions in mind heightened student awareness of not only the amount of disciplinary activity among the relevant disciplines, but also the *differing* insights produced by these disciplines. By asking these three questions of each relevant discipline, the student was able to reduce the former list of seven to five disciplines that were *most* relevant to the problem of human cloning. These are listed here along with an explanation for their selection:

- Biology: The cursory literature search found that more biologists are writing about human cloning than are scholars from any other discipline. This is understandable because human cloning is itself a biological procedure. Students also found that biologists are advancing some of the most important theories on human cloning and are expressing the greatest diversity of opinion on this issue.

- Bioethics: Essays written by bioethicists contain important bridging concepts and methods. The essays may appear to be similar to those written by philosophers but differ from them in one important respect: They are science based.

- Philosophy: Though essays written by philosophers appear to overlap those written by bioethicists, there are important differences. For one thing, the essays are not science based, but are grounded in humanistic ethics, thus offering a perspective that contrasts sharply with that of bioethicists on this issue. For another, philosophers tend to exclude important bridging concepts and methods that bioethicists tend to include.

- Religious Studies: Religion and the world's major faith traditions are among the most powerful influences in our society today. This explains, for example, why U.S. congressional hearings on "hot button" social issues such as human cloning typically include taking testimony from representatives of the major faith traditions. Therefore, including the perspective of religion seemed appropriate given the amount of attention religious studies scholars have devoted to this issue, the popular interest in their views, and the need to understand value systems that are faith based rather than empirically based.

- Law: Though the amount of legal scholarship on the issue is far smaller than that from the other disciplines, law offers insights that approach the issue from a unique perspective and is therefore pertinent.

In the end, course constraints required limiting the number of disciplines to three. The decision to consider biology, philosophy (i.e., humanistic ethics), and religion as "most" relevant to the problem of human cloning was made based on the criteria already noted. Whether these criteria or others are used to differentiate between disciplines that initially appear to be relevant and those that are in fact most relevant, the essential thing is to develop some means by which to identify and justify the disciplines ultimately used and make this decision-making process explicit. Later STEPS in the IRP will validate whether the disciplines selected are in fact the most relevant.

Chapter Summary

STEP 3 of the interdisciplinary research process involves taking three actions: (1) selecting disciplines that are potentially relevant to the problem because it falls within their research domains, (2) mapping the problem to identify its various disciplinary parts, and (3) reducing the number of disciplines to those that are most relevant. Students are well advised to think through the problem and use both the phenomena and perspective approaches to identify disciplines potentially relevant to the problem. The STEP of identifying relevant disciplines requires, among other things, that students have a clear understanding of the overall behavioral pattern of the problem they are studying. Students are urged to map the problem to reveal its disciplinary parts and causal linkages. Perceiving linkages and cause-and-effect relationships necessary to deal with complex problems is not possible using a traditional single-discipline approach.

It is in dealing with complex real-world problems that the interdisciplinary research process proves its analytical power and demonstrates its unmatched ability to construct a more comprehensive understanding. A simple research map, for example, can help students new to interdisciplinary research visualize the process from beginning to end. More advanced students working with more complex problems can benefit from using

(Continued)

(Continued)

concept, theory, and system maps to break down the problem into its constituent parts and see how the parts relate to the whole. By asking the three prescribed questions of each potentially relevant discipline, students should have little difficulty in identifying those that are most relevant. This process serves three practical purposes: (1) It deepens the student's understanding of the problem, (2) it may reveal the need to refine or restate the research question, and (3) it will make the full-scale literature search more productive.

Notes

1. The mental representation of the internal organization of a problem or system may be expressed in a text format or depicted by a visual-spatial technique called concept or knowledge mapping (Czuchry & Dansereau, 1996, pp. 91–96). The knowledge map is composed mainly of nodes of various shapes, sizes, and colors with meaningful links that are labeled to designate the relationships among the nodes. Succinct statements are incorporated in each node, and the links are identified by a letter abbreviation. The links may indicate characteristics, components, consequences, direction of action, outcomes, predictions, subsets, or subtopics; therefore, a legend is included to explain the meaning of the labels for the links. The spatial arrangements of the nodes on the map may be hierarchical, radial (spider), chain, flowchart, or even multidimensional; some map designers have positioned the nodes as trees, ladders, bridges, rockets, or other symbolic shapes for the theme of the map. The most significant purpose of the map's configuration is to depict relationships among the various aspects of the concepts in each node; consequently, the nodes may also be arranged to designate inclusions, exclusions, overlapping concepts, or chronological sequences (C. Atha-Weldon, personal communication, March 2005). Szostak (2004) speaks of mapping the causal links among relevant phenomena to show which phenomena are implicated in a particular research question and how they are related. The researcher might even map which theories are implicated along different links. A useful base from which to sketch such a map when doing work in the social sciences or humanities is his list of phenomena that appears in Chapter 2 as Table 2.4. Table 2.3 is useful in identifying natural science phenomena.

2. Machiel Keestra (2012) presents an example of a particular application of systems thinking. He initiates a promising discussion of the advantages of focusing on mechanisms whereby scholars can engage in intervention, stimulation, or activation experiments with components or operations of individual mechanisms, and look for consequences on that level in a way they usually cannot when testing theories about humans.

3. "When one finds nonlinear relationships, they are normally in reinforcing (i.e., positive feedback) loops, not balancing loops. But not all reinforcing/positive feedback loops involve a nonlinear relationship. Positive feedback loops may merely reinforce or strengthen a relationship (in which case no nonlinear relationship may be involved), but some of them can produce new (not just enhanced) outcomes because the intensified relationship has reached a tipping point (what complex theorists call a bifurcation point) or threshold (in which case a nonlinear relationship is probably involved" (William H. Newell, personal communication, January 8, 2011).

Exercises

Potentially Relevant Disciplines

4.1   Identify disciplines *potentially* interested in the following questions:

- Should the international space station continue to be funded?
- Should the prison camp for alleged terrorists at Guantánamo Bay be closed?
- Should novels with racially pejorative words be used in literature courses in public schools?

Think Systems

4.2   Systems thinking promotes interdisciplinary learning and facilitates the interdisciplinary research process. Create a system map to describe the effects of a factory closure on a local economy.

Mapping

4.3   This chapter has argued that you should map the problem as part of the process of identifying relevant disciplines. Identify the kind of map that would best aid your understanding of each of the following problems, and explain:

- The causes of homelessness
- Why couples divorce
- How to create a new business

Getting Perspective on the Problem

4.4   Use the examples presented in Tables 4.2 through 4.5 to show how the disciplines and their perspectives can illuminate aspects of the problem of teens dropping out of high school.

Image by Free-Photos from Pixabay

5

CONDUCTING THE LITERATURE SEARCH

LEARNING OUTCOMES

By the end of this chapter, you will be able to

- Explain the meaning of "literature search"
- Identify the reasons for conducting the literature search
- Identify the special challenges confronting interdisciplinarians
- Explain how to conduct the *initial* literature search
- Explain how to conduct the *full-scale* literature search

GUIDING QUESTIONS

Does it make any difference *when* you conduct an interdisciplinary literature search?

In what ways is conducting an interdisciplinary literature search different from a disciplinary literature search?

What challenges do you face in conducting an interdisciplinary literature search, and how can you overcome these?

CHAPTER OBJECTIVES

Basic to any research effort is the systematic gathering of information about the problem: the literature search. The integrated model of the interdisciplinary research process (IRP) introduced in Chapter 3 places the literature search at STEP 4. However, this placement is somewhat arbitrary because the search for information is a dynamic and uneven enterprise that spans multiple STEPS of the IRP. While there is no "right time" to take this STEP for each and every research project, the literature search must begin early on and is often conducted in phases, beginning with (or even preparatory to) STEP 1, defining the problem.

After defining *literature search* in the context of interdisciplinary studies, we present reasons for conducting a *systematic* literature search. Our primary focus is on how to conduct a library-based interdisciplinary literature search, which we divide into two substeps: the *initial* search and the *full-scale* search. We discuss these in detail, the unique challenges they present to interdisciplinarians, and strategies for overcoming them. Once the search process is completed, you can proceed to constructing the annotated bibliography (if required) and writing the interdisciplinary literature review (if required), although these subjects are beyond the scope of this chapter and this book. Detailed descriptions of the process and challenges the annotated bibliography and literature review pose to interdisciplinarians are found in Newell (2007b).

⟳ **DRAWING ON DISCIPLINARY INSIGHTS**

1. Define the problem or state the research question.

2. Justify using an interdisciplinary approach.

3. Identify relevant disciplines.

4. **Conduct the literature search.**

   ○ **Understand the meaning of the literature search.**

   ○ **Appreciate the reasons for conducting the literature search.**

   ○ **Special challenges confronting interdisciplinarians.**

   ○ **The *initial* literature search**

   ○ **The *full-scale* literature search**

5. Develop adequacy in each relevant discipline.

6. Analyze the problem and evaluate each insight or theory.

MEANING OF LITERATURE SEARCH

The process of gathering scholarly information on a given topic is the domain of the **literature search**, although the term commonly used in the natural and the social sciences is *literature review* (Reshef, 2008, p. 491).[1] This search carefully examines previous research in journals, books, and conference papers to see how other researchers have addressed the problem. The literature search also serves to demonstrate adequacy in the disciplinary literature on the problem, how the current project is connected to prior research, and to summarize what is known about the problem.

REASONS FOR CONDUCTING THE LITERATURE SEARCH

Before performing this STEP, it is useful to reflect on the reasons for conducting the literature search.

Reason #1: *To save time and effort.* Discovering what is already known about the problem at the outset will prevent unwitting duplication of work that has already been done. It may also guide you to take the project in a new direction.

Reason #2: *To discover what scholarly knowledge has been produced on the topic by different disciplines.* **Scholarly knowledge** is knowledge that has been vetted by a discipline's community of scholars through its peer review process. **Peer review** means subjecting an author's scholarly paper or book manuscript to the scrutiny of experts in the field who evaluate it according to academic standards that are viewed as fair and rigorous by the discipline's members. Scholarly knowledge includes journal articles and books (published by university, academic, or commercial presses). Conference papers are often not peer reviewed; but if they are, they are held to a far less exacting standard. Consequently, conference papers are less reliable than journal articles but more reliable than websites, at least on average. The disciplinary nature of much of this literature is easy to identify. However, the growing number of interdisciplines and their journals is complicating the task of neatly dividing scholarly sources by discipline. Students should exercise caution when using online literature because it may not have been peer reviewed. You should consult librarians and disciplinary experts to assess the credibility of any online document.

Reason #3: *To narrow the topic and sharpen the focus of the research question.* For example, you may be interested in the topic of terrorism but quickly discover that the topic is too broad and the amount of literature on it too vast to be approached during the time allotted. This does not mean that you should abandon the topic. Rather, you will have to narrow the topic by focusing on one aspect of terrorism such as a particular form of terrorism, or terrorism in a particular region or historical period, or the causes of a particular form of terrorism. Narrowing the topic will make the literature search more manageable and rewarding. If, however, the topic is focused too narrowly, there may no longer be sufficient literature on the topic in two or more disciplines. This would likely mean that undergraduates could not pursue interdisciplinary analysis of the too-narrow topic, though graduate students, scholars, or interdisciplinary teams might be able to perform missing disciplinary analyses. The literature search will help you decide whether the topic is ripe for interdisciplinary inquiry.

Reason #4: *To explain how the problem developed over time.* Every problem has a history. Interdisciplinary writing often traces the historical development of a problem. It is useful to include this "historical development" information in the introduction to your paper.

Reason #5: *To identify prior disciplinary research and understand how the proposed interdisciplinary project may extend our understanding of the problem.* Reading the literature of each discipline that has addressed the problem reveals its perspective on the problem and deepens your understanding of its insights into the problem.

Reason #6: *To situate or contextualize the problem.* In general, this involves identifying the web of interrelationships in which the problem is embedded. These connections often surface in the process of applying systems thinking to the problem and by drawing a system map as described in Chapter 4.

Reason #7: *To develop "adequacy" in the relevant disciplines* (STEP 5 of the IRP and the subject of Chapter 6). Developing adequacy involves identifying the defining elements of each discipline's perspective and being alert to how the scholarship produced by the relevant disciplines often uses differing terminology (i.e., concepts) and differing explanations (i.e., theories) to describe similar problems. It is critical, therefore, for you to identify the relevant disciplines early on and compare their terminology, looking for differences and similarities in meaning. It is important to keep track of this information as you read.

Reason #8: *To verify that the disciplines identified at the end of STEP 3 as most interested in the problem are really relevant.* Narrowing the number of disciplines to those that are most relevant is possible only by conducting a literature search.

Reason #9: For graduate students or scholars who plan on doing their own research on the topic, the literature search highlights what sorts of research might be valuable: What theories and methods might be usefully applied to the question? What new relationships among phenomena might be investigated? Graduate students and scholars need to "read between the lines," thinking as much about what is missing as what is there.

SPECIAL CHALLENGES CONFRONTING INTERDISCIPLINARIANS

Though many research activities are common to both disciplinary and interdisciplinary research, special challenges confront interdisciplinarians in the literature search.

More Literature Must Be Searched

The problems studied by interdisciplinarians are more complex and often broader than those studied by single-discipline scholars. Thus, there is simply more literature to be searched. An interdisciplinary research project requires integration of insights and theories sourced in several disciplines. In contrast, a disciplinary research project requires searching the literature of only a single discipline. In fact, the term *literature search* is actually a misnomer because an interdisciplinary research project typically requires several separate literature searches, one for each of the potentially relevant disciplines (Newell, 2007b, p. 92).

Researchers Risk Being Seduced by What Disciplinary Experts Say

Students unfamiliar with interdisciplinarity and the research process risk becoming seduced by the existing literature. This seduction may lead to being overly impressed by a particular writer's approach *before first developing a clear understanding of how the problem should be approached in an interdisciplinary way.* This might mean concluding at the outset of the project that a particular author or a particular theory holds the key to fully understanding the problem. It may, but probably not. *The nature of interdisciplinary work is to suspend judgment and patiently go through the process, letting the process reveal if one's earlier assumption is in fact accurate.* (*Note:* Arvidson [2016] discusses how the steps in the IRP are similar to those recommended in the philosophy of phenomenology. He stresses how phenomenology also urges the researcher to suspend judgment.)

Library and Database Cataloging Methods Disadvantage Interdisciplinary Researchers

The current method of cataloging in libraries serves the interdisciplinarian poorly because it is organized along disciplinary lines rather than in terms of a comprehensive list of phenomena, concepts, theory types, methods, and so on. Library catalogs are not set up to connect different parts of the problem (or to connect related problems) studied by different disciplines. Nor are they set up to identify the same or similar problems that are given different labels by different disciplines. The present system of classification also makes it difficult to search by the *name* of a phenomenon. (*Note:* Martin [2017] discusses the challenges of interdisciplinary literature searches, with a focus on how librarians can assist these. Szostak, Gnoli, and Lopez-Huertas [2016] discuss how library classification could be improved to aid interdisciplinary search.)

Furthermore, different terminology is used in different disciplines to describe the same phenomenon. For example, the health sciences have been comfortable employing the word *disability* to describe limitations in physical capabilities. But psychologists and social scientists, aware that different people react differently to physical limitations, increasingly employ the word *impairment,* which indicates a challenge rather than an absolute impossibility. Sometimes, the same word can mean different things, such as the word *investment*: For an economist, this means spending on buildings or machines that can be used to increase production in the future; for an accountant, it means any expenditure made with the intention of earning a financial return.

Another pitfall is to use the terminology used by one discipline to search the literature of another discipline. To be sure, books may be classified with respect to multiple subjects, as in this example from Miami University's online catalog that references John Larner's 1999 book, *Marco Polo and the Discovery of the World.* Click on the title, and it lists Author(s), Location, Subjects, Formats, Material Type, Language, Audience, Published, LC Classification, Physical Description, and Table of Contents. Under Subjects, one finds

Travels of Marco Polo

Voyages and Travels

Travel, Medieval

Asia—Descriptions and travel—Early works to 1800

Polo, Marco—Travels of Marco Polo

Polo, Marco, 1254–1323? Travels of Marco Polo

If you were interested in Asian history, travel, voyages, or medieval studies, this book might be of interest. However, without consulting the table of contents or index, you could not tell that it deals with city planning, religion, horsemanship, the invention of paper money, and other issues (William H. Newell, personal communication, January 30, 2011).

Though some of these subjects span disciplines, this only partially alleviates the problem. One way to overcome this limitation is to identify which disciplines are likely to be interested in the problem, as explained in Chapter 4, and then use resources—bibliographies and guides, dictionaries, encyclopedias, handbooks, and databases—specific to each discipline. In this way, the search can proceed systematically and productively across the relevant disciplines. A fuller discussion of the current method of library and database cataloging and strategies for overcoming their limitations is the subject of the next section.

THE INITIAL LITERATURE SEARCH

The literature search is often conducted in two phases: the initial search and the full-scale search. The initial phase begins when selecting a topic and deciding if it is researchable in an interdisciplinary sense. If it is, then the full-scale search can get under way. This second phase precedes and/or coincides with STEP 5, developing adequacy in relevant disciplines (see Chapter 6).

When initiating a search, researchers have a responsibility to think through the problem in all of its aspects to the best of their ability. True, you cannot credibly think through a problem without knowing how others "see" it. However, focusing on only what disciplinary experts "see" risks limiting your own ability to "see" the problem in its entirety (Szostak, 2002, p. 106). *Interdisciplinary research involves, among other things, looking for what disciplinarians have failed to see.* You are wise to heed Newell's (2007b) advice: "Interdisciplinary scholars need to start broadly and narrow the focus towards more specialized sources as the topic takes shape" (p. 85). Read cursorily to see whether experts in the disciplines have published research on the problem. It is better to err on the side of inclusiveness at the very outset than to conclude prematurely that a discipline is not relevant. You are more likely to make a novel discovery if you read widely. You may also enhance your creativity by taking the time to follow up tangential questions that arise as you read.

NOTE TO READERS

Throughout the search process, (1) avoid being influenced by what disciplinary experts say and "see," and (2) suspend judgment about the problem. (Admittedly, this is hard to do, especially when working on an issue where your ethical stance or faith tradition informs your view.) Above all, if you begin the literature search with a firm opinion, you should strive not to allow it to influence the literature you read and deem relevant.

Searching Your Library's Collection

In general, begin your search with books and then consult the journal literature. However, when working in the natural sciences, parts of the social sciences, and some of the newer fields such as hospitality management, begin with journals and online publications because these, not books, are the primary modes of communicating research (Newell, 2007b, p. 85).

Books and Periodical Literature

There are various kinds of books: monographs, reports of major research studies, and anthologies on the problem by different experts. The usefulness of each type of book in interdisciplinary work depends on the problem being investigated, the disciplines involved, and the required depth of the research.

With time-sensitive topics, the material in books may be dated because books require more time to research and publish (about a year from the end of writing to publication). Even so, the book material often provides pertinent background information that is an important component in research.

Experts consistently report their findings first in periodical literature, including peer-reviewed scholarly journals such as *Social Science Quarterly* and semi-scholarly professional publications such as *American Demographics*. Journal articles usually are not classified according to the subject headings used for books nor contained in the main library catalog. There are a variety of journal databases that will allow searches of journals in one or more disciplines. Experts may also report their findings as papers delivered at professional meetings, dissertations, government documents, or policy reports. Encyclopedia articles may or may not be authoritative, depending on the scholarly credentials of the writer. Like Internet sources, they should be used with great care. Students may profitably consult Chart 5.1, "Types of Publications," in W. Lawrence Neuman's (2006) *Social Research Methods* (6th ed.), which identifies publication types, examples of each type, types of writer, and the purpose, strengths, and weaknesses of each type of publication.

The Two Systems for Classifying Literature

The Dewey Decimal System (DDC) is used primarily in public libraries and small academic libraries in the United States. As the older of two systems, it is grounded in less current disciplinary knowledge.[2] It is organized hierarchically by disciplinary subject matter (moving from the general to the more specific) into categories, subcategories, specialized topics, and levels of subtopics. The system is designed to facilitate browsing, one of the strategies for searching discussed below. (*Note:* Many European libraries—public and academic—utilize the Universal Decimal Classification. This was originally based on the DDC but has evolved separately for decades. It has recently made some efforts to develop schedules of terminology that can be applied across disciplines.)

The Library of Congress System (LCC) is used by large U.S. research and academic libraries. It has many more categories (by using letters instead of numbers at the most general levels) and is more reflective of the way knowledge has developed. The LCC is designed to serve the needs of the disciplines by organizing knowledge according to disciplinary categories enabling disciplinarians to easily find books on their specialty. Although many disciplinary books contain information that falls outside disciplinary lines, this development is not presently reflected in the LCC system. Therefore, students will need to either browse the physical shelves or search the online catalog for leads.

Interdisciplinary researchers need to identify not only books that address their topic, but also books that reference it cursorily. These latter books are important to the IRP because they may bring out contrasts in disciplinary perspectives that might not otherwise be apparent. These books and their lists of sources may also identify, or at least provide, hints regarding important linkages to other concepts, theories, or information relevant to the topic. The systems maps developed in Chapter 4 may be very useful in identifying phenomena and relationships to search for.

Fortunately, catalog entries for books assigned LCC call numbers often list more than one LCC subject heading to reflect the contents of the book. These subject headings, though far from exhaustive, at least reflect *some* of the topical diversity within the book (Searing, 1992, p. 14). Once the subject headings relevant to the problem or topic are identified, these subject headings may be used to conduct more advanced searches that will yield additional references on the topic.[3]

Subject and Keyword Searches

Searching is conducted by using *subject* searches and *keyword* searches. Searching subjects and keywords is most fruitful when you have a well-defined problem or question even though you are not yet certain if it is researchable. It is also useful for identifying gaps in the literature, verifying facts, and checking for accuracy of quotes and references (Foster, 2004).

Subject Searches. When searching library catalogs, begin by using subject searches rather than keyword searches (discussed below). Subject headings are assigned by librarians who specialize in specific disciplines and who look through books to decide what they are about. The designated subject headings bring together works whose authors use different terminology, jargon, or technical language. The results are generally relevant to the topic. Sadly, many library catalogues have come to stress keyword searching—because it is easier, albeit less effective—and thus students often have to first figure out how to perform a subject search. For both subject and keyword searches, students need to remember that terminology differs across disciplines.

Keyword Searches. Keyword searches are useful primarily within a discipline's literature. Since students often have to identify information that is from unfamiliar disciplines, keyword searching must be sensitive to the terminology used in different disciplines so as not to miss important works. Researchers frequently encounter the problem of disciplines using different keywords to describe some aspect of the problem. In the example of human cloning, keywords include "control," "legislation," and "impact" to describe the concept of power as it relates to the problem. Disciplinary handbooks, dictionaries, encyclopedias, introductory textbooks, and other sources are tools useful to developing familiarity with the terminology commonly used by each discipline. Keyword searches have the added benefit of getting researchers "to consider how different people look at these same concepts and think about how the concepts are related" (Newell, 2007b, p. 86). If a student is not finding enough works in a particular discipline, it may be useful to consult a thesaurus (as discussed below) for synonyms for the search terms the student has employed.

Searching Indexes, Databases, and Other Collections

Of particular interest to interdisciplinarians are the subject-oriented indexes, databases, and other collections that organize much of the disciplinary scholarship from which they must draw. Each of these entities has its own thesaurus (a book of words and their synonyms used by a particular discipline to provide something of a bridge between the terminologies of different disciplines) or classification system based on standardized terminology from that field. The controlled vocabulary of LCC subject headings connects different catalogs, databases, and indexes (whether disciplinary, cross-disciplinary, or comprehensive), whereas the classification schemes of thesauri connect different disciplines within a particular cross-disciplinary database or index.

In both cases, they provide a bridge between the terminologies of different disciplines. One can do a keyword search within a particular discipline, but one then needs to expand it through a subject-heading or classification search to find other disciplines and their terminology before conducting a keyword search in another discipline on the same topic. Many databases allow researchers to search a keyword within a particular field, such as a subject heading, when the exact subject heading is not initially known. This is a way to do a more precise search without the knowledge of exact controlled terminology. (Newell, 2007b, p. 87)

Indexes and databases useful for interdisciplinary research are listed in the Appendix. (*Note:* This chapter emphasizes subject searching, whereas the Appendix stresses keyword searching.)

Though many databases are still restricted to a particular discipline and are organized according to its jargon, new databases are increasingly crossing disciplinary lines, and their thesauri are offering controlled vocabulary that connects works from the contributing disciplines. Fortunately for interdisciplinarians, topical databases are now available that cut across disciplines, and referencing these provides a critical source of disciplinary scholarship upon which interdisciplinarians must draw. Topical databases are especially useful when searching for relevant journal articles.

As researchers move from discipline to discipline in search of different insights on the same topic, they need to check each discipline's thesaurus to find the term(s) to search for. Starting with a good keyword and Boolean search strategy (described in the Appendix) helps locate appropriate descriptors—as subject headings in indexes are often called. Newell (2007b) gives this example in the discipline of psychology, which has done a lot of work on "gender differences":

A keyword search for [gender differences] in a thesaurus will yield something like 10,000 hits. But a check of the PsycINFO thesaurus will reveal that it is not a valid subject heading and the use of "human sex differences" instead will produce more like 60,000 hits. (p. 87)

Interdisciplinarians can also "connect the scholarship of different disciplines by building on the connections discovered by previous scholars" (Newell, 2007b, p. 87). Once you have identified a key insight on the topic, you can use a citation index to identify a wide range of subsequent works citing that key work. Citation indexes such as *Science Citation Index* and *Arts & Humanities Citation Index* cover an entire academic area. A citation index will typically capture all relevant works unless works aren't cited in the first place because others don't find them. (*Note:* Because citation indexes often fail to capture citations in books, they are much less useful in the humanities than in the natural sciences.)

Newer interdisciplinary fields such as environment sustainability studies are best searched through a combination of subject headings and citation indexes. Since interdisciplinary studies explores topics that transcend the disciplines, interdisciplinary researchers end up going back and forth between the Dewey and LCC systems for classifying knowledge, specialized systems developed by the disciplines, and more generic systems developed by librarians and by commercial information providers, including the creators of general indexes such as EBSCO's Academic Search or Gale's Academic One File.

Searching on the Internet

Students searching online often work more quickly but less deeply. When faced with numerous pages of items ("retrieval sets"), they tend to select items from only the first few pages of results. The potential danger of this is to overlook an important source embedded deeper in the results. If a keyword search produces a very large retrieval set, it may be the result of using imprecise terminology. Using more precise terminology helps to reduce the size of the retrieval set.

Searching on the Internet is widespread across disciplines, with researchers in science and medicine almost preferring to search using electronic sources (Hemminger, Lu, Vaughn, & Adams, 2007). Search engines allow concurrent searches across a wide and diverse array of sources. However, the Internet will often guide one to nonscholarly sources: These are generally recent (missing classic works) and much cited (missing works that are narrowly focused). Internet searches face the same challenges as keyword searches discussed above.[4]

Strategies for Searching

Strategies to conduct the initial phase of the literature search include browsing, probing, skimming, evaluating, and deciding.

Browsing

Browsing through a body of assembled or accessible information is essential for interdisciplinary researchers working in new and rapidly developing fields (Palmer, 2010, p. 183). For students who are able to select their own topic or problem, browsing books and/or journal articles will likely begin as part of STEP 1.[5] However, *if both the topic or problem and the disciplines to be mined for insights have been preselected, students will skip the initial phase to engage in the full-scale literature search.*

Browsing printed material in physical libraries has the potential to result in serendipitous discovery because it tends to be broad and flexible, and to lead researchers to materials that they would otherwise not have found through direct searching (Palmer, Teffeau, & Pirmann, 2009, pp. 13–14). Researchers in the arts, humanities, area studies, and languages are more likely to consider browsing printed materials for serendipitous discoveries (Education for Change, 2002, p. 25).

Although browsing printed materials on library shelves continues to be important, browsing Web-based materials is increasingly common. The Web has had a profound impact on what and how we browse, and the rate at which we can move through a diverse array of digital sources. Users, according to one study, often engage in "bouncing" or "flicking," moving rapidly from one site to another and occasionally returning to explore material in depth (Nicholas, Huntington, Williams, & Dobrowolski, 2004). For example, the Web allows students to easily browse the tables of contents of journals and books. Web browsing can lead researchers to more conventional library resources that otherwise might not have been pursued.

Probing

Interdisciplinarians use probing to locate relevant information that falls outside their discipline or area of expertise when standard searching and browsing techniques are inadequate. Whereas browsing is often ad hoc in nature, probing is a more deliberate effort to locate information in disciplines thought to be relevant. "Researchers probe into peripheral areas outside their expertise," explains Palmer (2010), "to increase breadth of perspective, to generate new ideas, or to explore a wide range of types and sources of information" (p. 183).

To aid your probing, consult disciplinary research aids, including bibliographies, encyclopedias, dictionaries, handbooks, companions, and databases, *after* drawing up a list of potentially relevant disciplines. For example, an indispensable source of disciplinary references for students working in the social sciences is *Social Science Reference Sources* (Li, 2000).

NOTE TO READERS

As you probe the literature of unfamiliar disciplines and encounter new ideas, concepts, theories, and methods that you do not fully understand, you need to develop familiarity with the terminology to carry out your research. Resolving differences in vocabulary and terminology requires "translation work." This work is essential, says Palmer (2010), because "valid interdisciplinary research is necessarily based on a deep understanding of how concepts, methods, and results fit in the body of discourse and practice in which they were developed" (p. 183). Interdisciplinary research teams are continually engaged in translation work throughout the project. In fact, says Palmer, "learning enough about the perspectives and problems driving the interests of collaborators appears to be a key factor in the success of interdisciplinary research groups" (p. 183).

Skimming

Skimming is done in the initial phase of the project to (1) establish whether the topic or problem is researchable in an interdisciplinary sense and (2) develop a general understanding of the problem and its parts before consulting the more specialized journal literature in the full-scale literature search. You should start broadly and then narrow your search to more specialized sources during the full-scale search described below.

The action of skimming printed materials involves flipping through a volume in the library looking for key components beginning with the abstract (if it is a journal article), preface or introduction (if it is a book), or table of contents (if it is a book), then moving to section or chapter headings, lists, summary statements or conclusions, definitions, and illustrations. This process is applicable to digital documents where search features (such as the "Look Inside" function offered by Amazon.com for books) make it easier to pinpoint keywords, theories, and so on. Researchers are making increased use of journal abstracts in full-text databases (Nicholas, Huntington, & Jamali, 2007). However, students are cautioned not to substitute skimming for eventually reading the full publication. Scholars often begin with skimming the preliminary parts of a document such as the table of contents and introduction, but they then return to the document if it seems important to their research (Tenopir, King, Boyce, Grayson, & Paulson, 2005).

NOTE TO READERS

Skimming is useful for undergraduates, solo inter-disciplinarians, and graduate students who want to know how a particular discipline might address the topic or problem, and for interdisciplinary teams who often expect their disciplinary members to do new disciplinary research. A cursory search will provide an initial (though not conclusive) estimate of what and how much has been written about the topic, and enable one to develop a "feel" for the scholarly conversation. However, it is wise to defer a final decision about the feasibility of the topic until you complete the full-scale search wherein you have closely read all authors' insights to see if they are talking about the same thing.

Evaluating

As you skim books and journal articles (i.e., insights), you must make quick decisions about their relevance and utility. Here are some questions that you should ask about each book and journal article:

- Does it cover the topic as a whole or at least some aspect of it?

- Does it have something noteworthy to say about the topic or some aspect of it?

- Was it published recently?

- Is it peer reviewed (i.e., does it appear in an academic journal or as a book published by a university, academic, or commercial press that publishes scholarly work)?

- Has it been reviewed (if in book form)?

The more recent the publication, the more valuable are its bibliography and endnotes (or footnotes) because they provide authoritative connections to other work.

Deciding

The point of the initial literature search is to decide whether there is enough information on the problem to make it researchable in an interdisciplinary sense. Once you have reached your decision, note that it may not be final until you have conducted the full-scale search and read the relevant disciplinary insights closely.

Mistakes to Avoid When Beginning the Literature Search

Those new to interdisciplinary research commonly make two mistakes in the initial phase of the literature search. The first is failing to pay close attention to the disciplinary source of the books and articles being gathered. One unintended consequence of this oversight is to end up with a large number of insights drawn from one or two disciplines at the expense of insights from other equally important disciplines. Interdisciplinary research cannot succeed on this basis. *It is important to keep track of which discipline produced which insights.* If an author's disciplinary affiliation is not readily evident, a search of the writer's name may provide this information. Rigorous interdisciplinary research involves striking a balance between disciplinary depth and disciplinary breadth. This balance is seldom realized in undergraduate education, however, where students are typically limited to gathering a few insights from only two or three disciplines.

A second mistake is being unduly influenced by the *quantity* of insights that a discipline has generated. For example, if the initial search reveals that sociology has contributed only one or two essays on the problem, one may wrongly conclude that sociology is not as relevant to the investigation as other disciplines, each of which has produced several insights. *Quantity of material produced by a discipline should never decide disciplinary relevance.* Why? It may be that the one insight authored by a cultural anthropologist (and the *only* essay on the problem by any cultural anthropologist) contains information of such importance that an interdisciplinary understanding of the problem would not be complete without it. (*Note:* An advantage of multiple insights on the problem by a discipline is that you can then gauge whether a particular insight is accepted or controversial within that discipline.)

But what about when, for example, sociology has not addressed the specific topic, say, of rave music? Can one conclude that sociology has nothing to contribute to the discussion? Not necessarily. Sociologists may have addressed house music or punk rock or musical subcultures preceding rave, or investigated some other social phenomenon that influences rave music. Researchers should be open to possibilities that are not initially apparent. Many topics turn out to be subsets of a larger topic. Rather than giving up on sociology upon discovering that it has not examined rave, one should look at the larger context—in this case, what preceded rave. If a discipline has treated a more general category to which a topic or phenomenon is related, it would be quite surprising if that treatment does not have some direct applicability to the narrower topic.

Students are sometimes surprised to learn that a discipline that initially appears not to be relevant has produced insights into the problem after all. These discoveries are possible only by conducting a full-scale literature search. In this effort, students can again profitably use Table 2.4 (Chapter 2) to cross-check traditional disciplinary searches.

NOTE TO READERS

Graduate students and practitioners should look to see whether core concepts or theories from a particular discipline might usefully be applied to the topic, even though no one in the discipline has done so.

THE FULL-SCALE LITERATURE SEARCH

The initial search quickly evolves into the full-scale literature search once you have decided what the project should be about and that it is researchable. Conducting the full-scale search involves a more detailed application of the search strategies we have already discussed. It also involves identifying *all* relevant insights and theories on the topic.

The full-scale search is based on an assumption foundational to interdisciplinarity: that no community of scholars has been entirely wasting its time over the past decades or centuries. This assumption should motivate those who are genuinely interested in understanding a particular problem to discover what others have written about it. Interdisciplinary problem solving and scholarship involve integrating insights and theories from multiple disciplinary sources. These deserve to be brought together and organized into a coherent whole in a way that does not privilege any one perspective, theory, or viewpoint (Szostak, 2009, p. 9).

The full-scale literature search is as interested in disciplinary breadth (i.e., how many disciplines have written on the topic) as it is in disciplinary depth (i.e., the number, quality, and variety of disciplinary insights on the topic) (see Box 5.1). *Researchers at any level should always err on the side of inclusiveness during the full-scale search.* Insights that are only marginally relevant will be identified later on and then discarded. But *inclusive* does not mean "open-ended." Inclusive refers not to the quantity of disciplinary insights but to their *quality and diversity*. The challenge for the student is to keep these to a manageable number.

BOX 5.1

Conducting the full-scale literature search reflects the belief of many interdisciplinarians that there should be a symbiotic relationship between interdisciplinary research and disciplinary or specialized research. As Szostak (2009) notes, interdisciplinary integration is only possible because of the concerted efforts of thousands of scholars across all disciplines. Their insights provide both the disciplinary depth and the disciplinary breadth that makes interdisciplinary integration possible. Therefore, none of these insights can be taken for granted; each insight must be kept track of for signs of disciplinary perspective, including theories and methods employed, and then compared and contrasted with each other (Szostak, 2009, p. 7).

NOTE TO READERS

For graduate students and practitioners, the full-scale literature search is a way to immerse themselves in the scholarly conversations on the topic and achieve mastery of the problem in all of its complexity. For them and for members of research teams who already know what the project is about, the full-scale literature search provides baseline knowledge about the problem and allows team members who are going to specialize in only one part of the problem to become knowledgeable about other aspects with which they may be unfamiliar. It also serves to identify what is missing from existing research and thus ripe for further inquiry.

Disciplinary Sources of Knowledge

Conducting the full-scale literature search requires closely *reading* insights from multiple disciplinary literatures for specific content, as well as *organizing* this information.

Closely Read Multiple Disciplinary Literatures

Interdisciplinarians face a reading challenge that disciplinarians do not face. While disciplinarians are responsible for reading only their discipline's literature relevant to the problem, interdisciplinarians are responsible for reading insights in multiple disciplinary literatures. More particularly, they must pay attention to each discipline's perspective on the problem (in an overall sense) as well as keep track of each insight's theories, key concepts, and methods. As you closely read about each insight, identify the following:

- Name and discipline of the author

- Disciplinary perspective of the insight

- Thesis or argument of the insight and its place in the disciplinary literature

- Assumption underlying the insight

- Name of the theory

- Key concepts embedded in the insight

- Method (disciplinary) used

- Phenomena addressed and their relationships (This is invaluable for mapping the problem and organizing the insights.)

- Bias (ethical or ideological) of the insight

Organize Information

A literature search that involves working in only a few disciplines and reading only a handful of insights nevertheless entails gathering, organizing, and comprehending a considerable amount of information. To avoid becoming overwhelmed as the reading proceeds, students are encouraged to organize this information in some systematic way so that it can be easily accessed when performing subsequent STEPS of the IRP. *Experience has shown that the time spent in organizing and recording this critical information will be more than offset by the time saved in retrieving this same information as subsequent STEPS of the research process are performed.*

Table 5.1 demonstrates an effective way to organize the information related to each disciplinary insight. This table, created in Word or Excel, helps you visualize the progress of the search, make decisions about the need for further searching, identify which disciplines are most relevant to the problem, and select which insights should be closely read.

Academic papers and books are often organized around headings: Different chapters or sections describe the theories and methods explored, and the phenomena/data engaged. You should be aware of several types of statements you may encounter, including statements of opinion and statements of motivation. Authors may be guided by ethical precepts, and they may or may not be explicit about these. You should keep track of any ethical statements you encounter; we will discuss in Chapter 10 how to address ethical conflicts that may generate conflicts in insights. (*Note:* For detailed advice on critical reading of texts, see Repko, Szostak, and Buchberger [2020]).

TABLE 5.1 ● Data Table of Things to Track Concerning Each Insight								
Author	Disciplinary Perspective	Thesis	Assumption	Theory Name	Key Concept(s)	Method	Phenomena Addressed	Bias
Post	Psychology (cognitive)	"Political violence is not instrumental but an end in itself. The cause becomes the rationale for acts of terrorism the terrorist is compelled to commit." (Post, 1998, p. 35).	Humans organize their mental life through psychological constructs.	Terrorist psycho-logic	Special logic (Post, 1998, p. 25)	Case study	Individual human agents	Suicide terrorists are irrational.

Note: If constructed using Excel, the table can be easily expanded horizontally to include additional information about any one insight as well as vertically by adding as many insights as necessary. The utility of this table for interdisciplinary work on the undergraduate level will be increasingly evident as the interdisciplinary research process unfolds. It is best to use the author's own words to eliminate the possibility of skewing the writer's meaning, which may occur when paraphrasing. Interdisciplines, schools of thought, and the applied fields should be treated in the same way as traditional disciplines.

Consult Disciplinary Experts

Interdisciplinarians regularly consult disciplinary experts. The expert on the topic (or on some aspect of it) is able to provide authoritative feedback on the feasibility of the project, inform the researcher on current research activity under way or nearing completion, and direct the researcher to the most important literature, including conference papers and published or soon-to-be-published journal articles and books. Students working on a topic of their choice are well advised to consult faculty who may provide valuable insights concerning the project's feasibility, references to key publications, and practical advice.

Nondisciplinary Sources of Knowledge

Interdisciplinarians do not assume that *all* relevant knowledge has been generated by the disciplines. They know that relevant information sometimes comes from sources that are nondisciplinary. This knowledge has not been produced by trained disciplinary scholars, nor has it been vetted by the disciplines. While it may not be of interest to the disciplines or may be overlooked by them, it is nonetheless relevant to the inquiry. Such knowledge may include oral histories, eyewitness testimonies, artifacts, and artistic creations. As we saw in Chapter 1, transdisciplinary research seeks actively to involve people from outside the academy. These other sources of knowledge are useful or even necessary in a particular context or to think inclusively about a specific concern.

Interdisciplinarians are acutely aware that such knowledge has dramatically different standing in the academic world than knowledge that has stood the test of expert scrutiny. *All knowledge is not equally valid.* Under certain circumstances, these other sources of knowledge may achieve credibility in the academy and even find their way into the literatures of the disciplines. In women's studies, for example, testimonial or "lived experience" plays a crucial role. In native studies, "traditional knowledge preserved over centuries through oral tradition and interpreted by elders is central" (Vickers, 1998, p. 23). The willingness to consider using nondisciplinary knowledge sources is based on the assumption that knowledge accumulates and that people learn from and build upon what others have done (Neuman, 2006, p. 111). Today's knowledge is the product of yesterday's research, and tomorrow's knowledge will be based on today's research.

While knowledge produced by the disciplines, compared to these other sources of knowledge, is generally considered the proper focus of the modern academy, Richard M. Carp (2001) urges interdisciplinarians not to limit their search for relevant information to the disciplines, for the simple reason that their knowledge formations are incomplete (p. 98). Interdisciplinarians, he argues, should be more imaginative, more inquiring, and more self-reflective about what knowledge they are willing to use (p. 84). Historians, sociologists, anthropologists, and other disciplinary experts are constantly mining these diverse sources of knowledge for their own disciplinary purposes and often go on to publish their findings. In this way, nonscholarly knowledge finds its way into the academy. Oral histories of migrant workers are a good example of the kind of nonscholarly knowledge that is gathered and presented in a way that disciplinary scholars can accept. Investigating the high cost of health care may include the testimony of health care providers as part of one's study. (*Note:* Interdisciplinary students can attempt to gather and analyze nonscholarly knowledge themselves but should not use it until it has been vetted by experts. Once vetted, it may be used, provided that it is clearly identified and used in a scholarly way. *As a general rule, students should be skeptical of insights that have not been carefully tested by experts.*)

Chapter Summary

The literature search is itself a process within the larger IRP and spans the early STEPS of that process, beginning with STEP 1. This chapter defines the term *literature search* in an interdisciplinary sense, presents reasons for conducting the literature search, and discusses the special challenges confronting interdisciplinarians. The chapter's primary focus is on how to conduct a library-based interdisciplinary literature search. It divides the search process (somewhat arbitrarily) into two substeps: the initial search

(Continued)

(Continued)

and the full-scale search. The initial search begins as one is selecting a topic and deciding whether it is researchable in an interdisciplinary sense. Once this decision is made, the full-scale literature search can get under way. A successful search will confirm that the topic is researchable and is indeed appropriate to interdisciplinary inquiry. Important to later STEPS in the research process, the successful full-scale literature search will have identified the most relevant disciplinary perspectives and their insights, though final confirmation will have to await the completion of STEPS 7 and 8 (see Chapters 9 and 10). The chapter warns against limiting one's reading to familiar disciplines, while ignoring unfamiliar ones. The chapter urges researchers to categorize insights by disciplinary perspective, which is a research practice distinctive to interdisciplinary studies. The chapter also urges researchers to devise some method of organizing this information so that it can be easily retrieved during later STEPS of the IRP.

If the literature search is hurried, it may not reveal all the relevant disciplines. Those unfamiliar with the interdisciplinary research process may be tempted to end the literature search prematurely after finding a handful of sources produced by a few disciplines or a few experts. But subsequent STEPS in the research process will expose incomplete work done in earlier STEPS. The next chapter advances the research process by explaining how to develop adequacy in the relevant disciplines.

Notes

1. A thorough discussion of the literature review from a social science perspective is in Neuman (2006) and in Hart (1998). *Literature review* is an umbrella term that refers to specialized reviews including context review, historical review, integrative review, methodology review, self-study review, and theoretical review (Neuman, 2006, p. 112). Social scientists typically conduct the literature review at the outset of their research. Also see L. G. Ackerson (2007), a leading engineering and science librarian and scholar whose particular areas of expertise are "user information seeking behaviors" and "information resources for interdisciplinary research," who prefers the term *literature search* (p. vii). A literature *review* typically includes explaining how the disciplines have approached the problem over time—how disciplinary scholars tried to improve upon previous research by developing concepts, theories, and methods of investigation to better understand the problem—but how these have failed to provide either a comprehensive understanding of the problem or a satisfactory solution to it, thus necessitating an interdisciplinary approach.

2. Both LCC and DDC were first developed in the nineteenth century and have struggled to adapt as disciplines evolve to take on new topics and new interdisciplines emerge.

3. The best treatment of this topic from an interdisciplinary perspective is Chapter 6 in Fiscella and Kimmel (1999).

4. An excellent discussion of the problem of sources, reliability of sources, and finding and evaluating sources on the Internet is Chapter 5 of Booth, Colomb, and Williams (2003) and Chapter 11 of Repko et al. (2020).

5. The disciplines vary in the importance that they attach to books compared to peer-reviewed journal articles. Scholars in the humanities are as likely to publish their research in book form as they are to publish in peer-reviewed journals. This is less so in the social sciences and far less so in the natural sciences.

Exercises

Special Challenges

5.1 Concerning the problem or topic that you are investigating, what challenges have you encountered in your initial searching that are discussed in "Special Challenges Confronting Interdisciplinarians"? Is there a challenge that you have encountered that the discussion overlooks?

Initial Searching

5.2 Constructing a quality building depends on two things: using quality materials and having a blueprint or map of its parts and their interrelationship. This applies to constructing the building's foundation. Mapping the problem and stating it (or framing the research question) are critical to constructing the foundation of your research project. Concerning your project, what materials do you need to gather that will enable you to decide whether the topic is researchable in an interdisciplinary sense?

5.3 What subject headings and keywords are you using to search for literature on your topic? Are these more helpful when searching your university's databases or when using commercial search engines such as Google or Bing?

5.4 How have the strategies of browsing, probing, skimming, and evaluating helped you to decide if your topic is researchable in an interdisciplinary sense? After reflecting on "Mistakes to Avoid When Beginning the Literature Search," what must you do to make the topic researchable in an interdisciplinary sense?

Going Full Scale

5.5 How have you addressed the organizational challenges that your project involves? As you proceed with the full-scale search, is your organizational plan able to accommodate the growing number of insights and related bits of information?

5.6 Select the most recently published insight (either a book or a journal article) and read it for the specific content listed under the subheading "Organize Information" and demonstrated in Table 5.1. Then, scrutinize the sources/bibliography and notes/endnotes to make connections to other sources related to your topic.

5.7 After mapping your problem and conducting the full-scale literature search, have you discovered a gap in existing research that might be filled if you were to include nondisciplinary sources of knowledge?

©iStockphoto.com/fizkes

6

DEVELOPING ADEQUACY IN RELEVANT DISCIPLINES

LEARNING OUTCOMES

By the end of this chapter, you will be able to

- Explain what it means to develop adequacy in relevant disciplines
- Explain how to develop adequacy in relevant theories
- Explain how to develop adequacy in disciplinary methods
- Demonstrate how to use and evaluate methods in basic research
- Demonstrate how to provide in-text evidence of disciplinary adequacy

GUIDING QUESTIONS

How much do you need to know about a discipline to be able to effectively evaluate its insights?

How do you develop an adequate appreciation of the disciplines that you draw upon?

How do you indicate in your writing that you have an adequate appreciation of each discipline?

How can you evaluate the use of disciplinary theories and methods?

CHAPTER OBJECTIVES

This chapter explains how those new to interdisciplinarity can develop adequacy in disciplines relevant to the problem. Adequacy involves deciding *how much* and *what kind* of knowledge is required from each discipline's insights. It also involves identifying relevant theories and understanding disciplinary methods. The chapter examines the interdisciplinary position on methods, the quantitative versus qualitative methods debate, and how a discipline's preferred methods correlate to its preferred theories. The chapter discusses how advanced undergraduates and graduate students can incorporate their own basic research with the interdisciplinary process. It concludes by explaining the importance of providing in-text evidence of disciplinary adequacy.

 A. DRAWING ON DISCIPLINARY INSIGHTS

1. Define the problem or state the research question.

2. Justify using an interdisciplinary approach.

3. Identify relevant disciplines.

(Continued)

(Continued)

4. Conduct the literature search.

5. **Develop adequacy in each relevant discipline.**
   o **Explain the meaning of adequacy.**
   o **Develop adequacy in theories.**
   o **Develop adequacy in disciplinary methods.**
   o **Use and evaluate disciplinary methods in basic research.**
   o **Provide in-text evidence of disciplinary adequacy.**

6. Analyze the problem and evaluate each insight or theory.

THE MEANING OF ADEQUACY

Adequacy (in an interdisciplinary sense) is an understanding of each relevant discipline's cognitive map, sufficient to identify its perspective on the problem, epistemology, assumptions, concepts, theories, and methods to understand and evaluate its insights concerning the problem.

Adequacy Calls for Knowing the Goal of the Research Project

A rule of thumb is that the more ambitious the goal, the more knowledge (i.e., disciplinary depth and breadth) is required. If your goal is to integrate only a handful of insights or theories, then the disciplinary knowledge required will be limited, and the understanding constructed will be partial. This is characteristic of undergraduate work. If, however, your goal is to achieve integration and construct an understanding of the problem that is more comprehensive than what experts have achieved thus far, then more knowledge will be required. And if your goal is to integrate your basic research with existing research on the same problem, then even more knowledge will be required. The specific knowledge called for in this last case is discussed later under the heading "Decide Which Disciplinary Methods to Use in Conducting Basic Research."

Adequacy Involves Borrowing From Each Relevant Discipline

Interdisciplinarians borrow insights from disciplines. To draw those insights, you must also develop some adequacy in the discipline's concepts, theories, and assumptions. What is borrowed are each discipline's insights into the research question, the concepts through which they are expressed, the theories on which they are based, and the assumptions in which they are grounded. Borrowing also calls for a careful evaluation of the insights that are borrowed in terms of their credibility (Are they peer reviewed?) and relevancy (Do they illuminate some part of the problem?).

We saw in Chapter 2 that disciplinary researchers will be guided in their research by the perspective of their discipline. It follows that we can only appreciate and evaluate disciplinary insights if we study these in the context of disciplinary perspective. As we will see in the next chapter, we are then guided to ask whether the insights are biased by the perspective. Sadly, many supposedly interdisciplinary researchers skip the critical step of gaining adequacy in disciplines and thus fail to evaluate insights in the context of perspectives. They then unwittingly absorb disciplinary biases and are limited in their ability to develop a more comprehensive understanding of the problem.

Disciplinary scholars have long worried that interdisciplinary research is necessarily shallow, for interdisciplinary researchers cannot possess the depth of understanding of each discipline they draw upon that is possessed by disciplinary scholars. The argument of this chapter and book is that interdisciplinary scholars can develop enough disciplinary adequacy to appreciate, evaluate, and integrate disciplinary insights. But the task is not so simple that the interdisciplinary scholar can just read an article or two in a discipline without reflecting on the nature of that discipline. That sort of "hit-and-run" scholarship is indeed shallow.

Adequacy Involves Understanding Which Disciplinary Elements Are Applicable to the Problem

Developing adequacy calls for deciding which elements of a discipline's perspective (phenomena, epistemology, assumptions, concepts, theories, and methods) are most applicable to the problem. The information provided in Chapter 2 should help you do this. As you read each disciplinary insight, keep track of what elements it contains. In some cases, epistemology and assumptions may be of critical importance, but authors seldom mention these explicitly. (*Note:* If far more disciplinary depth is required, consult the discipline-specific research aids referenced in the tables in Chapter 2.)

NOTE TO READERS

For graduate students and even solo researchers, *how much* will likely involve critiquing the perspectives and elements of the relevant disciplines, and identifying and examining linkages among their insights and theories.[1]

Adequacy Involves Appreciating Debates Within Disciplines

We have stressed above the importance of appreciating disciplinary perspective. The interdisciplinary researcher should recognize that disciplinary scholars will disagree. (See, for example, Table 6.2, which details different theories within economics regarding occupational discrimination.) Indeed, scholarship of all types advances through disagreement: Scholars marshal evidence in favor of competing hypotheses, and over time, there is usually some tendency toward consensus (that often absorbs elements of different hypotheses). If you read only one article from a discipline, you will have a limited sense of disciplinary debates. If you read a handful of articles, you will likely gain a sense of what disciplinary scholars agree about and what they disagree about. You should be careful of using an insight that is widely questioned within the discipline it comes from, unless it seems that disciplinary perspective is biased against that insight.

By reading widely in a discipline with respect to the research question, the interdisciplinary researcher absorbs the discipline's evaluation of its insights. The researcher can then evaluate these insights with respect to the discipline's perspective (see Chapter 7), something that disciplinary scholars are unlikely to do. (The two types of evaluation—disciplinary and interdisciplinary—are complementary.) The interdisciplinary scholar thus has the best possible sense of the strengths and limitations of each insight.

The Degree of Needed Adequacy Varies

The depth and breadth required for undergraduates is usually quite modest and depends on the length and sophistication of the study undertaken. *A rule of thumb is the fewer the disciplines, the more feasible the requirements of developing adequacy.* This is borne out in the example of an undergraduate student paper on the causes of freshwater scarcity in Texas. The three disciplines that the student found most relevant to the project included Earth science, biology, and political science. The student's prior coursework in Earth science and biology provided the necessary disciplinary "depth" for the science component of the project. *Adequacy* in this instance meant that the student had a working knowledge of the major theories, key concepts, and research methods (i.e., how data are collected and used) of these two disciplines. However, the student lacked "depth" in political science, which was needed to develop the policy component of the paper and provide the necessary disciplinary "breadth" to fully understand the problem. Achieving adequacy in political science for this student involved developing a working knowledge of the theories, key concepts, and research methods that informed state regulations concerning freshwater resource management.

By contrast, William Dietrich's (1995) interdisciplinary study of the Columbia River system involved far greater depth and breadth of disciplinary knowledge. Dietrich, a science reporter for *The Seattle Times,* received the Pulitzer Prize in 1990 for coverage of the Exxon Valdez oil spill. To study the vast and complex Columbia River system comprehensively required Dietrich to develop adequacy in several disciplines and disciplinary specialties. These included environmental science (an interdisciplinary field), chemistry, physics, political science, Indigenous American history and culture (an interdisciplinary field), and economics.

Students researching much more modest problems than Dietrich's can achieve adequacy in relevant disciplines to varying degrees, as demonstrated in the following examples.

Example #1 required the least depth and breadth in relevant disciplines. A sophomore-level course introducing the field of interdisciplinary studies required students to identify relevant disciplines and their perspectives on the preselected topic of human cloning. Though none of the students had previously researched this particular topic, all had access to Table 2.2 in Chapter 2 on disciplinary perspectives. From it, they were able to identify two potentially relevant disciplines, the number required for the assignment. Applying these disciplinary perspectives to the topic, however, proved more challenging as it involved reading two peer-reviewed articles from each discipline and, on the basis of these, ferreting out three defining elements: key concepts, assumptions, and theories. Throughout the exercise, the instructor served as a facilitator and coach.

Example #2 required greater depth and breadth in relevant disciplines. In this junior-level problem-based course on the interdisciplinary research process, students were allowed to research one of three preselected topics: the causes of suicide terrorism, the controversy over euthanasia, or illegal immigration. Students were required to map the problem to help them visualize its complexity and link its parts to particular disciplines. They were asked to consult Table 2.4 (Chapter 2). Also useful was Table 2.5, "Epistemologies of the Natural Sciences." These resources helped them to verify their findings and possibly discover linkages to other disciplines. In addition, students were required to select the three disciplines that they considered most relevant to the problem. For these students, demonstrating adequacy involved identifying a minimum of two peer-reviewed insights on the problem from each discipline, and constructing a data table consisting of each insight's key concepts, assumption, theory, and methodology.

Example #3 required still greater depth and breadth in relevant disciplines. Students entered this senior-level interdisciplinary capstone course with an approved research proposal on a problem or question that related to their professional or academic goal. They had also developed disciplinary depth in two of the three disciplines pertaining to their chosen problem and had, in most cases, completed one disciplinary research methods course in addition to completing the interdisciplinary research process and theory course. For these students, demonstrating adequacy involved mapping the problem, consulting "Szostak's Categories" (Table 2.4), identifying a minimum of three theories on the problem from each relevant discipline (including interdisciplines and schools of thought), closely reading the relevant disciplinary insights, and populating a data table (recommended in Chapter 5) with information drawn from them.

To Sum Up

Disciplinary adequacy has two key (and complementary) elements:

- Understanding (aspects of) the perspectives of each discipline drawn upon
- Understanding debates within these disciplines (and thus how insights have been evaluated within the discipline)

At every level, from introductory to midlevel to senior project, students, like professionals, must develop adequacy in those disciplines (including subdisciplines, interdisciplines, or schools of thought) that are relevant to the problem. The depth and breadth varies considerably between levels, depending on the complexity of the problem, the goal of the research, and the availability of collaborators and their role. Undergraduates do not need to start out with much specialized knowledge to engage in interdisciplinary research.

If the problem requires sophisticated manipulation of data, or the mastery of highly technical language, or knowledge whose mastery requires formal coursework, then more depth and reliance on disciplinary experts is required. But if the problem can be illuminated adequately using a handful of introductory-level concepts and theories from each discipline, and modest information is readily and simply acquired, then a solo interdisciplinary researcher or even a first-year undergraduate student can handle it. Fortunately, it is possible to get some useful initial understanding of most complex problems using a small number of relatively basic concepts and theories from each discipline (Newell, 2007a, p. 253).

NOTE TO READERS

Interdisciplinary research, though challenging, is manageable at all academic levels. What is required is for students to follow a research process that brings them from initial idea to identifying relevant disciplinary perspectives, to critically analyzing their insights to creating common ground between them, and then to developing a more comprehensive understanding of the problem. This and later chapters will provide numerous student and professional examples of interdisciplinary research oriented toward the natural sciences, the social sciences, and the humanities.

DEVELOP ADEQUACY IN THEORIES

Some undergraduate students may find theories intimidating, but their concern is usually alleviated when they understand what theories are and what they are used for: Theories help scholars to understand some aspect of the natural or human world. They explain the behavior of certain phenomena and how parts of a system interact and why. Theories are tested by data and research, and seek to explain the available evidence. Disciplinary experts use theories to produce insights concerning a specific problem that would otherwise not be possible to achieve.

The Reason to Understand Theories

Researchers need a basic understanding of selected theories for the practical reason that theory is a major source of disciplinary insights. *It is virtually impossible to conduct research in most disciplines on any topic and not have to deal at some level with theory.* This is especially true for insights from the natural and social sciences, and even some of the humanities disciplines such as art history and literature. Since theory is so fundamental to disciplinary scholarship, *developing adequacy in relevant disciplines must include knowing about the theories relevant to the problem under study.* At a minimum, you should be able to identify the theories while reading the relevant insights. Advanced undergraduates typically work with existing theories; graduate students, solo interdisciplinarians, and especially interdisciplinary teams may well develop new theories.

Concepts and How They Relate to Theory

Each discipline has created a large number of concepts that constitute its technical jargon or terminology. Introductory disciplinary textbooks are excellent sources of concepts and their definitions. Though each discipline has its own specialized vocabulary, it is common to find that a concept in one discipline is also found in the vocabulary of another discipline, but with a somewhat different meaning. For example, the concept of "rational" in sociology may refer to values and behavior that are normative for the group or for society at large, whereas for religion, the same concept may be conditioned by one's belief in and behavior governed by the sacred writings of a faith tradition.

Interdisciplinarians are interested in concepts because when they are modified, they can often serve as the basis for creating common ground and integrating insights. Concepts can facilitate making general connections across disciplinary boundaries. For example, the concept of *role* is widely used. Business studies the role of the consumer; sociology studies the role of the individual in social structure; history studies a person's role in some event or process. Other widely used concepts include *area* and *gender*.

How do concepts relate to theory? Concepts are the most elementary "building blocks" of any theory.[2] Some concepts are found in only a single theory, but many are found in a wide range of theories. For example, the concepts of *class, socioeconomic status,* and *social stratification* are found in a wide range of sociological theories. As noted previously, concepts most often describe one or more phenomena or causal links embraced by a theory, but some concepts may address other attributes of a theory.[3]

How to Proceed

Identifying all major theories relevant to the problem is essential to maintaining scholarly rigor and producing an interdisciplinary understanding that is comprehensive. Just as it is necessary to identify all disciplines potentially relevant to the problem before selecting those that are most relevant, so too is it necessary to identify all relevant theories before selecting those that are most relevant. *The way to proceed is to identify the major theories relevant to the problem in a single discipline and then repeat this process in serial fashion with the other relevant disciplines until all the important theories are identified.* Authors will generally state which theories they are utilizing, unless their discipline is strongly wedded to only one theory in addressing a particular problem.

First, Identify Theories Within a Single Discipline

STEP 4 (Chapter 5) urges categorizing theories by discipline. Taking time to do this makes it easier to link, as the following writers do, each relevant discipline with a specific set of theories. The publications in which these theories appear should be read (or reread) with this question in mind: "What theory does each writer advance to explain the problem I am investigating?" This question is applicable to almost any topic, including these:

- The increase in teenage obesity
- Public funding of professional sports stadiums
- Affordability of prescription drugs
- The development of visual perspective in Renaissance art
- Freshwater scarcity in drought-stricken areas such as California
- Undocumented immigration

The following examples are of theories advanced by the same discipline on a particular topic. These are drawn from published work and student projects.

From the Natural Sciences. Smolinski* (2005), *Freshwater Scarcity in Texas.* Smolinski finds multiple theories within the discipline of Earth science relevant to the problem of freshwater scarcity in Texas, as shown in Table 6.1.

TABLE 6.1 ● Theories From a Single Discipline on the Causes of Freshwater Scarcity in Texas	
Relevant Discipline	**Theories From a Single Discipline on the Causes of Freshwater Scarcity in Texas**
Earth Science	1. Global Warming Theory 2. Overexploitation Theory 3. Infiltration Theory

Individual Earth scientists embrace one of three primary theories to explain the growing scarcity of freshwater in any locale: global warming, overexploitation, or infiltration. Deciding that these are the most relevant theories was possible only after Smolinski had grounded himself in the published Earth science literature. To ensure that his research was current and complete, Smolinski worked with an Earth science professor and consulted with professionals employed in the public and private sectors for additional and confirming insights.

From the Social Sciences. Fischer (1988), "On the Need for Integrating Occupational Sex Discrimination Theory on the Basis of Causal Variables." In introducing his interdisciplinary essay on occupational sex discrimination (OSD), Fischer writes, "A review of the literature in the fields of psychology, sociology, economics, philosophy and history reveals a wide variety of explanations of OSD, each reflecting the relevant 'looking glass' of the particular discipline (or school of thought)" (p. 22). This statement is appropriate to an interdisciplinary essay for two reasons: (1) It informs the reader that the researcher has conducted an in-depth literature search, and (2) it identifies the disciplines most relevant to the problem.

Fischer found that four disciplines and one school of thought (Marxism) had produced important insights into the problem. From these, he began with economics, identifying four important economic theories on OSD, as shown in Table 6.2.

TABLE 6.2 ● Theories From a Single Discipline on the Causes of Occupational Sex Discrimination (OSD)	
Relevant Discipline	**Theories From a Single Discipline on the Causes of OSD**
Economics	1. Monopsony Exploitation Theory 2. Human Capital Theory 3. Statistical Discrimination Theory 4. Prejudice Theory

TABLE 6.3 ● Theories From a Single Discipline on the Meaning of the *Rime*	
Relevant Discipline	**Theories From a Single Discipline on the Meaning of the *Rime***
Literature	1. Reader-Response Criticism
	2. Marxist Criticism
	3. The New Historicism
	4. Psychoanalytic Criticism
	5. Deconstruction

From the Humanities. Fry (1999), *Samuel Taylor Coleridge: The Rime of the Ancient Mariner.* Paul H. Fry identifies five literary theories that are used to interpret the meaning of Samuel Taylor Coleridge's complex and ambiguous classic romantic poem, *Rime of the Ancient Mariner.* These theoretical/analytical approaches, noted in some cases as "criticism," are identified in Table 6.3.

Summary and Analysis. In each case, the set of theories chosen by the author represents significant explanations of the problem. Each theory makes certain assumptions, expresses itself with certain concepts, and encourages the use of particular methods. Interdisciplinarians often have to work with a set of disciplinary theories that offer conflicting explanations of the problem. The temptation to avoid in this situation is to reduce the number of these theories prematurely so as to minimize conflict.

Second, Identify Theories Within Each of the Other Relevant Disciplines

After identifying the relevant theories within a single discipline, repeat the process for the other sets of theories as illustrated in these examples of student and professional work.

From the Natural Sciences. Smolinski[*] (2005), *Freshwater Scarcity in Texas.* Smolinski applied the same research process that he used to identify and understand the important Earth science theories to the disciplines of biology and political science, as shown in Table 6.4.

As students take additional steps in the interdisciplinary research process, they will modify and expand this table to keep track of the rapidly proliferating pieces of information that normally accumulate as the research process proceeds.

From the Social Sciences. Fischer (1988), "On the Need for Integrating Occupational Sex Discrimination Theory on the Basis of Causal Variables." Fischer examines the literature from each of the other relevant disciplines to identify other important theories on the causes of OSD, as shown in Table 6.5.

TABLE 6.4 ● Theories From the Relevant Disciplines on the Causes of Freshwater Scarcity in Texas	
Relevant Discipline	**Theories From a Single Discipline on the Causes of Freshwater Scarcity in Texas**
Earth Science	1. Global Warming Theory
Biology	2. Overexploitation Theory
Political Science	3. Infiltration Theory
	4. Single-System Theory
	5. Encroachment Theory
	6. Market Theory
	7. Border Theory

TABLE 6.5 ● Theories From the Relevant Disciplines on the Causes of Occupational Sex Discrimination (OSD)	
Discipline or School of Thought	**Theories From the Relevant Disciplines on the Causes of OSD**
Economics	1. Monopsony Exploitation Theory
Psychology	2. Human Capital Theory
Sociology	3. Statistical Discrimination Theory
History	4. Prejudice Theory
Marxism[a]	5. Male Dominance Theory
	6. Sex Role Orientation Theory
	7. Institutional Theory
	8. Class Conflict Theory

a. Marxism is a school of thought.

Summary and Analysis. These tables enable the researcher to keep track of the rapidly proliferating pieces of information that normally accumulate as the interdisciplinary research process (IRP) proceeds. In each case, performing additional STEPS in the research process will involve revisiting these lists, perhaps modifying or expanding them as necessary.

It is not unusual for a community of scholars to advance two or more theories concerning a particular problem. When this occurs, researchers would be wise to adopt Fischer's (1988) strategy of first identifying the discipline advancing multiple theories before dealing with the other disciplines, each of which may advance only one or two theories. Why is this? Theories from the same discipline are often easier to integrate than are theories from different disciplines, as will become evident in STEP 8.

When to Use a Deductive Approach to Theory Selection (for Advanced Students and Practitioners)

In these examples, interdisciplinarians were working on problems that had already attracted considerable attention from disciplinary scholars. They conducted their research *inductively*, that is, by reading widely in the literature and noting which theories were used. But what about working on a problem on which little scholarly research has been done, or where only *certain types* of theories have been advanced? Seeing such a gap in the literature, you may wish to advance a theory explaining the problem that disciplinary experts have overlooked. Unless you are already aware of an appropriate theory, selecting a theory from among several *types* or categories of theories will require using a *deductive* approach. This involves (1) selecting an appropriate *theory type* (taking care that it not be the same type that disciplinarians have already used), and then (2) selecting a theory from within the type.

To aid in making this decision, Szostak (2004) provides a "Typology of Theory Types." He suggests that we begin the narrowing process by asking of each theory the five "W" questions:

- *Who (agency)?* Who is the agent? Answer: Agents can be intentional (that is, capable of thinking and deciding) or nonintentional, and be individuals, groups, social institutions, nation states, and the international community.

- *What (action)?* What does the agent do? Answer: Agents may act, react, or express attitudes.

- *Why (decision making)?* How does the agent decide? Answer: Nonintentional agents do not decide, (and thus we can only attribute their actions to their inherent nature) but intentional agents have recourse to five types of decision making (rational, rule based, value based, tradition based, and intuitive).

- *When (time path)?* What time path does the causal process follow? Answer: Causal processes may be in equilibrium, cyclical, unidirectional, or unpredictable.

- *Where (generalize)?* How generalizable is the theory? Answer: Generalizability can be evaluated on a nomothetic/ideographic continuum where nomothetic represents generalizability and ideographic represents specificity.

Asking these questions has the added benefit of revealing (or confirming) what has been missed in disciplinary research. They can thus be of use even when researchers are not developing their own theories, but only evaluating the theories of others. If, for example, all of the theories you have identified focus on individual actions, but you think group behaviors are important, then you are guided to wonder what the individual-level theories are missing. Also, *answering* the five "W" questions will reveal source(s) of conflict between theory types. A theory that emphasizes groups will likely reach a different conclusion from one that emphasizes individuals. Table 6.6 illustrates how these questions can be applied to several important theories.

TABLE 6.6 ◆ Szostak's Typology of Selected Theories					
Questions to Ask of Each Theory					
Theory	**Who? Agents**	**What? Action**	**Why? Decision Making**	**When? Time Path**	**Where? Generalize**
Most natural science, outside of biological science	Nonintentional agents	Passive	No active decision-making process	Various	Various
Evolutionary Biology	Nonintentional generally individuals	Active	Inherent	Not the same equilibrium	Nomothetic
Evolutionary Human Science	Intentional individual (group)	Active	Various	Not the same (any equilibrium)	Nomothetic
Complexity (Describes systems of interaction among phenomena; applied across natural and human sciences)	Catastrophe, chaos	Active and passive	Not strictly rational but must involve adaptive elements	Varies by version	Generally nomothetic
Action (Including theories of praxis)	Intentional individual (relationship)	Intersection of action and attitude	Often rational, but may be subconscious and unpredictable	Various	Generally idiographic
Systems (Recognition that patterns in social life are not accidental)	Various	Action and attitude	Various; emphasize constraints	Various	Generally nomothetic
Psychoanalytic	Intentional individual	Attitudes	Intuition; others possible	Various	Implicit nomothetic
Symbolic Interactionism	Intentional relationships	Attitudes	Various	Stochastic	Idiographic; some nomothetic
Rational Choice	Individual	Action	Rational	Usually equilibrium	Nomothetic
Phenomenology	Relationships (individuals)	Attitudes (actions)	Various	Various	Various

Source: Szostak, R. (2004). *Classifying science: Phenomena, data, theory, method, practice* (pp. 82–94). Dordrecht: Springer. With kind permission from Springer Science+Business Media.

Understanding Why Theories Conflict Is Preparatory to Creating Common Ground and Performing Integration. Table 6.6 shows how several common theories can be classified using the five "W" questions. A key point to take away from the table is that theories differ along each of these dimensions, and may thus each be better suited to addressing either different research questions or different aspects of a single research question (there might, for example, be a place for both individual and group actions in a particular situation).

(*Note:* The term **nomothetic** that appears in the right column refers to a theory that is applicable to a broad range of phenomena. **Nomothetic theory** posits a general relationship among two or more phenomena, and nomothetic researchers are concerned with showing a broad applicability. **Ideographic** refers to a theory that is applicable only to a narrow range of phenomena and under a constrained set of circumstances. **Ideographic theory** posits a relationship only under specified conditions, and ideographic researchers wish to explain the relevance of a theoretical proposition in a constrained set of circumstances [Szostak, 2004, pp. 68, 108].)

DEVELOP ADEQUACY IN DISCIPLINARY METHODS

The disciplinary research methods discussed here are not to be confused with the IRP itself. *The IRP is an overarching research process that subsumes disciplinary methods.* Developing adequacy in disciplinary methods varies greatly among the audiences addressed by this book. Undergraduate students must be familiar with the methods used by authors and understand how these methods may skew the authors' insights and theories. Some graduate students and solo interdisciplinarians may apply disciplinary methods themselves, while others need only be cognizant of the methods employed in order to identify, evaluate, and examine linkages among the insights and theories of the contributing disciplines. Interdisciplinary teams will generally employ disciplinary methods. This section defines *disciplinary method* and discusses how to develop adequacy in disciplinary methods.

Disciplinary Method Defined

To review, **disciplinary method** refers to the particular procedure or process or technique used by a discipline's practitioners to conduct, organize, and present research. Method implies an orderly and logical way of doing something. In the natural and social sciences especially, methods are the means by which to obtain evidence of how some aspect of the natural or human world functions (Szostak, 2004, p. 100).[4]

Methods Used in the Natural Sciences, the Social Sciences, and the Humanities

Fortunately, the number of methods used by the disciplines is quite small compared to the number of theories they favor. Most methods used by the major disciplines in the natural sciences, the social sciences, and the humanities fall into one of the following categories.

The Natural Sciences

The natural sciences generally emphasize quantitative research strategies:[5]

- Experiments (usually in a laboratory setting)
- Mathematical models
- Classification (of natural phenomena)
- Mapping
- Statistical analysis
- Careful examination of physical objects (as when geologists study rocks)

The Social Sciences

The social sciences use both quantitative and qualitative research strategies:[6]

- Experiments (often in an applied setting)
- Mathematical models
- Statistical analysis
- Surveys (qualitative)
- Interviews (qualitative)
- Ethnography/unobtrusive measures (qualitative)
- Physical traces (as in archaeology or paleontology)
- Experience/intuition (as used in interpretation of data) (qualitative)
- Classification (of human phenomena)
- Triangulation or mixed methods[7]

The Humanities

The humanities typically emphasize qualitative research strategies:[8]

- Textual analysis (content analysis, discourse analysis, and historiography)

- Hermeneutics/semiotics (study of symbols and their meaning)

- Experience/intuition (as used in interpretation/appreciation of creative works)

- Classification (of periods, schools of thought, etc.) (Berg, 2004, p. 4; Szostak, 2004, pp. 66–130)

- Cultural analysis (a combination of textual and semiotic analysis)

The Interdisciplinary Position on Methods Used in Disciplinary Research

The interdisciplinary position on disciplinary methods is that there are many methods, each with different strengths and limitations, and that no one method or overall approach should be privileged over any other in interdisciplinary work. *Interdisciplinarians should not be bound by the theory–method combinations that disciplinarians find convenient.* This view follows from the belief that each discipline relevant to a problem has something to contribute to understanding the problem. The interdisciplinary position is mainstream, in that there is now philosophical, if not scientific, consensus that no single method, or broad scientific approach such as positivism, animates knowledge formation today (Szostak, 2004, p. 100).

Adequacy in Relevant Disciplines Must Include Understanding Disciplinary Research Methods

Adequacy in disciplines must include understanding disciplinary research methods for this reason: *Since the evidence on which insights are based is derived from the application of methods, analysis of these insights must involve a familiarity with the potential limitations of those methods.*

Generally speaking, most undergraduate research involves integrating insights and theories drawn from published research, though some interdisciplinary courses and senior projects do conduct fieldwork. At this level, the main point in paying attention to methods is to understand how disciplines can and do choose methods that are good at (and biased toward) their favored theories, which, in turn, influence the insights and theories produced. Undergraduates can utilize their understanding of the potential limitations of particular methods (see below) in analyzing insights produced using these. Graduate students and scholars can contemplate how different methods might yield different insights. They might pursue such methods themselves.

Adequacy in Relevant Disciplines Includes Knowing the Interdisciplinary Position on the Quantitative Versus Qualitative Methods Debate

Adequacy in disciplines includes knowing the issues involved in the quantitative versus qualitative methods debate and the interdisciplinary position on it. Historically, disciplinary scholars have been divided over the value of qualitative versus quantitative methods. **Quantitative research strategies** emphasize evidence that can be quantified, such as the number of atoms in a molecule, the flow rate of water in a river, or the amount of energy derived from a windmill. **Qualitative research strategies** focus on the what, how, when, and where of a thing—its essence and its ambiance. Qualitative research, then, refers to meanings, concepts, definitions, characteristics, metaphors, symbols, and descriptions of things or people that are not measured and expressed numerically (Berg, 2004, pp. 2–3). Qualitative research is often more useful for a new problem (or research question) or to frame an old problem or question in a new way. Whereas most of quantitative research relies on numbers, qualitative research tends to rely on words, images, and descriptions.[9]

The quantitative versus qualitative debate is largely over. Today, most authors on research methodology stress mixing methods rather than distinguishing between methods (Hall & Hall, 1996, p. 35). Practitioners recognize that the relative importance of the two broad approaches may, nevertheless, vary according to the characteristics of the research problem. Interdisciplinary practitioners and students should accept both quantitative and qualitative approaches because each has been long used, accepted, urged, and useful (Tashakkori & Teddlie, 1998, p. 11). This methodological inclusiveness now characterizes many of the leading books on methods in the social sciences, including Bruce L. Berg's (2004) *Qualitative Research Methods,* which is written for students as much as for scholars (see Box 6.1). He urges researchers "to consider the merits of both quantitative and qualitative research strategies" (p. 3).[10] This is sound advice for interdisciplinary students who are concerned to identify the perspectives of various writers on a problem.

BOX 6.1

There are two common misconceptions about qualitative research. The first is that reliance on numbers results in a more certain and more valid result than qualitative research can provide. Those working in the social sciences are well aware of the tendency to give quantitative orientations more respect. However, qualitative methods are not only fruitful, but they can also provide greater depth of understanding than can be achieved by relying on quantitative methods alone. Though some qualitative research projects have been poorly done, says Berg (2004), qualitative approaches shouldn't be dismissed just because some studies failed to apply them properly. He adds that qualitative methods can and should be extremely systematic and have the ability to be reproduced by subsequent researchers. "Replicability and reproducibility, after all, are central to the creation

(Continued)

(Continued)

and testing of theories and their acceptance by scientific communities" (Berg, 2004, p. 7). *Interdisciplinarians, no less than disciplinarians, should be concerned that their research stands the test of having subsequent researchers examine the same problem using the same disciplines, insights, concepts, theories, and methods and achieve the same results.*

The second misconception is the tendency of some to associate qualitative research with the single technique of participant observation, while others extend their understanding of qualitative research to include interviewing as well (Berg, 2004, pp. 2–3). However, qualitative research strategies also include methods such as "observation of experimental natural settings, photographic techniques (including digital recording), historical analysis (historiography), document and textual analysis, sociometry, sociodrama and similar ethnomethodological experimentation, ethnographic research, and a number of unobtrusive techniques" (Berg, 2004, p. 3).[11] In practice, researchers pursuing quantitative methods often report qualitative data (for example, explaining the process used to collect statistical data), while qualitative researchers often quantify ("most interviewees said. . .").

If humans were studied using just quantitative methods, the danger would arise that conclusions—although mathematically precise—might fail to fit reality or, worse, distort that reality. Qualitative methods provide a way to evaluate and understand unquantifiable facts about actual people or artifacts left by them, such as art, literature, poetry, photographs, letters, newspaper accounts, diaries, and so on. Qualitative techniques explore how people structure their daily lives, learn, and make sense of themselves and others. Researchers using qualitative methods are thus able to understand and give meaning to humans and their activities.[12]

NOTE TO READERS

Graduate students in particular should be aware that there is a vast literature on "mixed methods research" (see Hesse-Biber & Johnson, 2015, for a good overview). This literature discusses both advantages of and strategies for employing multiple methods in a single research project. There is, not surprisingly, a large overlap between this literature and the IRP (see Szostak [2015a] for a discussion). The mixed methods literature mostly stresses the advantages of using both quantitative and qualitative methods, but its techniques can and have been applied to using multiple quantitative or qualitative methods. It has tended to pay less attention to "disciplinary perspective" than we have in this book, but it has nevertheless appreciated that different methods are grounded in a complex of theories, epistemologies, and concepts. The motive for discussing mixed methods research is a point central to this chapter: a recognition that different methods have different strengths and limitations. We can thus urge interdisciplinary researchers to employ mixed methods research methodology when performing research, just as we can urge students evaluating insights generated by others to seek insights generated by different methods.

USE AND EVALUATE DISCIPLINARY METHODS IN BASIC RESEARCH

The term *basic research* refers to scientific investigation using one of the dozen methods described above for collecting and analyzing data to verify existing hypotheses, theories, or ideas or to propose new hypotheses, theories, or ideas. The main motivation for interdisciplinarians (whether working solo or in teams) to conduct basic research is to address potential linkages between phenomena studied by different disciplines—linkages that often escape the attention of disciplinary researchers. A second motivation is to combat bias in the methods that disciplinary researchers may choose to apply: The interdisciplinary researcher may then apply a method preferred in a different discipline to a causal link already examined by disciplinary scholars. Interdisciplinarians will then integrate the findings of their basic research with that conducted by disciplinarians. The results of that integrated product are different from, but complementary to, the results of research in the disciplines.

Interdisciplinarians involved in research that has a basic research component have to decide how to select a method(s) that is appropriate to the problem and that will best provide new insights that can be integrated with existing insights to generate a superior understanding. *Using a disciplinary method to conduct basic research as part of an interdisciplinary study should not be equated with or substituted for the research model described in this book, but should be used in conjunction with it. The IRP subsumes whichever disciplinary method(s) is used.* (*Note:* Even if one is not performing basic research using disciplinary methods, the information presented in the following discussion is useful to evaluate disciplinary insights in terms of the methods employed.)

Most disciplinary researchers in the social sciences have at least one research strategy or methodological technique they feel most comfortable using, which "often becomes their favorite or only approach to research" (Berg, 2004, p. 4). This is likely true for researchers in the natural sciences, as well as the humanities. Interdisciplinarians, however, should not emulate disciplinarians in this regard. *There are important outcome-shaping implications of selecting any research method, and interdisciplinarians must be aware of these implications when they make decisions about which research methods to use.*

Identify Strengths and Limitations of Different Methods

We applied the five "W" questions to identify different types of disciplinary theory in Table 6.6. We can use the same questions plus five more to identify the strengths and limitations of different disciplinary methods.

- How many agents can be investigated? Some methods focus on one or two in detail while others can embrace millions.

- Can the four key elements of a causal argument be investigated? (Researchers look for correlation between cause and effect, temporality [cause occurs before effect], identification of intermediate variables/processes between cause and effect, and seek ways to rule out alternative explanations.)

- Does the method allow for induction? That is, does it allow the data to suggest new hypotheses? Some methods are designed to test hypotheses rather than suggest new ones.

- Can agents be followed through time?

- Can agents be followed through space?

Table 6.7 applies the five "W" questions plus the questions above to 10 of 12 methods (the others are mapmaking and semiotics) commonly employed by scholars. It shows that different methods have different strengths and limitations. It can thus be used to suggest methods that might yield alternative understandings of any problem and provides advice on which methods are best suited to a particular project. (*Note:* Table 6.7 supports two key observations: [1] Some methods are better suited to address particular research questions than are others, and [2] any complex research question will benefit from the application of multiple methods for a complex question will surely involve different sorts of agents, causation, and so on.)

Consider, for example, the limitations of experiments. Though well suited for exploring specific causal relationships among natural objects (such as a chemical reaction), this method is less well suited for studying the decisions of intentional agents because they may behave differently in a controlled setting than in the real world. If considering the utility of conducting experiments on human subjects, you can reasonably ask whether participants are likely to behave "realistically" (i.e., as they would in other settings). And if considering the utility of experiments on non-intentional agents, you should be aware

TABLE 6.7 ● Typology of Strengths and Limitations of Methods

Criteria	Participant Observation	Physical Traces	Statistical Analysis	Survey	Textual Analysis
Type of Agent	All	All; but group only in natural experiment	Intentional individuals; relationships Indirect	Intentional individuals; others indirect	All
Number Investigated	All	Few	Few	One	All
Type of Causation	Action (evolutionary)	Passive, Action	Attitude; acts indirectly	Attitude	All
Criteria for Identifying a Causal Relationship	Aids each, but limited	Potentially all four	Might provide insight on each	Some insight on correlation, temporality	All; limited with respect to intermediate, alternatives
Decision-Making Process	Indirect insight	Some	Some insight; biased	Yes; may mislead	Some insight
Induction?	Little	Some	If open	Yes; bias	Little
Generalizability	Both	Both	Idiographic	Idiographic	Both
Spatiality	Some	Constrained	From memory	From memory	Difficult to model
Time Path	No insight	Little insight	Little insight	Little insight	Emphasize equilibrium
Temporality	Some	Constrained	From memory	From memory	Simplifies
Type of Agent	Intentional individual; relationships groups?	All; groups and relationship indirect	All; groups and relationship indirect	Intentional individuals; groups indirect	Intentional individuals; others indirect
Number	Few; one group	Few	Many/all	Many	One/few
Type of Causation	Action (attitude)	Passive, action	Action, attitude	Attitude; acts indirectly	Attitude, action
Criteria for Identifying a Causal Relationship	All, but rarely done	Some insight to all four	Correlation and temporality well; others maybe	Some insight on correlation	Some insight on all
Decision-Making Process	All	No	No	Little	Some insight; biased

(Continued)

TABLE 6.7 ● (Continued)

Criteria	Participant Observation	Physical Traces	Statistical Analysis	Survey	Textual Analysis
Induction?	Much	Much	Some	Very little	Much
Generalizability	Idiographic; nomothetic from many studies	Idiographic; nomothetic from many studies	Both	Both	Idiographic; nomothetic. from many studies
Spatiality	Very good; some limits	Possibly infer	Limited	Rarely	Possible
Time Path	Some insight	Some insight	Emphasize equilibrium	Little insight	Some insight
Temporality	Very good up to months	Possibly infer	Static, often frequent	Longitudinal somewhat	Possible

Source: Adapted from Szostak, R. (2004). *Classifying science: Phenomena, data, theory, method, practice* (pp. 138–139). Dordrecht: Springer.

Note: The "criteria" reflect the 10 questions listed in the text above.

that natural processes might be influenced by environmental factors that may prevent a reaction from occurring.

Another limitation of experiments is that they are not inherently inductive (they are designed to test a hypothesis). A mistake in experimental design may sometimes yield surprising results and suggest a new hypothesis (as when Fleming discovered the effects of penicillin on bacteria by accident). Still, if you wish to generate a new hypothesis, then you may wish to consider more inductive methods such as textual analysis or observation.

Decide Which Disciplinary Methods to Use in Conducting Basic Research: An Example

When discussing theories earlier, we urged you to keep track of the relevant theories. We urge you to do the same with methods. Compared to theories, methods are much easier to track because there are only about a dozen or so used by the disciplines. However, be aware that broad methods such as textual analysis employ several distinct techniques that you may also wish to track.

Once you have narrowed the number of possible disciplinary methods to a few, decide which of these should be used. Consider, for example, this student research project, the focus of which is the declining salmon populations in the Columbia and Snake River systems. A topic, as noted earlier, may take any number of research directions. In this instance, the student narrowed the topic to two possible avenues of inquiry:

Option A.   Explain the 80% reduction of the salmon populations that has occurred since the system of dams was built.

Option B.   Examine the current state of the salmon populations in the Columbia and Snake Rivers.

The student decided that the disciplines most relevant to both options were biology, economics, and history (the assignment limited the number of disciplines to three). Consulting Table 6.8, "Research Methods Typically Used by Each Discipline," makes it easy to connect each discipline to its preferred research method(s).

The next task was to decide which of these methods is most appropriate for each avenue of inquiry. Note that each option defines the topic in a way that influences the choice of methods. Option A seeks to explain how the problem developed and who or what was responsible. Table 6.9 shows that the methods most appropriate to Option A are mapping (biology), statistical analysis and critical thinking (economics), and identification and interpretation of historical documents (history).

Option B seeks a more comprehensive understanding of the current state of the salmon population in the Columbia and Snake Rivers. Table 6.10 shows which methods are

TABLE 6.8 ● Research Methods Typically Used by Each Discipline	
Relevant Disciplines	**Research Methods Typically Used**
Biology	• Experiments • Mathematical models • Classification of natural phenomena • Mapping • Simulations (computer)
Economics	• Mathematical models • Statistical analysis of empirical data
History	• Identification of primary source material from the past in the form of documents, records, letters, interviews, oral history, archeology, etc., or secondary sources in the form of books and articles • Critical analysis in the form of interpretation of historical documents into a picture of past events or the quality of human and other life within a particular time and place

TABLE 6.9 ● Possible Methods to Use in Option A		
Relevant Disciplines	**Research Methods**	**Potential Candidate for Inclusion and Justification**
Biology	Experiments	No. The focus of the topic is on what happened in the past.
	Mathematical models	No. It is inappropriate for a historical explanation of how the problem developed.
	Classification of natural phenomena	No. There is nothing concerning the topic that requires classification.
	Simulations (computer)	No. Computer simulations are not appropriate for explaining past decisions.
Economics	Mathematical models	No. It is inappropriate for a historical explanation of how the problem developed.
	Statistical analysis of empirical data	Yes, provided that the statistical information was used to justify the original decision to build the dams.
	Critical thinking	Yes.
History	Identification of primary source materials	Yes. Those supporting and opposing the building of the dams could be interviewed. Scientific studies and past and present government hearings could be analyzed.
	Critical analysis in the form of interpretation of historical documents	Yes. The historical documents must be interpreted to answer the research questions: "How did the problem develop, and who was responsible?"

potential candidates for inclusion in this option. The Option B version of the topic results in a greatly expanded number of possible methods available to the interdisciplinarian.

Deciding which disciplinary research methods to use depends, primarily, on whether you are going to conduct basic research on the problem. It also depends on the disciplines interested in it, on the theories producing insights concerning it, and on the time and resource constraints. The researcher will decide which methods are most relevant based on the requirements or limitations of the interdisciplinary project.

How a Discipline's Preferred Methods Correlate to Its Preferred Theories

The fact that Table 6.7 relies in part on the same five "W" questions that are employed in Table 6.6 establishes empirically that certain methods will be better suited for

Relevant Disciplines	Research Methods	Potential Candidate for Inclusion and Justification
TABLE 6.10 ⬤ Possible Methods to Use in Option B		
Biology	Experiments	Possibly. Experiments recently concluded or in progress may illuminate how the system of dams is impacting various salmon populations.
	Mathematical models	Possibly. Modeling seasonal migrations of salmon under varying conditions would be useful.
	Classification of natural phenomena	No. There is nothing concerning the topic that requires classification.
	Mapping	Possibly. This method can show the progressive impact of past decisions on salmon populations.
	Simulations (computer)	Possibly. Computer simulations are useful to predict various outcomes under variable conditions.
Economics	Mathematical models	Possibly. For reasons stated.
	Statistical analysis of empirical data	Yes. An abundance of recent or current statistical data are likely available.
	Critical thinking	Yes.
History	Identification of primary source materials	Yes. Those presently supporting or opposing the dams could be interviewed. Scientific studies and past and present government hearings should be analyzed to provide background information and immediate context.
	Critical analysis in the form of interpretation of historical documents	Yes. Documents and sources must be interpreted to answer the research question: "What is the current state of the salmon populations in the Columbia and Snake Rivers?"

investigating certain theories: those that provide similar answers to the five "W" questions. For example, mathematical modeling and statistical analysis are not particularly well suited to examining human decision making if humans follow diverse (or random) decision-making strategies. But if we assume that humans make decisions rationally, then we need to only know their preferences (what they are trying to achieve) and their options: We can determine what decision they will make without studying how they go about making it. If we assume certain preferences (say, that individuals wish to maximize their income), then we can model or estimate statistically the decisions they will make. By relying on rational choice theory, then, economists are able to employ these quantitative methods that would otherwise be poorly suited to examining human decisions. If researchers wish to theorize nonrational decision making—that individuals are influenced by culture or peer pressure, for example—then they may need to employ different methods: Sociologists would use surveys and interviews and observation, and behavioral economists (who do not assume rationality) would use experiments.

Disciplinarians do not randomly choose theories and methods, but choose methods that are particularly good at investigating their favored theories. They also choose to investigate phenomena that are well suited to their theories and methods. (Economists have thus tended to pay little attention to culture, which is not directly implicated in rational choice theory and is hard to measure quantitatively.) Since each discipline tends to value its own theories and methods and subject matter, a huge barrier to interdisciplinary understanding emerges. The sociologist who claims that his interviews show an important cultural influence on individual decisions can easily be dismissed by the economist for using the wrong method and wrong theory to study the wrong phenomena. As always, be mindful of the power of disciplinary perspective.

Knowing that a discipline has likely chosen a method that makes its theory look particularly good, it is useful to speculate on what other methods might indicate if they were applied to the particular theory. Graduate students or scholars wishing to apply disciplinary methods themselves are guided to consider methods not usually employed to investigate a particular theory. As Berg (2004) warns us, researchers who mistakenly think of choice of method as having little or no connection to choice of theory "fail to recognize that methods impose certain perspectives on reality" (p. 4). We must be careful, therefore, not to take disciplinary perspective for granted and pursue a method simply because that is the method that the discipline employs to investigate that theory. (Fischer [1988] makes the related point that we should not ignore a method that may have gone out of favor in a discipline.)

Berg (2004) illustrates this challenge in the example of researchers deciding to canvass a neighborhood and arrange interviews with residents to discuss their views of some social problem. Their decision to use this method of data collection, he says, means that they have already made a theoretical assumption, namely, that reality is fairly constant and stable. Similarly, when researchers make direct observations of events, they assume reality is deeply affected by the actions of all participants, including themselves. Thus, each method—interview and unobtrusive participant observation—reveals a slightly different facet of the same social problem.

If one of the disciplines is from one of the "harder" (i.e., quantitatively oriented) social sciences such as psychology, then methods such as experiments and statistical analysis will likely have been applied by disciplinary researchers. But qualitative methods may be appropriate and may yield quite different insights. On the other hand, if one of the disciplines is from the humanities, then methods such as semiotics or a type of textual discourse analysis might have been applied, but other methods might prove useful. For *some* scientific questions, one method clearly excels: experiments. Experiments are unrivaled for the analysis of nonintentional agents. Even for scientific questions, though, experiments are fallible. Interdisciplinarians working on a science-oriented topic should supplement experimental evidence with evidence from other methods (Szostak, 2004, pp. 27–28).

Methods and Epistemology

Choice of methods is also closely associated with choice of epistemology. Those disciplines that favor quantitative methods tend toward an epistemological viewpoint that we can achieve precise and unchanging understandings of the world. Favoring of qualitative methods, especially in the humanities (but also in many social sciences), is often associated with epistemological attitudes that are more skeptical and that argue that different people will inevitably achieve different understandings of the world. This epistemological distinction has widened with the emergence of postmodernism and related epistemologies in recent decades.

As we saw in Chapter 2, interdisciplinary researchers should strive to respect different epistemologies just as they should respect different methods. A "both/and" approach to modernism and postmodernism (and by extension to various types of each of these), and the methods associated with these, is thus urged.

The Concept of Triangulation in Research Methodology

Triangulation or drawing upon multiple lines of insight is commonly associated with surveying, mapmaking, navigation, and military practices. Triangulation can be formally defined as identifying a point (on a map or on the ground) by drawing triangles to it from known points. Surveyors can be more accurate by taking sightings from multiple positions. Triangulation provides a useful metaphor for interdisciplinary research. Every method provides a different line of sight directed toward the same point or research problem. By combining several lines of sight, researchers can produce an integrated picture of the problem and have more ways to verify theoretical concepts.[13]

Triangulation of research methodology involves using multiple data-gathering techniques to investigate the same problem/system/process. In this way, findings can be cross-checked, validated, and confirmed. The important feature of triangulation, explains Berg (2004), "is not the simple combination of different kinds of data but the attempt to relate them so as to counteract the threats to validity identified in each" (p. 5). The term *validity* refers to "how well the measurement we use actually represents the true condition of interest— the thing we are trying to measure" (Remler & van Ryzin, 2011, p. 106).[14] In other words, since different methods have different biases, we can use multiple methods to counteract bias.

Regarding the question of how many methods should be used in an interdisciplinary research project, the answer is "It depends on the problem, how it is stated, and how ambitious the investigation is." Ultimately, the number of disciplinary research methods used in a given project depends on many factors, not the least of which is the scope and complexity of the problem and whether or not one decides to integrate one's own basic research with that of disciplinary experts.

PROVIDE IN-TEXT EVIDENCE OF DISCIPLINARY ADEQUACY

Students need to provide in-text evidence that they have developed adequacy in the disciplines they are using. **In-text evidence of disciplinary adequacy** may be expressed in many ways, such as statements about the disciplinary elements that pertain to the problem, the disciplinary affiliation of leading theorists, and the disciplinary methods used. It is easy to weave this evidence into the narrative if this confirming information has been collected and organized in a retrievable way. It may be particularly easy to display adequacy, while doing the analysis in STEP 6 (Chapter 7), wherein the student is expected to place the insights in the context of perspective, especially theory and method.

Certainly, adequacy requires the use of the most current and authoritative scholarship pertaining to the problem. Interdisciplinary work tends to focus on complex and real-world problems, or on intellectual problems that are not necessarily real world. Some problems are time sensitive. In the case of human cloning, for example, scholarship is being produced from multiple disciplines and interdisciplines in response to procedural breakthroughs, scandals, and legislative attempts to control this reproductive technology. Each of these developments raises new questions and prompts a new round of scholarly comment, often rendering earlier analysis less useful or even obsolete. When working with time-sensitive topics, the most recent scholarship must be consulted.

There are two practical reasons for providing in-text evidence of disciplinary adequacy. First, it demonstrates academic rigor. Interdisciplinary students bear a heavier responsibility than disciplinary students do in their research because interdisciplinarians have to establish adequacy in two or more disciplines. Therefore, students should be concerned to counter possible criticisms of superficiality by doing what more and more professional interdisciplinary writers are doing: Identify those disciplines that pertain to the problem, justify using an interdisciplinary approach, and weave into the narrative an explanation for using certain methods and applying certain theories.

Second, in-text evidence of disciplinary adequacy highlights the distinctive character of the research project compared to that of disciplinary research. *Paying attention to the interdisciplinary research process involves not just moving through its various STEPS; it also involves being self-consciously interdisciplinary.* This means reflecting on your biases (disciplinary and personal), serving as an honest broker when confronting conflicting viewpoints, and all the while keeping in view the end product that prompted the research in the first place.

Chapter Summary

Adequacy means knowing enough information about each discipline to have a basic understanding of how it approaches, illuminates, and characterizes the problem. There are two key elements:

- Appreciating disciplinary perspective in general, and disciplinary theories and methods in particular
- Appreciating disciplinary debates about the problem being investigated, and thus disciplinary evaluation of particular insights

For those conducting basic research with the intention of integrating results with published disciplinary insights on the same problem, additional considerations apply. For one thing, the way they state the problem will greatly influence which method(s) they are likely to use. A consequence of this decision is that interdisciplinarians may have to restate the problem or reframe the research question. We stressed that whatever *disciplinary method is used to conduct basic research as part of an interdisciplinary study, that method should not be equated with or substituted for the IRP but should be used in conjunction with it.* In the end, interdisciplinarians must make the difficult decision concerning which theories and which methods are most appropriate. Interdisciplinarians have not only the freedom to make these decisions, but also the responsibility to make them transparent. Whether or not they choose to use tables to store and juxtapose data as has been urged, it is important to develop some method to keep track of the information as it is gathered so that nothing important is lost or overlooked when performing subsequent STEPS.

Clearly, interdisciplinarians have to engage in far more preparatory work than do disciplinarians. This need not be onerous if done systematically; it must be done in order to perform the later STEPS of the IRP that involve integration. STEP 6 is deciding *whether* and *how* the disciplinary elements that have been chosen in STEP 5 (i.e., the insights and their assumptions and theories) adequately illuminate the problem.

Notes

1. Some interdisciplinarians, such as Hal Foster (1998), believe that to be interdisciplinary one must be "disciplinary first," meaning professionally "grounded in one discipline, preferably two" (p. 162).

2. Mieke Bal (2002) argues that concepts constitute interdisciplinary method in the humanities and that their use is inclusive of social science methodology (p. 5).

3. There are some philosophers who would argue, though, that concepts are theory based: You need to understand the theory before you can understand the concept. In concept theory, this approach is called "theory theory." See Szostak, R. (2013). Communicating complex concepts. In M. O'Rourke, S. Crowley, S. D. Eigenbrode, & J. D. Wulfhorst (Eds.), *Enhancing communication and collaboration in interdisciplinary research* (pp. 34–55). Thousand Oaks, CA: Sage; and Szostak, R. (2015a). Interdisciplinary and transdisciplinary approaches to multimethod and mixed methods research. In S. N. Hesse-Biber & R. B. Johnson (Eds.), *The Oxford handbook of multimethod and mixed methods research inquiry* (pp. 128–143). Oxford, UK: Oxford University Press.

(Continued)

(Continued)

4. Szostak (2004), in *Classifying Science: Phenomena, Data, Theory, Method, Practice,* is careful to distinguish between methods and "techniques or tools, such as experimental design or instrumentation, or particular statistical packages" (p. 100). Tools and techniques and so on are a subset of methods. His chapter on classifying methods is indispensable reading for students.

5. Introductory textbooks on the scientific method include Stephen S. Carey's (2003) *A Beginner's Guide to Scientific Method* (2nd ed.) and Hugh G. Gauch, Jr.'s (2002) *Scientific Method in Practice.*

6. Textbooks on social science research methods include Linda E. Dorsten and Lawrence Hotchkiss's (2005) *Research Methods and Society: Foundations of Social Inquiry,* W. Lawrence Neuman's (2006) *Social Research Methods: Qualitative and Quantitative Approaches,* John Gerring's (2001) *Social Science Methodology,* and Chava Frankfort-Nachmias and David Nachmias's (2008) *Research Methods in the Social Sciences* (7th ed.). For research methods of social science-oriented fields, see Frank E. Hagan's (2005) *Essentials of Research Methods in Criminal Justice and Criminology* and William Wiersma and Stephen G. Jurs's (2005) *Research Methods in Education: An Introduction.*

7. Bruce L. Berg (2004) says that triangulation was first used in the social sciences "as a metaphor describing a form of *multiple operationalism* or *convergent validation,*" meaning "multiple data collection technologies designed to measure a single concept or construct" (p. 5). For many social scientists, triangulation is usually restricted to three data-gathering techniques to investigate the same phenomenon. Alan Bryman (2004) provides extensive discussion of mixed methods in Chapters 21 and 22 of *Social Research Methods.*

8. Though written for the social sciences, Berg's (2004) *Qualitative Research Methods for the Social Sciences* may be used profitably in the humanities. In contrast to the social sciences, there are only a few books on commonly used research methods in the humanities. These include Catherine Marshall and Gretchen B. Rossman's (2006) *Designing Qualitative Research,* John W. Creswell's (1997) *Qualitative Inquiry and Research Design: Choosing Among Five Traditions,* and Matthew B. Miles and Michael Huberman's (1994) *Qualitative Data Analysis: An Expanded Sourcebook.* Examples of excellent research methods textbooks used in particular humanities disciplines include those by Laurie Schneider Adams (1996), *The Methodologies of Art: An Introduction;* Hong Xio (2005), *Research Methods for English Studies;* Martha Howell and Walter Prevenier (2001), *From Reliable Sources: An Introduction to Historical Methods;* and James J. Scheurich (1997), *Research Method in the Postmodern.*

9. Useful books on qualitative research in all of its aspects include those by John W. Creswell (1997), *Qualitative Inquiry and Research Design: Choosing Among Five Traditions;* John W. Creswell (2002), *Research Design: Qualitative, Quantitative, and Mixed Methods Approaches;* and Norman K. Denzin and Yvonna S. Lincoln (Eds.) (2005), *The SAGE Handbook of Qualitative Research.*

10. Berg (2004) notes that qualitative methodologies have not predominated in the social sciences (p. 2). His chapter "Mixed Methods Procedures" provides a comprehensive summary of this approach and a survey of the literature on the subject.

11. In terms of the methods that will be discussed below, photography can be considered a kind of observation, historical research generally involves textual analysis, but sometimes interviews or archaeological investigation, and ethnographic experimentation blends experiment and observation.

12. The general purpose of qualitative research derives from the theoretical perspective of symbolic interaction that is one of several theoretical schools of thought associated with the social sciences. It focuses on subjective understandings and the perceptions of and about people, symbols, and objects. Human behavior depends on learning rather than on biological instinct. Humans communicate what we learn through symbols, the most common of these being language. The core task of symbolic interactionists as researchers, Berg (2004) explains, is to "capture the essence of this process for interpreting or attaching meaning to various symbols" (p. 8). By contrast, positivists use empirical methodologies borrowed from the natural sciences to investigate phenomena. Their concern is to provide rigorous, reliable, and verifiably large amounts of data and the statistical testing of empirical hypotheses. Qualitative researchers, on the other hand, are primarily interested in individuals and their so-called life-worlds. "Life-worlds include emotions, motivations, symbols, and their meanings, empathy, and other subjective aspects associated with naturally evolving lives of individuals and groups" (p. 11).

13. Marilyn Stember (1991), among others, also makes this point in "Advancing the Social Sciences Through the Interdisciplinary Enterprise." On triangulation, see pp. 151–153 in R. Szostak (2004), *Classifying Science: Phenomena, Data, Theory, Method, Practice.*

14. For a thorough discussion of "validity" and the difficulties involved in establishing whether a measure is valid or not, see pp. 106–115 in D. K. Remler and G. G. Van Ryzin (2011), *Research Methods in Practice: Strategies for Description and Causation.*

Exercises

Decisions

6.1   What decisions are involved in developing adequacy in disciplines relevant to the problem?

Theories

6.2   Crime statistics are a moving target, sometimes increasing due to certain factors, and at other times decreasing due to other factors. Find out what the crime statistics are in your town or city, and compare the statistics spanning a 10-year time span. Draw upon or formulate two theories that explain the rise or fall of crime (or particular types of crime) in your community, making certain that each theory is from a different disciplinary perspective.

(Continued)

(Continued)

6.3   Related questions are these: What theory or theories do local officials use to explain either the rise or the decrease in crime? Does each theory reflect a particular disciplinary perspective? Do the various theories advanced explain the increase or decrease in crime comprehensively?

Gaps

6.4   Let's say you are researching a problem that has generated considerable attention from disciplinary scholars and that they have advanced various theories to explain the behavior of a certain phenomenon or the cause of some behavior. How do you discover if there is a gap in the research? How would you go about advancing a theory to explain the problem (e.g., its cause or the behavior) that disciplinary experts have overlooked?

Methods

6.5   Concerning the above example of crime in your community, would using qualitative or quantitative methods be more helpful in formulating theories to explain the cause of crime, and thus create policies to combat crime?

6.6   What assumption(s) would you be making if you preferred either qualitative or quantitative methods of data collection?

Adequacy

6.7   Concerning the topic you are researching, how can you provide "in-text" evidence of disciplinary adequacy?

©iStockphoto.com/fizkes

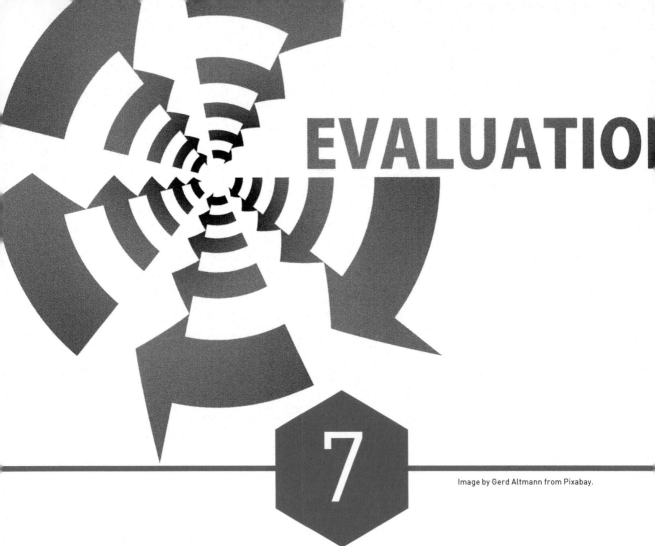

EVALUATION

7

Image by Gerd Altmann from Pixabay.

ANALYZING THE PROBLEM AND EVALUATING INSIGHTS

LEARNING OUTCOMES

By the end of this chapter, you will be able to

- Analyze the problem from each discipline's perspective
- Evaluate the insights produced by each discipline

GUIDING QUESTIONS

How do you evaluate disciplinary insights?

How do you gain familiarity with disciplinary perspective in order to do so?

In particular, how do you evaluate the theories, methods, data, and phenomena associated with disciplinary insights?

How does interdisciplinary analysis add to the evaluation of insights performed within disciplines?

CHAPTER OBJECTIVES

The disciplines provide different lenses or perspectives (in a general sense) for viewing the same problem and illuminating its parts. The challenge of STEP 6 is to view the problem through the perspective of each relevant discipline and then evaluate each discipline's insights concerning the problem to reveal their strengths and limitations. The movement is from the general (each discipline's perspective on the problem) to the particular (each insight concerning the problem). The chapter presents various strategies for evaluating insights. This work of analyzing perspectives and evaluating insights in terms of their strengths and limitations is foundational to preparing insights for integration, which is the focus of Part III.

 A. DRAWING ON DISCIPLINARY INSIGHTS

1. Define the problem or state the research question.
2. Justify using an interdisciplinary approach.
3. Identify relevant disciplines.
4. Conduct the literature search.
5. Develop adequacy in each relevant discipline.
6. **Analyze the problem and evaluate each insight or theory.**
   - **Analyze the problem from each discipline's perspective.**
   - **Evaluate the insights produced by each discipline.**

ANALYZE THE PROBLEM FROM EACH DISCIPLINE'S PERSPECTIVE

Analyzing the problem from each disciplinary perspective involves moving from one discipline to another and shifting from one perspective to another. Newell (2007a) describes this process of "moving" and "shifting":

> We must take off one set of disciplinary lenses and put on another set in its place as each discipline is examined. A possible initial effect of doing this is "intellectual vertigo" until one's brain can refocus. Experienced interdisciplinarians have developed the mental flexibility that enables them to shift easily from one disciplinary perspective to another. They do this in much the same way that multilingual persons shift easily from English to French to German without really having to think hard about what they are doing. (p. 255)

Analyzing the problem requires viewing it through the lens of each disciplinary perspective *primarily* in terms of its insights and theories. For example, if you want to construct a *disciplinary* understanding of the *Rime of the Ancient Mariner,* you would naturally mine the discipline of English literature for insights and interpretive theories. However, if you want to construct an integrated and more comprehensive understanding of the *Rime,* you would search for insights generated by authors in disciplinary and interdisciplinary fields that approach the text from different theoretical positions such as reader response, Marxist criticism, new historicism, psychoanalytic criticism, cultural analysis, and deconstruction (e.g., Fry, 1999). (*Note:* When working in the humanities, one typically works with a mix of disciplinary and interdisciplinary fields.)

How to Analyze a Problem From Each Discipline's Perspective

The problem of acid rain was a classroom exercise designed to illustrate disciplinary perspective taking. Table 7.1 views the problem from the perspective of each relevant discipline stated in terms of an overarching *what* or *how* question that can be asked of any problem.

The overarching question posed by each relevant discipline or interdiscipline reflects its perspective (in an overall sense), as noted in Table 2.2 in Chapter 2 (excluding law, environmental studies, and science and technology studies, which were not examined). One benefit of asking overarching *what* or *how* questions framed in each discipline's perspective is that these questions can reveal disciplinary bias. Table 7.1 reveals disciplinary bias in terms of the questions asked; the disciplines may also be biased in the answers they provide.

TABLE 7.1 ● Disciplines and Their Perspectives Stated in Terms of Overarching Questions Asked About the Problem of Acid Rain	
Discipline and Interdiscipline	**Perspective Stated in Terms of the Kinds of Overarching Questions Asked**
Physics	What are the fundamental physical principles underlying electrical power production that lead to acid rain?
Engineering	How does the design of the power generation process lead to acid rain?
Chemistry	What molecular changes lie behind the creation of acid rain and its effects?
Biology	How does acid rain affect flora and fauna?
Economics	What public policies could encourage firms to produce less acid rain?
Political Science	How could those public policies be adopted and implemented?
Law	How could those policies be enforced?
Environmental Studies	How is the problem of acid rain part of a complex environmental system?
Science and Technology Studies	How is the problem of acid rain a reflection of the relationship between scientific and technical innovations and social, political, and cultural values of society?

Note: Environmental studies and science and technology studies are interdisciplines.

The very premise of interdisciplinary studies is that disciplines can rarely explain all aspects of a complex problem. When dealing with a problem that is complex, such as acid rain, you can discover how complex it really is by mapping it as Figure 7.1 does, and by connecting the problem to the perspective of each relevant discipline, as Table 7.2 does. The table was constructed by taking each discipline's general perspective (provided in Table 2.2 in Chapter 2) and applying it to the specific problem of acid rain.

For example, the general perspective of economics, as stated in Table 2.2 (Chapter 2), is that it "emphasizes the study of market interactions, with the individual functioning as a separate, autonomous, rational entity, and perceives groups (even societies) as the sum of individuals within them." In other words, almost any problem is viewed as a "rational" (and therefore predictable and quantifiable) working out of "market interactions." As applied to the complex problem of acid rain in the United States,

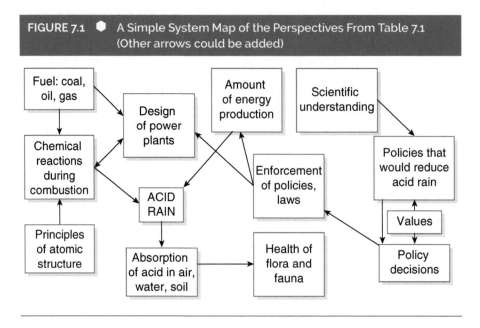

FIGURE 7.1 ● A Simple System Map of the Perspectives From Table 7.1 (Other arrows could be added)

therefore, economics sees it as a market problem resulting from decisions made by rational actors. (*Note:* Economists are increasingly flexible regarding rationality. Markets do not put a price on pollution; economists often recommend that governments do so.)

But the focus of our interest is not on the problem of acid rain in its entirety (i.e., its causes *and* effects) but *only* on its causes. The system map of Figure 7.1 reveals the various causes of the problem including the power plants that produce the electricity; their power source, which is predominantly coal; the chemical properties of power plant emissions; federal and state regulations; and so forth. So, the perspective of economics applied to the problem of acid rain is that it is caused by the behavior of that portion of the economic system that drives decisions about the use of coal in power plants. This "telescoping down" strategy of applying a general disciplinary perspective to a particular complex problem or part of it is not possible without first mapping the problem and understanding how its parts interact with each other. Table 7.2 shows how each discipline's unique perspective allows it to focus narrowly on only one aspect of the complex problem of acid rain.

Examples of Analyzing Problems From Disciplinary Perspectives

Analyzing problems from various disciplinary perspectives is demonstrated in several multithreaded examples from published and student work. The categorization of these

TABLE 7.2 ● Disciplinary Perspectives Relevant to the Problem of Acid Rain	
Relevant Discipline or Interdiscipline	**Perspectives on Acid Rain in a General Sense**
Physics	May see acid rain as a consequence of basic thermodynamic principles underlying the operation of an electricity-generating power plant
Engineering	May see acid rain as a power plant design problem
Chemistry	May see acid rain as the result of a series of chemical processes
Biology	May see acid rain as posing a biological problem for downwind flora and fauna
Economics	May see acid rain as the result of the behavior of that portion of the economic system that drives decisions about the use of coal in power plants
Political Science	May see acid rain as a regulatory problem
Law	May see the destructive effects of acid rain on property as a question of who is responsible and thus liable
Environmental Studies	May see acid rain as complex physical, chemical, and biological interaction caused by human activity
Science and Technology Studies	May see acid rain as an example of how embedded social, political, and cultural values affect scientific research and technological innovation and how these, in turn, affect society, politics, and culture

Note: Environmental studies and science and technology studies are interdisciplines.

examples refers to the area of the researcher's training and the orientation of the topic, more than to the disciplines from which insights are drawn.

From the Natural Sciences. Smolinski[*] (2005), *Freshwater Scarcity in Texas.* Smolinski has already identified the disciplines most relevant to the complex problem of freshwater scarcity in Texas in STEP 3 (Chapter 4). These are listed in the left-hand column in Table 7.3. Using the information on disciplinary perspective provided in Table 2.2 (Chapter 2) and information gathered in the literature search, he is able to apply the general perspective of each relevant discipline to the specific problem of freshwater scarcity in Texas and then state these narrowed perspectives in general terms. This is possible even though each discipline embraces two or more conflicting theories explaining why the problem exists and is offering possible solutions.

TABLE 7.3 ● Disciplinary Perspectives Relevant to the Problem of Freshwater Scarcity in Texas	
Discipline Most Relevant to the Problem	**Perspective on Problem Stated in General Terms**
Earth Science	Sees the problem as a consequence of human interference in natural geological systems
Biology	Sees the problem as a consequence of human degradation of the Earth's major systems: atmosphere, biosphere, geosphere, and hydrosphere
Political Science	Sees the problem as a reflection of institutional and interest group factors

These general statements of each discipline's perspective applied to the problem provide a glimpse into how it will be possible to eventually integrate conflicting theories within each discipline.

From the Social Sciences. Fischer (1988), "On the Need for Integrating Occupational Sex Discrimination Theory on the Basis of Causal Variables." Fischer is grappling with a complex social problem involving several disciplines—and a school of thought (i.e., Marxism)—and their perspectives. While it is unlikely that undergraduate or even graduate students would be required to work with as many disciplines and theories as Fischer does, his approach is instructive because of the way he simplifies the process for us by briefly describing each perspective in a concise narrative. Once again, the general perspective of each relevant discipline is applied to the problem, and Fischer states these narrowed perspectives in concise statements, as shown in Table 7.4.

TABLE 7.4 ● Disciplinary Perspectives on the Problem of Occupational Sex Discrimination (OSD)	
Discipline or School of Thought Most Relevant to the Problem	**Perspective on Problem Stated in General Terms**
Economics	OSD is caused by rational economic decision making on the part of males and females (Fischer, 1988, pp. 27–31).
History	OSD is caused and perpetuated by longstanding institutional forces (pp. 32–34).
Sociology	OSD is caused by a process of female socialization different from men which, in turn, is directly reflected in their occupational structures (p. 34).

Discipline or School of Thought Most Relevant to the Problem	Perspective on Problem Stated in General Terms
Psychology	OSD is caused and perpetuated by males to maintain the traditional male–female division of labor (pp. 35–36).
Marxism	OSD is a necessary act in preserving the institutions of capitalism (p. 32).

Note: Marxism is a school of thought.

From the Social Sciences. Delph[*] (2005), *An Integrative Approach to the Elimination of the "Perfect Crime."* In her senior project, Delph uses one interdiscipline (criminal justice) and two subdisciplines (forensic science and forensic psychology) in the study of how "perfect crimes" (i.e., the criminal is not caught) can be eliminated or, at least, greatly reduced. An interdiscipline brings together aspects of other disciplines to focus on a complex phenomenon such as crime. As an interdiscipline, criminal justice brings together the subdisciplines of forensic science (biology) and forensic psychology (psychology) along with elements of sociology that focus on criminal behavior. Criminal justice generates its own theories and uses a variety of research methods borrowed from biology, psychology, and sociology. Early on, Delph decided to limit her study to the interdiscipline and subdisciplines listed in Table 7.5. Their perspectives, as they apply to the problem, are stated concisely in general terms.

TABLE 7.5 ● Disciplinary Perspectives Relevant to the Problem of Eliminating the "Perfect Crime"	
Interdiscipline or Subdiscipline Most Relevant to the Problem	Perspective on Problem Stated in General Terms
Criminal Justice	Sees the persistence of "perfect (i.e., unsolvable) crimes" as a result of inefficient investigatory processes
Forensic Science	Sees the persistence of "perfect crimes" as a result of improper application of forensic science to criminal investigations
Forensic Psychology	Sees the persistence of "perfect crimes" as a result of insufficient attention to the art of criminal profiling

Note: Forensic science and forensic psychology are subdisciplines.

TABLE 7.6 ● Disciplinary Perspectives on the Meaning of a Graffito	
Discipline or Interdiscipline	**Perspective on Problem Stated in General Terms**
Art History	Sees the graffito as an autographic art object reflective of a period of Dutch culture and thus providing a window into that culture
History	Sees the graffito as a product of Dutch history and a window into Dutch culture
Linguistics	Sees the graffito as self-referential
Literature	Sees the graffito as allographic literature
Philosophy	Sees the graffito as an epistemological argument
Psychology	Sees the graffito as an expression of psychic mourning
Cultural Analysis	Sees the graffito as a "text-image" that embodies the program of cultural analysis

Note: Cultural analysis is an interdiscipline.

From the Humanities. Bal (1999), "Introduction," *The Practice of Cultural Analysis: Exposing Interdisciplinary Interpretation.* Bal, befitting an interdisciplinary approach, is able to state, in general terms, each relevant discipline's overall perspective on the meaning of the art object, a graffito, which is her focus of study. A table such as Table 7.6 is useful in keeping track of the several perspectives involved in the analysis. Juxtaposing these perspectives in this way enables the researcher to more readily identify possible areas of overlap and points of conflict among them, which is called for in STEP 6.

From the Humanities. Silver* (2005), *Composing Race and Gender: The Appropriation of Social Identity in Fiction.* Silver studies the ethics of appropriation of social identities by fiction writers, a topic that had been practically ignored, especially by fiction writers themselves. "Appropriation" refers to the practice commonly used by actors, filmmakers, and fiction writers of assuming another person's social identity. Appropriation, or "putting oneself in another's shoes," helps the artist or author to gain a more personal understanding of those people who are being appropriated. Among the examples of appropriation, she cites is that of John Howard Griffin who wrote *Black Like Me* to call attention to racial injustice. By appropriating the skin of a black man, Griffin became the first known white person to experience directly the kind of discriminatory treatment known only to black people. Silver explores this topic from the perspectives of three disciplines: sociology,

Chapter 7 ■ Analyzing the Problem and Evaluating Insights 191

TABLE 7.7 ● Disciplinary Perspectives on the Problem of the Appropriation of Social Identity in Fiction	
Discipline Most Relevant to the Problem	**Perspective on Problem Stated in General Terms**
Sociology	Sees the problem arising from socially constructed power dynamics and guilt feelings that impact race and gender relations
Psychology	Sees the problem arising from writers attempting empathetic arousal of their audience
Literature	Sees the problem as arising from authors' need to achieve a sense of authenticity

psychology, and literature. She applies the general perspective of each discipline to the problem of appropriation, as shown in Table 7.7.

Benefits of Analyzing Problems From Disciplinary Perspectives

Analyzing a problem from the perspectives of only a few disciplines may reveal that the problem is too broad and needs to be narrowed further. If so, STEP 1 will have to be revisited and the statement of the problem reworded to reflect the narrowed focus (see Chapter 3). Analyzing the problem of acid rain reveals that it is a very broad problem that requires consulting quite a few disciplines if a truly comprehensive understanding of its causes is to be achieved. Deciding which part of the problem to focus on would most likely result in reducing the number of disciplines (telescoping down) to those that are the most relevant to the narrowed focus. *The progression of the research is from defining the problem to identifying its parts to identifying the disciplines that specialize in those parts.*

Applying each discipline's general perspective to the problem and stating this perspective concisely in writing is worthwhile for four reasons:

1. Telescoping down forces us to think deductively, to move from the general to the particular.

2. Applying a general disciplinary perspective to a particular complex problem, or part of it, requires that we map the problem and understand how its parts interact with each other. We may well identify gaps between disciplinary perspectives.

3. The action of writing each discipline's narrowed perspective in a concise statement is itself an integrative act in this sense: If a discipline has produced

two or more conflicting insights or theories about the causes of a problem, these conflicting insights or theories will still probably share the discipline's general perspective, as well as the narrowed statement of that perspective.

4. Focusing so intensely on perspective enables us to verify whether or not the disciplines selected earlier are as relevant to the problem as we thought they were.

NOTE TO READERS

Creating tables to capture this information is worthwhile activity because the juxtaposition of disciplinary perspectives reveals similarities and differences that otherwise might be overlooked. It is easier to compare things that are close together than to compare things that are far apart. Juxtaposing the perspectives of the relevant disciplines makes it easier to see the narrowness of each perspective and the necessity of pushing on with the interdisciplinary research process (IRP). *Integration cannot proceed unless we first identify similarities and differences (and sometimes gaps) between perspectives and their insights.* At a minimum, this strategy will enable you to perform STEP 7 more efficiently, which involves identifying conflicts between insights or theories.

Sometimes a discipline has an obvious perspective on a problem but no published insights. In this circumstance, undergraduate researchers will have to move on, but graduate researchers may consider filling the gap.

EVALUATE THE INSIGHTS PRODUCED BY EACH DISCIPLINE

Having analyzed the problem through the lens of each discipline's perspective, we can now move from perspective to insights. You should have identified these insights by the end of the literature search in STEP 4 and confirmed their relevance in STEP 5.

Evaluating insights involves using different strategies organized under the following headings: (1) disciplinary perspective (in a general sense), (2) the theories used in generating insights, (3) the data used as evidence for insights, (4) the methods employed, and (5) the phenomena embraced by insights.

While you should understand how each strategy works in isolation, you will often find that evaluation proceeds best when these strategies, illustrated below, are used in concert. Evaluating insights requires close reading to detect the author's understanding, explanation, or argument.

Strategy 1: Evaluate Insights Using Disciplinary Perspective

The purpose of evaluating the insights produced by the relevant disciplines is to see how each author's understanding of the problem may be skewed. The term **skewed understanding** means the degree to which the insight reflects the biases inherent in the discipline's perspective, and thus the way the author understands the problem. Skewed understanding results from the author's deliberate decision or unconscious predisposition to omit certain information that pertains to a problem. (See Box 7.1 on scholarly bias.)

BOX 7.1

Individual scholars, interdisciplinary as well as disciplinary, often bring their personal biases or settled prejudices to certain problems. For example, sociology, since its founding, has been more liberal than is economics, and "studies" of any sort (e.g., women's, environmental, and even religious) tend to be more liberal than are their disciplinary counterparts. It bears emphasizing that interdisciplinarians value diversity of perspective and seek out conflicting viewpoints with which they disagree. They can live with ideological diversity and tension. They also value difference and ambiguity. The trap that they should avoid falling into is drawing insights and theories disproportionately from disciplines or schools of thought with which they agree or are more familiar, whether this is done consciously or unconsciously (Newell, 2007a, p. 252).

Throughout the entire IRP, the goal must be kept in mind: to construct a more *comprehensive* understanding of the problem. The understanding cannot be truly comprehensive if it excludes certain views or is dominated by certain views. If the scales of scholarship are prejudiced, then the rigor, comprehensiveness, and intellectual integrity of the project will be seriously, or even fatally, compromised.

As with previous STEPS, performing STEP 6 *may* involve revisiting earlier STEPS. Here, it is important to connect each part of the problem with those disciplines that are interested in it and that have produced important insights and theories on it. Performing STEP 6 thoroughly and with integrity will enable one to identify which perspectives are the most compelling and which ones are (possibly) missing.

Researchers are well advised to query the epistemological, metaphysical, ethical, and ideological elements of a discipline's perspective (the other elements of disciplinary perspective will be treated below). A discipline that is pessimistic of the potential for scholarly understanding, or that doubts the existence of an external reality, will likely produce more ambiguous insights than does a discipline that is confident of progressing toward precise understandings of a fixed reality. A discipline that thinks a particular outcome is good will likely study it differently than does a discipline that thinks it bad. Last but not least, policy implications may influence scholarly conclusions (generally subconsciously).

In the following example, Repko (2012) identifies the major authors from the discipline of psychology who are considered experts on the causes of suicide terrorism based on how frequently they are cited by other authors: Jerrold M. Post, Albert Bandura, and Ariel Merari. Table 7.8 shows Repko's evaluation of each insight in terms of (a) its perspective on the problem, (b) its underlying assumption, and (c) its strength and limitation.

TABLE 7.8 ● Evaluating Insights From Psychology			
Author	**Insight's Perspective**	**Assumption**	**Strength/Limitation of Insight**
Post	Terrorists reason logically but employ what he calls "a special logic" or "psycho-logic."	Terrorists are born, not made.	Strength: It challenges other research that shows that most suicide attackers are psychologically normal. Limitation: It fails to account for external influences such as culture.
Bandura	Terrorists rationalize their acts of violence using various techniques.	Terrorists are made, not born.	Strength: It addresses causal factors external to the individual attacker including political factors, the influence of culture on one's sense of identity, and the influence of sacred beliefs. Limitation: It is silent concerning a person's cognitive predisposition and personality traits that may influence decision making.
Merari	Terrorists need a specific mindset to carry out suicide attacks that is shaped by various factors.	Terrorists are made, not born.	Strength: It takes into account personality factors, especially the psychological impact of a broken family background, but attributes primary responsibility to recruiting organizations and their charismatic trainers. Limitation: It fails to address the causal factors of politics, culture, and religion that research outside of psychology has shown to be highly influential in the development of a suicide attacker.

In *The Perspective of Each Insight*, Repko (2012) queries the disciplinary perspective of psychology:

> Psychology typically sees human behavior as reflecting the cognitive constructs individuals develop to organize their mental activity. Psychologists also study inherent mental mechanisms, both generic predispositions as well as individual differences. Psychology is confident of progressing toward understanding this phenomenon, and has produced insights marked by precise (though at times conflicting) understandings of *individual* terrorist behavior. As applied to suicide terrorism, psychologists generally understand the problem in terms of individuals whose behavior is the product of mental constructs and cognitive restructuring. Psychologists attempt to identify the possible motivations behind a person's decision to join a terrorist group and commit acts of shocking violence, and are consciously aware that their findings have policy implications. They tend to agree that there is no single terrorist mindset, a finding that greatly complicates attempts to profile terrorists groups and leaders on a more systematic and accurate basis. (p. 130)

While evaluating insights in terms of perspective is important everywhere, it may be particularly important in the humanities where a variety of perspectives toward works of art or literature can be identified.

From the Humanities. Bal (1999), "Introduction," *The Practice of Cultural Analysis: Exposing Interdisciplinary Interpretation.* In this example, Bal draws on the perspectives of various critical approaches to objects and texts to develop the most comprehensive understanding of the graffito possible. These approaches and their perspectives stated in general terms include Reader Response, Marxist Criticism, New Historicism, Psychoanalytic Criticism, Deconstruction, and Cultural Analysis as are shown in Table 7.9.

TABLE 7.9 ● Critical Approaches and Their Perspectives Stated in General Terms	
Critical Approach	**Perspective**
Reader Response	The reader of the graffito is to cite direct references in the text to show that the world of the text corresponds to the one in which the reader is situated.
Marxist Criticism	The graffito is a product of work and, as such, is to be understood as the product of a complex web of social and economic relationships as well as the prevailing ideology to which the majority of people uncritically subscribe.

(Continued)

TABLE 7.9 ● (Continued)	
Critical Approach	**Perspective**
New Historicism	The critic/reader of the graffito is to be highly conscious of and even discuss preconceived notions before situating the text in its historical/literary context.
Psychoanalytic Criticism	The graffito, as an invention of the mind, provides a psychological study of the writer, and of the reader.
Deconstruction	The graffito is an ambiguous text consisting of words from the many discourses that inform it, the meaning of which is ultimately indeterminate.
Cultural Analysis	The graffito heightens one's awareness of one's situatedness in the present, the social and cultural present from which one looks at the graffito.

This humanities example of Bal's graffito illustrates how to handle a large number of insights and theories. The possible insights and theories *from just one discipline* have to be multiplied by the number of disciplines relevant to the topic. In the Bal example, this means six disciplinary literatures multiplied by the number of insights and theories generated by each discipline relating to the graffito—a formidable challenge even for graduate students and practitioners. One solution is to focus not so much on each critical approach, but on how the perspective of each approach reveals new meaning when applied to a particular object and/or text as Bal does. In this example, the next task (see below) is to apply each perspective to the graffito to expose its meaning.

The Assumptions of Each Insight

These assumptions are not broad philosophical assumptions, but rather the narrower assumptions that an author might actually make in a publication. Identifying these is often challenging because authors do not always make their assumptions (whether broad or narrow) explicit.

Even so, it is possible to discover authors' assumptions from their disciplinary perspective by applying that perspective to the problem. In Table 7.8, none of the psychology authors stated their assumptions, so Repko applied the disciplinary assumptions of psychology to each author's approach to the problem of suicide terrorism. This process did not produce a uniform result as demonstrated by two very different assumptions noted in the center column. But underlying these different assumptions, the larger and more basic assumptions of the discipline are evident, and this, as we shall see in Chapter 11, can serve as the basis for integrating the insights produced by psychology.

The Strengths and Limitations of Each Insight

You can identify the strength(s) and limitation(s) of each insight and what it assumes to be true about the problem only by reading it carefully and then comparing it to the others. Newell (2007a) observes, "Each discipline has its own distinctive strengths," and "the flip side of those strengths is often its distinctive limitations" (p. 254).

> A discipline such as psychology that is strong in understanding individuals is thereby weak in understanding groups; its focus on parts means that its view of wholes is blurry. A discipline such as sociology that focuses [primarily] on groups doesn't see individuals clearly; indeed, at the extreme it sees individuals as epiphenomenal—as little more than the product of their society. Empirically based disciplines in the social and natural sciences cannot see those aspects of human reality that are spiritual or imaginative, and their focus on behavior that is lawful, rule-based, or patterned leads them to overlook human behavior that is idiosyncratic, individualistic, capricious and messy, or to lump it into unexplained variance. Humanists, on the other hand, are attracted to those aspects and tend to grow restive with a focus on behavior that is predictable, feeling that it misses the most interesting features of human existence. (p. 254)

We should stress here that interdisciplinary scholars bring two key strategies to the task of evaluating disciplinary insights: evaluating these in the context of disciplinary perspective, and comparing the insights from different disciplines. Disciplinary scholars are unlikely to employ either strategy.

Strategy 2: Evaluate Theories Used in Generating Insights

A second strategy is to evaluate the theories that produced the relevant insights, identifying their strengths and limitations. There are two ways to evaluate theories. One is to state the theory, detect its assumption(s), and identify its explanatory strength(s) and limitation(s). The other is to ask the five "W" questions introduced in Chapter 6 to evaluate the appropriateness of each theory to the problem. We begin with the first approach.

State the Theory, Detect Its Assumptions, and Identify Its Explanatory Strengths and Limitations

In the natural sciences and social sciences, authors typically use theories to explain cause or behavior; these theories also reflect their disciplinary perspective. So, when working in the natural or social sciences, you should expect to work with theories, as well as the insights they produce. The progression of evaluation, then, is from disciplinary perspective to disciplinary theories. In order to evaluate insights that are produced by theories, you must evaluate the theories themselves, as van der Lecq and Fischer do in the following examples.

From the Natural Sciences. van der Lecq (2012), "An Interdisciplinary Approach to the Evolutionary Origin of Language." The focus question of van der Lecq's research is "What was the primary function for which human language was selected?" After identifying all relevant disciplines and their insights, she evaluates the ability of their theories to answer the research question, as shown in Table 7.10.

TABLE 7.10 ● Disciplinary Theories and Their Strengths and Limitations in Explaining the Evolutionary Origin of Language

Discipline or Subdiscipline	Theory	Strength of Theory	Limitations of Theory
Biology, Anthropology	"Grooming and gossip theory" (Social Brain Hypothesis)	Dunbar (1996): The theory explains why language emerged and why only with humans and posits that the primary function of language is social rather than instrumental. (*Note:* Instrumental theories assume that language evolved for exchange of technical information, such as explaining how to make tools or coordinating hunts.) (p. 201)	1. We cannot really be certain whether what is true for us (that we talk mainly about social issues) was true of our ancestors. 2. It doesn't explain why language is much more complex nowadays than is necessary for social bonding. 3. It is too general, in the sense that it does not specify what humans are talking about while bonding. (p. 201)
Cognitive Science, Biology	Political Hypothesis	Dessalles: "Language is a biological trait that cannot have been a *mere* product of culture" (p. 202).	1. The theory's assumption that altruistic behavior contradicts Darwinian evolution is doubtful because altruistic behavior is natural among primates and perfectly compatible with Darwinian evolution (p. 203). 2. The theory "lacks explanatory power because it is hard to believe that we are motivated to use language in order to gain status all the time." 3. The theory's strong emphasis on the contrast between biology and culture is outdated (p. 203).

Discipline or Subdiscipline	Theory	Strength of Theory	Limitations of Theory
Biology, Anthropology, Evolutionary Biology	Niche Construction	Odling-Smee: The theory "addresses the importance of human behavior and cultural processes in human evolution. It is also an interdisciplinary approach that crosses the boundaries between natural (physical) and cultural explanations of the evolution of language." It is "based on the assumption that language is a cultural phenomenon and that it emerged gradually" (p. 208).	1. Criteria for determining the validity of competing theories has yet to be developed. 2. The theory is based on the assumption that language is a cultural phenomenon and that it emerged gradually (p. 208).

Source: Van der Lecq, R. (2012). Why we talk: An interdisciplinary approach to the evolutionary origin of language. In A. F. Repko, W. H. Newell, & R. Szostak (Eds.), *Case studies in Interdisciplinary Research* (pp. 191–223). Thousand Oaks, CA: Sage.

From the Social Sciences. Fischer (1988), "On the Need for Integrating Occupational Sex Discrimination Theory on the Basis of Causal Variables." Fischer analyzes the problem of occupational sex discrimination (OSD) from the perspective of each of the relevant disciplines in terms of the theories that each discipline advances to explain the problem. Note that the insights of each theory appearing in Table 7.11 reflects the unique and narrow perspective that produced it, and shows the explanatory strengths and limitations of each cluster of theories. Fischer obtained this information by closely reading (or rereading in some cases) the various theories he had earlier identified as most relevant to the problem of OSD.

Simply the fact that a theory has explanatory limitations should not disqualify it from being used. However, these limitations *do* mean that the insights are skewed by the way the theory understands the problem, and interdisciplinarians should acknowledge this, as Fischer does.

When there are multiple theories within a discipline, the interdisciplinary researcher can learn much about the strengths and weaknesses of these by reading the disciplinary literature. The interdisciplinary researcher can add to disciplinary analyses by comparing theories from different disciplines. For example, when a theory from cognitive science suggests that altruism is not evolutionarily sound, familiarity with evolutionary theory can suggest otherwise (see Table 7.11).

TABLE 7.11 ● Disciplinary Theories and Their Strengths and Limitations in Explaining the Problem of OSD

Discipline or School of Thought	Theory	Insight of Theory	Explanatory Strength of Theory	Explanatory Limitation of Theory
Economics	1. Monopsony Exploitation	1. OSD is caused by rational decision making that focuses on the demand for labor.	1. It explains economic motivation of employers.	1. Economists assume that individuals are rational and self-interested, and therefore fail to account for prejudice, sex role discrimination, and "tastes" adverse to hiring women.
	2. Human Capital	2. OSD is caused by rational decision making that focuses on the supply of labor.	2. It explains economic motivation of employers.	2. Economic motivation fails to account for prejudice, sex role socialization, and "tastes" adverse to hiring women.
	3. Statistical Discrimination	3. OSD is caused by rational decision making that focuses on the higher turnover "costs" associated with female employees.	3. It explains economic motivation of employers.	3. It assumes that individuals are rational and self-interested, and therefore fails to account for prejudice, sex role discrimination, and "tastes" adverse to hiring women.
	4. Prejudice	4. OSD is the result of some employers indulging their own sexual prejudices.	4. It is open to factors extending beyond economic motivation.	4. It assumes that individuals are rational and self-interested, and therefore fails to account for prejudice, sex role discrimination, and "tastes" adverse to hiring women.

Discipline or School of Thought	Theory	Insight of Theory	Explanatory Strength of Theory	Explanatory Limitation of Theory
History	1. Institutional Development	1. OSD is caused and perpetuated by longstanding institutional forces.	1. It identifies historical trends that may have produced the problem. 2. It places the problem in a broad context.	1. It is unable to analyze the behavior of groups. 2. It is unable to account for psychological motivation of individuals.
Sociology	1. Sex Role Orientation	1. OSD is caused by a process of "female socialization different from men which, in turn, is directly related in their occupational structure" [p. 34].	1. It identifies conflict among social groups and institutions. 2. It accounts for differences between adult men and women in terms of how they were raised.	1. Its focus on groups fails to account for individual behavior motivated by complex psychological factors or genetic predisposition.
Psychology	1. Male Dominance	1. OSD is caused and perpetuated by males to maintain the traditional male–female division of labor.	1. It explains individual behavior and decision-making processes.	1. It is unable to study group behavior.
Marxism	1. Class Conflict	1. OSD is a necessary act in preserving the institutions of capitalism.	1. It explains macro trends and developments.	1. Economic considerations fail to explain behavior of all groups or individuals.

Note: Marxism is a school of thought.

Ask the Five "W" Questions to
Evaluate the Appropriateness of Each Theory

To determine the appropriateness of each theory to the problem at hand, the reader may ask of it the five "W" questions.

- *Agency:* "Who is the agent?" If each of the theories operates at the level of individuals, but one feels that group processes are important to the problem, then you need to ask how the insight might change if group agency were considered. A theory on group agency would need to be included with the theories to be integrated. Note that the theories investigated by Fischer above variously address intentional individuals, intentional groups, and nonintentional agents (institutions).

- *Action:* "What does the agent do?" The decision in the Fischer example is whether or not to engage in OSD. The first few theories surveyed by Fischer emphasize actions: the decisions that employers or workers make. Later theories are more focused on explaining attitudes. Note that prejudice theory assumes certain attitudes in explaining actions: It can thus readily be integrated with theories that seek to explain how prejudices arise.

- *Decision Making:* "How does the agent decide?" The example illustrates how to query the decision-making processes that are addressed by each of the theories. Concerning the economic theories, Fischer (1988) comments, "'Economic man' supposedly makes economic decisions in a predictable and exact way, always acts intentionally and deliberately, never acts impulsively or altruistically, knows the consequences of her or his actions, and acts to maximize economic benefit to herself or himself" (p. 24). The value of asking this question is this: If a theory posits a form of decision making, but one thinks other forms of decision making are vital to the problem, then one needs to reflect on how the insight might be changed by integrating insights from theories that posit different types of decision making. The theories surveyed by Fischer capture elements of tradition-based and value-based decisions, and perhaps also intuitive (for prejudice and socialization may be largely subconscious processes).

- *Time path:* "When?" Fischer's focus is on explaining a particular (and longstanding) outcome. The theories here all tend to envision a similar sort of causal process (generating an equilibrium in which OSD is perpetuated). Is it worth looking for theories that might explain a move toward/away from OSD?

- *Generalizability:* "Where?" Rational choice theories are highly generalizable whereas theories emphasizing particular attitudes or traditions or institutions *may* only apply to certain societies. The researcher can reflect on whether OSD differs markedly across societies.

Since disciplines tend to prefer a small number of theories that are similar in many ways, disciplinary scholars are unlikely to ask all of these questions about a theory in their discipline. In asking these questions, then, the interdisciplinary researcher can add to disciplinary evaluation of a particular theory.

NOTE TO READERS

Here is an instance of where one "integrates as one goes," performing some integration while looking ahead to the integrative phase of the IRP. Insights from the discipline of economics that presume individuals are rational and self-interested may need to be reassessed when the problem involves social, religious, or cultural behavior (as in the case of OSD) that is based on other motivations as well. "In using skewed insights," says Newell (2007a), "interdisciplinarians need to maintain some psychological distance from the disciplinary perspectives on which they draw, borrowing *from* them without completely buying *into* them [italics added]" (p. 254).

After surveying the four major economics theories in his study of OSD, Fischer (1988) concludes that economists define the problem largely (but not exclusively) in terms of the economic motivation of employers. He correctly concludes that each of these economic theories is appropriate to the problem even though each "offers a highly restrictive and incomplete explanation of the causes of OSD." Therefore, it is necessary to "broaden the analysis to include other prominent theories of OSD" (p. 31). This phrasing alerts the reader to the overall limitations of the economic theories and explains why Fischer finds it necessary to examine theories advanced by the other relevant disciplines.

The strategy of evaluating insights by evaluating the theories that produce them, as illustrated in the van der Lecq and Fischer examples, is applicable to any situation where one has identified two or more theories from two or more disciplines.

However, merely evaluating theories from multiple disciplines does not constitute full interdisciplinarity, but only constitutes multidisciplinarity. Full interdisciplinarity is achieved only by continuing the IRP and actually integrating the most important theories as the basis for constructing a more comprehensive understanding of the problem. This is the subject of Part III of the book.

Strategy 3: Evaluate Insights
Using the Evidence (Data) for Insights

A third strategy for evaluating insights is to focus on the data that authors use as evidence for their insights. Interdisciplinarians must be keenly aware that the data presented by disciplinary experts may also be skewed. Each discipline has an epistemology, or way of knowing, and it collects, organizes, and presents data in a certain way that is consistent with this knowing. To say that the data presented by disciplinary experts may be "skewed" is not to allege that the data are falsified or sloppily gathered or presented in a biased way (though each of these sometimes occurs). Rather, it is to say that experts *may* omit or fail to collect certain kinds of data for various reasons. This is because experts are interested in certain kinds of questions and amass data to answer these questions without consciously realizing that they may be excluding other data that would, if included in the study, modify or even contradict the study's findings. Part of the task in identifying conflicts in insights (Chapter 9) is to identify and evaluate the different kinds of evidence used by each discipline or theory to support its insights. What one discipline counts as evidence may be discounted or considered inappropriate by another. Therefore, *the interdisciplinarian should be alert to possible conflicts arising over the different kinds of evidence used by each discipline or theory.*

Data are, of course, closely connected to the methods employed by researchers (which will be discussed below). As noted earlier, the sciences and the harder social sciences employ the methods of experiments, models, and statistics, all of which constitute what are thought to be convincing evidence. Different disciplines consider evidence in terms of what makes one piece of knowledge persuasive and other pieces of knowledge not persuasive. In women's studies, for example, the testimonial (i.e., "lived experience") is considered persuasive, and "in native studies, traditional knowledge preserved over centuries through an oral tradition and interpreted by Elders is central" (Vickers, 1998, p. 23). Neither kind of evidence would be considered valid by those adhering to a positivist/empiricist epistemology and using quantitative evidence.

Historians, on the other hand, count a wide array of artifacts as evidence, including diaries, oral testimony, and official documents, none of which are accepted or considered appropriate by the sciences. Evidence for literary criticism consists of the imaginative application of theory—including the six theories discussed in Table 7.9—to the text. For Bal (1999), what counts as evidence is the discovery *in the text*, using the technique of close reading, of traces of each theory's imprint. These may be summarized in the meaning of the verb *to expose*. One example makes the point. A close reading of the word *note*, with which the text of the graffito begins, "shows" a speech act in its purest form: direct address to the reader in the present but loaded with "pastness" (pp. 7–8). The student working with a theory (or series of theories) must be sufficiently grounded in the theory to identify the evidence that the writer advances to support it.

In reading and thinking about the sources you gather, you should ask these two questions:

- What counts as evidence in this discipline?

- What kind of evidence is this disciplinary author omitting that would shed additional light on the problem?

Examples of How Supportive
Evidence Reflects Disciplinary Perspective

The close connection between disciplinary perspective and the kind of supportive evidence typically used by practitioners in the discipline is illustrated by examining essays by scholars from two disciplines and a profession that grapple with this question: Should schools adopt computer-assisted education? [This is a debate, we might note, that continues decades later, though computers have become increasingly common in classrooms.]

Essay #1. Discipline: Communications/Information Technology. Clifford Stoll (1999) argues in *High-Tech Heretic: Why Computers Don't Belong in the Classroom and Other Reflections by a Computer Contrarian* that schools should not adopt computer-assisted education. His expertise in the field of information technology extends to the business aspect of it, and this is reflected in the kind of evidence he presents to support his case: the hidden financial costs of computers, reference to supportive essays in the disciplinary journal *Education Technology News,* examples of schools having to make hard choices between making needed repairs and buying technology, and careful examination of the mythical cost savings derived from automating education administration.

Essay #2. Discipline: Psychology (Learning Theory). The National Research Council (NRC) is the research arm of the National Academy of Sciences, a private, nonprofit scholarly society that advises the federal government in scientific and technical matters. Its influential study, *How People Learn: Brain, Mind, Experience, and School,* argues that computer-assisted education can enhance learning (Bradsford, Brown, & Cocking, 1999). The supportive evidence used by the NRC includes references to state-of-the-art learning software and several experimental projects such as GLOBE, which gathered data from students in over 2,000 schools in 34 countries (Bradsford et al., 1999).

Essay #3. Profession: Education. In 1999, the Alliance for Childhood, a partnership of individuals and organizations, issued the report *Fool's Gold: A Critical Look at Computers in Childhood* that subsequently appeared in a leading education journal. The report argues that computer-assisted education does not benefit young children. This view, a matter of heated debate within the teaching profession, was nevertheless included in the Education Department's own 1999 study of nine troubled schools in high-poverty areas, as well as in extensive references to studies by leading education experts, including Stanford professor (of education) Larry Cuban, theorist John Dewey, Austrian innovator Rudolf Steiner, and MIT professor Sherry Turkle (Alliance for Childhood, 1999).

Reflecting on These Examples

These examples show how each discipline or profession amasses and presents evidence that reflects its epistemology. However, in all three cases, the experts omit evidence that they consider outside the scope of their discipline or profession. "Facts," then, are not always what they appear to be. They reflect what the discipline and its community of experts are interested in.

It is easy for students and the public to be seduced by data produced by "experts" on the problem, mistakenly assuming that the data must surely be "correct" and objective because they came from an authoritative source. The lesson here is that the reader must evaluate the kind of evidence used by disciplinary authors, understand how they use that evidence, and ask whether other types of data might possibly alter the insight.

Yet again the interdisciplinary researcher employs two key strategies in evaluating the data employed within a particular insight. First, one asks about disciplinary perspective, and how this might constrain what kind of data are considered appropriate. Second, one compares insights from different disciplines to ask which data might be usefully employed in evaluating an insight from another discipline.

Strategy 4: Evaluate Insights Using the Methods Authors Employ

A fourth strategy that is useful in evaluating insights is to focus on the methods their authors employ. Chapter 6 introduced the dozen or so distinct methods employed by scientists (often in combination) to conduct research and produce new knowledge. The focus here is on the importance of recognizing how these methods may be skewed. As with theories, says Szostak (2004), there are key questions that should be asked of any method used. These questions were introduced in Chapter 6, but it is useful to state them in greater detail here.

> *Question 1: Who is being studied?* Is the focus of the method on intentional agents or nonintentional agents? Some disciplines—and, therefore, their methods—pay more heed to individuals than to groups. As with theories, the interdisciplinarian should query how appropriate a particular method is to the type(s) of agency inherent in the problem at hand.

> *Question 2:* A subsidiary question here involves the number of agents that a method investigates. For example, surveys deal with numerous people, interviews deal with few, and observation (especially participant observation) deals with fewer. The problem with these methods, Szostak (2004) says, is that if any subset (from one to many) of the relevant population is examined, researchers will face questions of sampling: "Is the sample biased, or does it represent the average of the larger population, or perhaps the most common attributes of the larger population (which can be quite different from the average)?" (p. 104).

Question 3: What is being studied? Some methods study actions (and some especially reactions), while others study attitudes (Szostak, 2004, p. 105). For many research questions, we may need to understand both how people think and how they behave.

Question 4: What elements of a causal explanation can be addressed? Philosophers have appreciated that there are four key questions to ask about any causal relationships. Methods differ considerably in how well they address each of these. Does the method allow us to establish a correlation between cause and effect? Does the method allow us to establish that the cause precedes the effect? Does the method allow us to identify variables or processes that might operate between cause and effect? Finally, does the method allow us to dismiss alternative explanations? (*Note:* In all cases, the word *cause* is used in its most general sense to refer to any sort of influence that one phenomenon might exert on another. Note also that the interdisciplinary scholar should always be open to the possibility of multiple influences generating a particular result.)

Question 5: What sorts of decision making can the method examine? As in the case of theory, the focus of this question is on the decision-making process used by agents, including the passive decision making of nonintentional agents. Particular methods, observes Szostak (2004), "prove to be best suited to answering one type of question, but methods are differentially applicable to different types of decision making" (p. 107).

Question 6: Does the method have inductive potential? Some methods test specific hypotheses and thus usually have limited capacity to suggest alternative hypotheses.

Question 7: What sort of causal process can the method investigate? Can it explore equilibria, change in a particular direction, cycles, or stochastic processes?

Question 8: Where is the place or setting of the phenomenon or process to be studied? Phenomena can be analyzed in one place or as they are in motion. Analysis can be performed in a natural setting or in an artificial setting. The latter allows the researcher to control variables and thus isolate a particular cause–effect relationship but at the risk that agents behave differently than they would in a natural setting (Szostak, 2004, p. 108).

Question 9: Are the results of the method highly generalizable? Is it easy to imagine that the results obtained from the method would be achieved in a different setting? This may be less likely with interviews than with an experiment involving a chemical reaction.

Question 10: When (at what time) can the phenomena be studied? Szostak (2004) asks, do researchers analyze a set of phenomena at one point in time, continuously through time, or at several discrete points in time? Do they analyze all phenomena at the same time(s) or at different times? The advantage of continuous time is that the researcher can study the process of change, but analysis at particular points of time is easier. (p. 109)[1]

In Chapter 6, we provided Table 6.7 that briefly summarizes answers to these 10 key questions. However, you may find it easier to navigate Table 7.12 that provides a textual summary of some of the strengths and limitations of selected methods. These summaries are not exhaustive and should be used only as a starting point for further study. Students wishing more detail should consult Chapter 4 of Szostak (2004).

If an insight about a problem that you believe reflects group processes is provided by a method that investigates individuals, then you can legitimately wonder whether other methods might be more appropriate, and whether they would provide support for the insight in question. The same sort of analysis is possible for each of the 10 questions listed above. The challenge here is to use these questions to reflect on the nature of the problem they are investigating and compare this to the methods employed to investigate it. The reward is not to accept without question the methods employed by researchers, to appreciate that no method is perfect, and to reflect on whether particular insights are driven by the limitations of the method employed.

As with theories, disciplinary researchers are likely to stress the strengths of their favored methods. While the disciplinary literature will examine whether a particular author employed a method appropriately, it is far less likely to ask whether the method was appropriate to the question being investigated. In asking the 10 questions above, we are closely examining a particular element of disciplinary perspective. As with other aspects of disciplinary perspective, we can be aided by comparing across disciplines: What did the method employed by one author illuminate that the method employed by another missed?

Strategy 5: Evaluate Insights Using the Phenomena Embraced by Insights

A fifth strategy used to evaluate the appropriateness of an insight is to focus on the phenomena that each discipline considers within its research domain. The authors' insights are skewed by the way they define the problem (i.e., those parts of the problem they overlook). Their insights are also skewed in the way that they look at what they *do* see. This is due to the phenomena or the behavior they choose to investigate. Overall, their choice of phenomena influences their choice of method, which in turn influences their choice of theory. Focusing on phenomena points up the importance of first mapping the problem (see Chapter 4) in order to see which parts of it are covered by the disciplines. The examples by Fischer and Bal illustrate, to varying degrees, how mapping the problem (either consciously or unconsciously) helped them to identify relevant disciplines, theories, and the insights they produced concerning the problem at hand.

TABLE 7.12 ◆ Strengths and Limitations of Disciplinary Methods (Not exhaustive)		
Method	Strengths	Limitations
Experiments	Experiments are primarily a deductive tool whereby the subject is manipulated in a particular way and results are measured. They are potentially highly reliable because "they can be easily repeated with all sorts of subtle changes to research design." Experiments are best at identifying simple cause–effect relationships, and "can illuminate some aspects of decision making, such as the degree to which people are swayed by the views of others" [pp. 119–121].	Analysis of group behavior is generally, but not always, unfeasible. Since experiments involve control and manipulation, advocates of qualitative research, including feminists, have often been hostile to them. "With intentional agents, researchers must worry about signaling the desired result to subjects" [pp. 118, 120–121].
Surveys	Focus is on the individual level with results often broken down on a group basis. Results "speak to average tendencies of group members rather than group processes themselves" [p. 135]. They can point to important differences among group members. They can provide quantitative data on attitudes at a point in time [p. 135].	They may contain too few causal variables and usually speak only directly to relationships [but network analysis can identify relationships by surveying people with whom they interact] [p. 135].
Statistical analysis [secondary data analysis]	This most popular single method in social science involves analysis of statistics collected by others, including those generated by surveys or experiments. Data are often available for huge numbers of people, can be aggregated to show group tendencies as well as differences within groups, and are available for both nonintentional and intentional agents [p. 131]. This method can establish correlations extremely well.	"Secondary data cannot provide detailed insight into how any intentional agent forms attitudes." When a correlation is established, researchers must use judgment and rely on theory in inferring causation [Silverman, 2000, pp. 6–9]. McKim [1997] states that statistical analysis should only be taken as evidence of a causal relationship if reinforced by plausible theory and direct evidence from other methods [p. 10]. Szostak [2004] notes that researchers "must also worry about the strength of a relationship; researchers often celebrate the 'statistical significance' of a result without taking the necessary—and inherently qualitative—step of asking if the relationship is important. They may thus too easily embrace a theory with limited explanatory power, or

(Continued)

TABLE 7.12 ● (Continued)

Method	Strengths	Limitations
		reject a theory with great explanatory power because their sample size was small and thus statistical evidence not established" [p. 131]. Researchers "must look at how data are recorded, and ask whether either those reporting or those recording had likely biases" [p. 134].
Content or textual analysis	This method includes "a variety of techniques . . . often grounded in different theoretical understandings of the meaning of a text." Texts speak most directly to the intentions of the author and "can provide valuable insights into the author's perception of groups, relationships, and nonintentional agents." Authors often reveal why they or others made decisions as they did, though they may be incorrect or purposely biased [pp. 136–137].	Theorists disagree whether, and to what degree, the core message of any text can be identified. Though texts can provide a diversity of interpretations, there are limitations to the insights that can be drawn from any text. "No text can be understood fully in isolation from other texts, since language is symbolic." A further limitation is that researchers can only build upon the information that the author (consciously or not) provides [pp. 136–137]. Szostak cautions researchers that authors may bias their understanding of events. "Content analysis, by which quantitative analysis is performed on how often particular ideas or phrases appear in a text, is one technique which can potentially identify intentions of which the author was not even aware" [p. 137].
Participant observation (PO) (including ethnographic fieldwork)	PO is the most common form of observational analysis, though discreet observation is also used. PO researchers focus on intentional agents and emphasize attitudes, but may also study actions as well as constraints/incentives. PO follows subjects over a period of time as they make decisions and often ask subjects to explain why they acted as they did. PO may be the best way to study certain or unique events and identify idiosyncrasies that prevent a rule from operating [pp. 127–128]. Palys (1997) notes that PO is almost always used in conjunction with interviews, surveys, and/or textual analysis, allowing researchers to compare what people say and do, and reducing the problem of researcher bias.	Only a small number of individuals and relationships can be studied at one time [p. 127]. Goldenberg (1992) notes that the very presence of an observer may cause participants to behave differently and feign different attitudes. He also says that many researchers believe that the method is better for exploration/induction than hypothesis testing [p. 322]. Though PO research is inherently inductive, it is possible for researchers to ignore evidence that conflicts with their desired conclusions [Szostak, 2004, p. 128]. Szostak notes that protocol analysis (whereby participants are asked to perform a task and describe verbally their thoughts while doing so) "effectively combines elements of PO and interviewing. It is more inductive than other types of PO. It is also more artificial, and thus raises questions of whether participants will both do and think as they would in a less artificial environment" [p. 129].

Method	Strengths	Limitations
Interview	Interviews are more costly than surveys and thus tend to involve fewer people. Interviews are good for identifying attitudes that encouraged certain actions (but often people do not know why they act as they do). Questions can elicit insights into constraints/incentives imposed by impersonal agents (p. 122). Narrative analysis overcomes problems of researchers biasing results through their questions by asking people to tell their own stories (p. 123).	Interviews can speak only indirectly to relationships and group processes. "They are limited in their ability to identify temporal priority . . . and dependent on the researcher asking appropriate questions" (p. 122). Interviewees may be misled about why they acted/thought as they did/do, either purposefully or through faulty memory (p. 122). Because interviews necessarily involve small numbers of people, "generalizations require integration of results across many studies" (p. 123).
Case study (a mixed method generally incorporating textual analysis, observation, and/ or interviews)	Case studies provide insight into a particular issue or theory in rich detail (while statistical analysis tends to seek patterns across numerous cases). Case studies can be quantitative and/or qualitative (p. 140).	"Researchers should be careful not just to report those observations that seem to lend themselves to generalizations; other information may encourage the development of alternative theories or the recognition of limits to existing theories" (p. 141).

Source: Szostak, R. (2004). Classifying science: Phenomena, data, theory, method, practice. Dordrecht: Springer. With kind permission from Springer Science+Business Media.

From the Social Sciences.   Fischer (1988), "On the Need for Integrating Occupational Sex Discrimination Theory on the Basis of Causal Variables." Fischer is explicit in identifying the strengths and limitations of relevant disciplines based on the phenomena that each one typically investigates. Rather than discuss all five disciplines that Fischer considers relevant to the problem of OSD, he limits his discussion to economics and history. As a trained economist, Fischer has a professional mastery of his discipline. The strength of economics, he says, is that it views the problem of OSD as having to do with economic behavior. But that strength is also its limitation. The problem with economics, he says, is that its theories and methods are skewed. At one time, political economists, as they called themselves, believed that social, cultural, psychological, and political factors were as much a part of their discipline as were economic factors. But since 1900, the desire of orthodox economists to transform the discipline into a science has resulted in their whittling down political economy to "economics proper," meaning that economists focus narrowly on the behavior of "economic man" (or woman— sex was not an issue) and the exclusion of normative issues (e.g., epistemological issues).

After mapping the problem (subconsciously in Fischer's case), Fischer asked which phenomena are excluded from certain insights. He concluded that the economic theories on OSD ignore many of the key causal dimensions of the problem (including particular institutions and cultural attitudes) and offer an "incomplete and unpersuasive explanation" of it (p. 26). This skewed approach, says Fischer, explains why a major sex discrimination case involving Sears Roebuck and Company and its employees used expert witnesses who were historians rather than economists. Historians, Fischer notes, "understand the importance of long-run institutional forces causing and perpetuating OSD, an area neglected by most economists" (p. 26).

When a discipline focuses on a complex problem such as occupational sex discrimination, it immediately redefines the problem in a way that allows the discipline to make use of its distinctive elements. The result is that authors offer powerful but limited (and sometimes skewed) theories and insights on the problem. This is not surprising because, as previously noted, disciplines specialize in different kinds of problems and on certain parts of the problem.

Fischer's explanation of how a discipline can so easily skew its approach to a problem is instructive for interdisciplinarians. By not mapping the problem to reveal its key parts, even professionals can easily buy into their discipline's skewed perspective. If this is true for professionals, it is likely to be equally true for students. This emphasizes the importance of mapping the problem and connecting each part to a discipline that studies that part. This also points up the need to develop adequacy in each relevant discipline (in this case, the phenomena that they typically study), yet to avoid being impressed with its skewed perspective, theories, and insights. *When evaluating any particular insight you should compare the phenomena implicated in the particular insight with those you think are implicated in the problem as you have mapped it, and then ask how addressing these other phenomena might alter the insight.*

Notably, Fischer, a trained economist who knows well the strengths and limitations of his discipline, is not a trained historian. Yet, he took the time to develop adequacy in history to the extent that he knows its perspective on the subject of OSD.

From the Humanities. Bal (1999), "Introduction," *The Practice of Cultural Analysis: Exposing Interdisciplinary Interpretation.* Bal uses cultural analysis because it is an interdisciplinary field with its own theory and method that enables the practitioner to explain more comprehensively than traditional disciplinary approaches can the enigmatic meaning of a complex object such as the graffito. For example, cultural analysis differs from a traditional historical approach in its focus on the viewer's situatedness in the present and its seeking to understand the past as part of the present in which the viewer is immersed (p. 1). History tends to view the past, whether person, event, or object, as the product of evolutionary trends and developments leading up to it, societal forces, and individual decisions, which it seeks to describe in rich detail. Using history as it is usually practiced, then, would skew one's understanding of the graffito by ignoring the silent assumptions that historians consciously or unconsciously impose on the past.

The strength of art history (which is increasingly interdisciplinary in orientation) is its ability to study an art object in terms of concepts such as quality, its visual appeal, how the artist fits into the canon of great artists, the social context of the object, the uses and misuses of stylistic analysis, and the relationship of art history and movements such as postmodernism and feminism (Fernie, 1995). Though there is much that art history can contribute to one's understanding of the graffito, there is much that it is not equipped to deliver. For one thing, the graffito is unique and the author is unknown, making it impossible to fit it into the canon of great art and artists. Nor can art history comment on the graffito's philosophically profound message, evaluate its poetic structure, or explain how it embodies the concept of "culture."

Philosophy and its subdiscipline of epistemology concern how we can know what is real and what is not real. Relying on epistemic philosophy in the case of the graffito, however, would skew our understanding of the text:

Note

I hold you dear

I have not

thought you up

The text contains a statement of nonfiction, "I have not thought you up." However, this statement appears to be contradicted by the address—"Note"—that changes a real person, the anonymous writer's beloved, into a self-referential description of the note; that is, a referential "dear" becomes a self-referential "Note" or short letter. This changes the note into a piece of fictional prose (Bal, 1999, p. 3). If the note is fiction, then epistemic philosophy is of little use to deepen our understanding of the text's meaning. Instead, we would turn to literature because it quite easily suspends ontological questions (i.e., concerning the kinds of things that have existence), and it has no trouble distinguishing between fiction and reality.

Literature's strengths in evaluating the graffito are its enabling one to approach it from different theoretical perspectives and situate it biographically, critically, and historically. These strengths, however, are balanced by literature's limitations: its inability to situate such an evaluation culturally, understand it as an art object, explain it semantically, and probe it epistemologically and ontologically.

The strength of psychology is to probe the recesses of a person's psyche, whether artist or viewer of the art object. However, psychology is not equipped to situate an art object in a cultural and historical context, examine its poetic form and narrative mode, or understand it as an aesthetic and ethical expression.

Figure 7.2 maps the forces that combined to create the graffito, including the author's personality, experiences, values, philosophical views, and understandings of literature

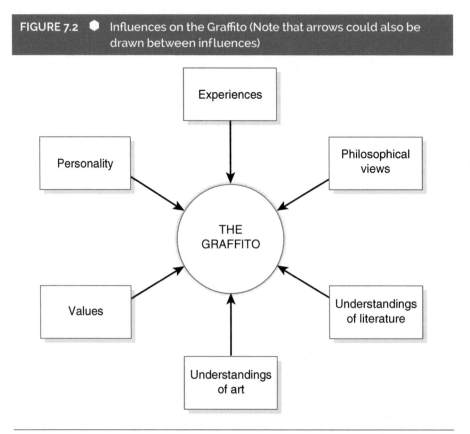

FIGURE 7.2 ● Influences on the Graffito (Note that arrows could also be drawn between influences)

and art. We would thus be guided, like Bal, to draw upon insights from the several disciplines that study these things.

Unfortunately, some authors who purport to engage in interdisciplinary work do not talk as self-consciously about their methodology as Bal does. One way to fathom an author's methodology is to read reviews of the writer's book(s). Another way is to read the author's writing on the topic closely, paying particular attention to the footnotes (or endnotes) that may contain important methodological clues. A third way is to consult disciplinary experts, who should be called upon in any event to confirm one's understanding.

CHECKLIST FOR EVALUATING PREVIOUS RESEARCH

Readers can profitably use this checklist to evaluate previous research (i.e., insights and theories) on the problem before identifying conflicts in insights and theories and locating their sources (Chapter 9):

- Reflect self-critically on the bias that you or the expert may be bringing to the problem. Szostak's (2002) insight is worth noting here: "While disciplines are an important source of bias, human nature, individual psychologies, and the diverse roles that people play in society are also sources of bias" (pp. 112–113).

- Identify the disciplinary background of each scholar (if this was not done during the literature search). The theory or method used by an author will generally reflect the general perspective of the discipline. One should question how this perspective influenced the questions that the author raised, the theory and method used, and the understanding produced.

- Determine whether some key phenomena (i.e., parts of the problem) were excluded from previous analysis and the impact of this (Szostak, 2002, p. 112).

- Identify the evidence that each scholar's discipline typically considers as valid. Analyze the evidence provided for each insight and ask whether different evidence would point in a different direction.

- Identify the epistemology that each scholar's discipline typically embraces. (The challenge here may be in discovering whether the author is assuming a modernist stance or is embracing one of the newer critical stances.) Ask to what extent this influenced the scholar's approach and thus the nature of the insight.

- Identify the theory that informs each scholar's insight (if it is theory based) and be familiar with its strengths and limitations. Ask how the scholar applied the theory to the problem.

- Identify the research method that each scholar uses and be familiar with its strengths and limitations. As with theory, the goal is to assess to what extent the insight reflects limitations of the method.

As we have noted multiple times along the way, you need to do two things in performing these various actions: reflect on (the elements of) disciplinary perspective, and compare across insights from different disciplines.

Chapter Summary

In order to analyze a problem and evaluate the insights and theories on it, you must appreciate the strengths and corresponding limitations of each discipline's perspective (in a general sense) and be aware that these strengths and limitations flow primarily from differences in phenomena studied and theories and methods used. Limitations in phenomena are best identified by mapping the problem. Limitations in theories and their insights are best identified by asking the five "W" questions. Limitations in methods are best identified by asking a second series of 10 questions. Only then is it possible to evaluate the insights and theories generated by each discipline's community of authors and assess the relevance of these to the problem.

In this STEP of the IRP, you may decide to replace one of the relevant disciplines with a discipline that seemed only marginally relevant in earlier STEPS. Fischer (1988), for example, had originally found that Marxism was only marginally relevant to explaining the causes of OSD compared to the more nuanced theories advanced by modern economists. At this juncture, you may realize the need to extend the literature search to learn more about what additional insights, concepts, or theories a particular discipline or school of thought has to offer. You may decide that the wording of the problem that once looked neutral (i.e., did not appear to privilege any one discipline) now seems too much indebted to the perspective of one of the relevant disciplines. Or you may realize that the very conception of the problem is overly reflective of a particular discipline.

At the end of STEP 6, you will have finally decided which parts of the problem to investigate and thus which disciplinary insights and theories are truly relevant to the problem. By completing this STEP and reflecting on previous work, you are now ready to engage in the integrative part of the IRP. This is the focus of Chapters 8 to 12 of Part III.

Note

1.   For a fuller discussion of these 10 questions, see Szostak (2004), Chapter 4.

Exercises

The Location of a Wind Farm

7.1 In response to public pressure to reduce its dependence on coal and switch to renewable energy sources, a public utility is proposing to build a large wind farm in shallow water within view of prime beachfront property, exclusive resort hotels, and private luxury condos whose taxes are a significant source of revenue to the local economy. Assuming that disciplinary experts from a nearby university will be asked to conduct studies on the proposal by the local governing council, what disciplinary perspectives should be included to enable public officials to make a fully informed decision to approve or not approve the construction of the wind farm? The problem should first be mapped in order to reveal the parts of the problem and identify relevant disciplines.

The Problem of Homelessness

7.2 Say you are participating in a service learning course that is focusing on the problem of homelessness in a particular area of town. Before going out into the "field," you are to write a brief background paper on the subject. Having had extensive coursework in business, you prefer to examine the problem from a business perspective.

    a.   Is this possible in an interdisciplinary course? Why, or why not?

    b.   How might you make use of your business background in a way that is appropriate to interdisciplinary inquiry?

    c.   How could you state the research problem in a way that does not privilege any one disciplinary perspective?

The Progression of Research

7.3 If you are limited to working with only two or three disciplines, which disciplinary perspectives would contribute to the fullest understanding of the problem, and why?

Personal Bias

7.4 You have deeply held beliefs concerning the issue of euthanasia and believe it is ethically permissible, or even desirable, in certain situations. How can you remain true to your deeply held convictions and yet engage in interdisciplinary inquiry concerning this issue when evaluating insights?

Limits of Perspectives

7.5 Select one of the above problems in 7.1 or 7.2 and identify the limits that three disciplinary perspectives impose on developing a more comprehensive understanding of it.

7.6 Given the limits of disciplinary perspectives and the fact that they are skewed, how does viewing a complex problem through multiple disciplinary lenses help us to understand the problem?

(Continued)

(Continued)

Evaluating Theories

7.7 How does Bal's work illustrate the importance of being flexible in developing research strategies that balance disciplinary depth with theoretical breadth?

How Supportive Evidence Reflects Disciplinary Perspective

7.8 In the above example of the proposed offshore wind farm project, what kinds of data/evidence would be considered valid by each contributing disciplinary expert? More generally, what are the ways that factual information may be skewed by the very nature of disciplinarity?

How Methods May Be Skewed

7.9 Select a method from Table 7.12 and discuss its strengths and limitations when applied to the example of OSD, the graffito, or the wind farm. How can asking the 10 questions of any disciplinary method provide important insights into a scholar's understanding of the problem?

Phenomena Embraced by Insights

7.10 Referencing the above problem of homelessness, what phenomena are omitted from the discussion of possible causes when only the disciplines of economics and psychology are considered? The problem should first be mapped to reveal the parts of the problem and identify relevant disciplines.

7.11 How is cultural analysis, though an interdisciplinary approach, limited in its approach to understanding an object such as the graffito?

INTEGRATING
INSIGHTS

PART III

INTEGRATION

8

Image by Gerd Altmann from Pixabay.

UNDERSTANDING INTEGRATION

LEARNING OUTCOMES

By the end of this chapter, you will be able to

- Explain what interdisciplinary integration is
- Discuss the generalist critique of integration
- Discuss the integrationist case
- Explain the importance of perspective taking and cultivating certain qualities of mind
- Describe the model of integration used in this book
- Answer the three fundamental questions concerning integration

GUIDING QUESTIONS

What do we mean by interdisciplinary integration?

(Why) is integration essential to interdisciplinarity?

How do you integrate disciplinary insights? (Note that this question animates the next chapters as well.)

What does the result of integration look like?

CHAPTER OBJECTIVES

This chapter introduces the second part of the interdisciplinary research process (IRP) and presents an overview of interdisciplinary integration. It explains what interdisciplinary integration is, discusses the controversy over the role that integration should play in interdisciplinary work, and emphasizes the critical importance of perspective taking and of cultivating certain qualities of mind. The chapter also introduces the model of integration used in this book in terms of what it integrates, how it integrates, and what the result of the integration looks like. It concludes by raising and answering three fundamental questions concerning integration.

 B. INTEGRATING DISCIPLINARY INSIGHTS

7. Identify conflicts between insights and their sources.
8. Create common ground between insights.
9. Construct a more comprehensive understanding.
10. Reflect on, test, and communicate the understanding.

WHAT INTERDISCIPLINARY INTEGRATION IS

Merely examining a behavior or object from different disciplinary perspectives does not *by itself* constitute interdisciplinary work. *Without integration, these different perspectives would lead to mere multidisciplinary work* (Hacking, 2004, p. 5). The reason is illustrated in a poem by American poet John Godfrey Saxe (1816–1887), based on an Indian fable about six blind men and an elephant. Each of the men thoroughly investigated a particular part of the massive beast, and each emphatically concluded that what he had "observed" was very much like a wall (its sides), a spear (its tusk), a snake (its swinging trunk), a tree (its leg), a fan (its ear), and a rope (its tail). The poem ends,

And so these men of Indostan

Disputed loud and long,

Each in his own opinion

Exceeding stiff and strong,

Though each was partly in the right,

And all were in the wrong! (Saxe, 1963)

The lesson is that simply having six different experts from six different disciplines examine an object will likely yield at least six different understandings of the object. This is the nature of multidisciplinarity. What is lacking, of course, is any attempt to integrate these conflicting insights, insofar as this is possible, into one composite description or image of the object so as to provide a more comprehensive understanding of it. This understanding is not "owned" by one discipline or insight. We extend this analogy to include a seventh blind person representing the interdisciplinarian who queries the other six about the object. That person then integrates the information provided by the six disciplinarians in an attempt to construct a more comprehensive understanding of the elephant.[1]

Interdisciplinarians generally agree on the centrality of integration to interdisciplinarity and the interdisciplinary research process,[2] and they are moving toward consensus about what integration should encompass. "Integration," asserts Klein (2010), "is widely viewed as the litmus test of interdisciplinarity" (p. 112). Though integration is not easy, it is possible, even for those new to the field.

A Definition of Integration

A good way to start defining a word is to look at its etymology or historical beginnings. The English word *integration* can be traced back to the Latin word *integrare*, meaning "to make whole." As a verb, *integrate* means "to unite or blend into a functioning whole."

Over the centuries, says Klein (2012), "the idea of integration was associated with holism, unity, and synthesis" (p. 284).

A synonym of integration is the noun *synthesis*. Because the terms *integration* and *synthesis* are so close in meaning, many practitioners use them interchangeably. This book uses *integration* because it appears in prominent definitions of interdisciplinary studies and in the definition we presented in Chapter 1. This chapter examines both the *process* and the *product* of integration.

From the literature, two important ideas about integration emerge that should be included in its definition. Practitioners repeatedly refer to interdisciplinary integration as a *process* as opposed to an *activity*. This is deliberate. Process conveys the notion of making gradual changes that lead toward a particular (but often unanticipated) result, whereas *activity* has the more limited meaning of vigorous or energetic action not necessarily related to achieving a goal.

Additionally, practitioners refer to the result of the integrative process as culminating in a new and more comprehensive understanding. This understanding is sometimes called "the integrative result," "the new whole," "the new meaning," "the integrative product," "the extended theoretical explanation," "the conceptual blend," or "the cognitive advancement." Though these terms have similar meanings, the one generally used in this book is "more comprehensive understanding." Achieving the new understanding requires combining expert insights from disciplines, subdisciplines, interdisciplines, and schools of thought, and sometimes from perspectives beyond the academy. *The premise of interdisciplinary studies is that the disciplines themselves are the necessary preconditions for and foundation of interdisciplinarity.*

Integration does not derive from a predetermined pattern or a universal template that is applicable beyond the specific problem, issue, question, or research project. Integration is something that we must create. The definition of interdisciplinary integration, introduced in Chapter 1, is repeated here:

Interdisciplinary integration is the cognitive process of critically evaluating disciplinary insights and creating common ground among them to construct a more comprehensive understanding. The understanding is the product or result of the integrative process.

Integration, therefore, is *the means* by which the end or goal of the research process is achieved. Klein (1996) states, "Synthesis connotes creation of an interdisciplinary outcome through a series of *integrative actions* [italics added]" (p. 212). These "integrative actions," decisions, or STEPS are the subject of Chapters 9 to 12.

Traits of Integration

Several traits attend the term *integration*:

- They convey the meaning of goal-driven behavior to solve a problem, answer a question, or resolve an issue.

- Central to integrative activity is critically evaluating disciplinary insights.

- The nature of the integrative process is creative combining or uniting.

- What is combined, united, or blended are the disciplinary concepts, assumptions, or theories used to generate their insights.

- The result of integration is valid only for a particular problem and context.

- The object of this process is the creative formation of something new, greater than, and different from the sum of its parts—a more comprehensive understanding. (*Note:* The distinctive characteristics of the understanding are discussed later in this chapter.)

THE CONTROVERSY CONCERNING INTEGRATION

Practitioners usually take one of two positions concerning the place of integration within interdisciplinary studies: "generalist" and "integrationist."

The Generalist Critique

Generalists are those who reject the notion that integration should be *the* defining feature of genuine interdisciplinary research and teaching. They understand interdisciplinarity loosely to mean "any form of dialog or interaction between two or more disciplines," while minimizing, obscuring, or rejecting altogether the role of integration (Moran, 2010, p. 14).

Some generalists see the terms *interdisciplinarity* and *integration* as synonymous with *teamwork*, as in team teaching or cross-disciplinary communication on research teams. Other generalists prefer to distinguish between types of interdisciplinarity by focusing primarily on the kinds of questions asked rather than on integration and leave "the question of integration open" (Lattuca, 2001, pp. 4, 78, 80). Still others go so far as to reject any definition of interdisciplinary studies that "necessarily places priority emphasis on the realization of synthesis [or integration] in the literal sense" (Richards, 1996, p. 114).

The core of the generalist case is that in most cases, integration is simply not achievable. Generalists identify at least four factors that complicate, retard, or even prevent integration: (1) disciplinary fragmentation, (2) unbreachable epistemological barriers, (3) conflicting perspectives and ideologies, and (4) a variety of possible results.

Disciplinary Fragmentation

Each discipline is fragmented, say generalists, and "the fragments, too, are fragmented" (Dogan & Pahre, 1990, p. 5). While specialization can advance knowledge, it can also isolate subfields from each other. Psychologist Elaine Hatfield and historian Richard Rapson (1996) point to hundreds of specialized subdisciplines, each "speaking their own languages, adopting their own definitions and methodologies, asking their separate questions, and rarely addressing one another" (p. viii). As the number of distinct specializations expands, it becomes harder to achieve a comprehensive understanding.

Epistemological Barriers

Generalists point to a second factor that makes integration problematic: unbreachable epistemological barriers between disciplines. The result is "incommensurability of concepts, different units of analysis, differences in worldviews, expectations, criteria, and value judgments" (Rogers, Scaife, & Rizzo, 2005, p. 268). Some types of knowledge are so qualitatively distinct, claims Richards (1996), that they will not permit integration under any circumstances. This problem is said to occur when, for example, one attempts to combine insights from the sciences (natural and/or social) that are analytical-deductive with those from the humanities, which are subjective-evaluative. In such cases, he asserts, "there would seem to be unbreachable epistemological barriers preventing genuine integration" (p. 122). These barriers also render developing a common vocabulary or basis for common communication between disciplines problematic. Consequently, generalists doubt the efficacy of seeking common ground "as a way to uniformly resolve all-important differences" because "in choosing some common ground over others, there can be competing versions of what is held in common" (Fuchsman, 2009, p. 79).

Conflicting Perspectives and Ideologies

Generalists see as a third obstacle to integration the conflicting perspectives and ideologies within disciplines and across disciplines. Hyland (2004) stresses conflict *within* disciplines. "Most disciplines," he says, "are characterized by several competing perspectives and embody often bitterly contested beliefs and values" (p. 11). Szostak (2002), on the other hand, stresses conflict *across* disciplines because authors typically use the theories and methods that their home disciplines favor (p. 111). This fact, explains biologist Ernst Mayr (1997), makes consensus within and across disciplines hard to achieve because "disagreeing scientists adhere to different underlying ideologies, making certain theories acceptable to one group which are impossible for another group" (p. 103).

A Variety of Possible Results

A fourth objection to integration is that when the conditions for interdisciplinarity exist, a variety of results are possible. There can be **full integration** where all relevant disciplinary insights have been integrated into a new, single, coherent, and comprehensive understanding or theory that is consistent with the available empirical evidence; **partial integration** when only some insights have been integrated and it applies to only some part(s) of the problem; *multiple integrations* when two or more investigations of the same problem and working with the same materials produce different understandings; or *no integration* when the problem cannot be addressed by drawing on two or more disciplinary literatures and the conditions for using an interdisciplinarity approach cannot be satisfied.

Generalists charge that an increasing number of interdisciplinary combinations will simply produce a plurality of more comprehensive understandings, or competing metatheories, and that these will result in theoretical confusion. Given these possible outcomes, epistemological barriers, and ideological preferences, one critic concludes that "the concept of integration as a single coherent entity no longer fully applies" (Fuchsman, 2009, pp. 79–80). Consequently, generalists see no justification for including integration in any definition of interdisciplinary studies.

As a result of the four arguments above, generalists often voice a preference for theory competition and alternative integrations. Generalists prefer offering readers a menu of competing theories, each of which responds to a different but overlapping set of questions, instead of presenting one integrative theory that combines elements of several competing theories. For example, some criminologists argue that criminology theories should remain separate and unequal and that "theory competition" is preferable to theory integration (Akers, 1994, p. 195).[3]

This preference is analogous to the position taken in the fine and performing arts, and often the humanities that study them, where scholars prefer setting up a range of alternative integrations for readers or viewers to consider rather than providing one. Rationales for doing this may include the desire to engage others in the issue, the feeling that it would be presumptuous to take a stand, the sense that declaring one integration "best" is premature, or the feeling that the best integration depends on the particularities of each situation (Newell, 2012, pp. 299–314). The strongest rationale for not pursuing integration further is the recognition that works of art and literature are inherently ambiguous.

The Integrationist Case

Integrationists regard integration as *the key distinguishing characteristic* of interdisciplinarity and *the goal* of fully interdisciplinary work. The core of their position is that integration is achievable and that researchers should come as close to achieving integration as possible given the problem under study and the disciplinary insights at their disposal. They point to the following developments that make integration achievable and desirable:

- Theories from cognitive psychology that are supportive of integration

- New models of the interdisciplinary research process that feature techniques demonstrated to achieve integration

- The publication of groundbreaking integrative work on a wide range of complex problems (see Box 8.1)

- The insistence on the centrality of integration by leading interdisciplinary and transdisciplinary organizations (see Box 8.2)

- The availability of ways to adjudicate between competing explanations and theories

Theories From Cognitive Psychology That Are Supportive of Integration

Integrationists assert that generalists are overlooking important theories developed by cognitive psychologists on common ground and cognitive interdisciplinarity that show that integration is a natural cognitive process. Indeed, the theories of Clark and Bromme (introduced in Chapter 1) and Svetlana Nikitina provide the theoretical premise underlying interdisciplinary integration. Digging deeper into these theories will advance our understanding of the concept of common ground and how it makes integration possible.

Clark's Theory. Central to Clark's theory of common ground is its emphasis on language in everyday interactions. He finds, for example, that all people take as common ground aspects of human nature such as physical senses, communal lexicons (i.e., sets of word conventions in individual communities), and cultural facts, norms, and procedures (1996, pp. 106–108). Clark explains that when it comes to coordinating a joint action, "people cannot rely on just any information they have about each other. They must establish just the right piece of common ground, and that depends on them finding a shared basis for that piece" (pp. 93, 99). As applied to interdisciplinary work, Clark's theory of common ground means that the interdisciplinarian should expect to find latent linguistic commonalities (i.e., common ground) between insights from different disciplines that often provide the basis for integration.

Bromme's Theory. Bromme (2000) has developed the theory of cognitive interdisciplinarity that he and others are applying to communication across academic disciplines, especially the natural sciences. A significant finding of Bromme is that in interdisciplinary communication, common ground is frequently "discovered" when the partners of cooperation—that is, the relevant disciplines—"find out that they use the same concepts with different meanings, or that they use different terms for approximately the same concepts" (p. 127).

Bromme's theory has direct applicability to the interdisciplinary research process developed by integrationists. This process calls for interdisciplinarians, whether they are developing a collaborative language or trying to integrate conflicting disciplinary insights, to *first* identify the concepts with different meanings or the theories providing different explanations *before* attempting to discover common ground. Once these are identified, the interdisciplinarian can then proceed with creating the "**common ground integrator**"—that is, the *concept, assumption, or theory*—by which these conflicting insights (whether disciplinary or stakeholder) can be integrated. We will explore various strategies for achieving common ground in later chapters.

Nikitina's Work on Interdisciplinary Cognition. In her pioneering work on interdisciplinary cognition, Nikitina (2005) sees integration as a cognitive process that occurs commonly and naturally. She presents two metaphors that describe this process. The first is a musical chord because each note contributes uniquely to a comprehensive whole that is something more than the sum of its parts. The second is color, which is "some sort of cooperative thing between our brain and the world." Color, she says, is "a psychophysical unity, informed both by physics and by individual perception, with neither perspective being definitive" (p. 406).

How These Theories Inform the Debate Over the Role of Integration

These theories inform the debate over the role of integration in interdisciplinary work in at least five ways:

1. They claim that the activity of establishing common ground is a normal and basic feature of human communication and, therefore, is *natural* and *achievable*. What integrationists assert is possible, cognitive psychologists have discovered that everyday people do routinely.

2. If it is possible for humans from differing social and other societal contexts to establish common ground and to communicate, then it should also be possible to establish common ground between disciplines, which are human constructs, as well as between stakeholders (i.e., those in society who have an interest in a societal issue, and might be included in a transdisciplinary research project or interviewed as part of an advanced interdisciplinary research project).

3. If common ground is natural and achievable, then so should be the result of integration—a more comprehensive understanding—because that is what results from common ground in everyday conversation.

4. And if integration is natural and achievable, then there is no reason to divorce integration from a definition of interdisciplinarity as generalists feel compelled to do.

5. Since cognitive psychologists assert people normally integrate without consciously thinking about this, integrationists feel safe in asserting that interdisciplinarians who are self-conscious about process should be able to integrative purposefully.

Integration, then, involves a normal set of cognitive activities that may be explained in terms of distinct and recurrent mental operations.

New Models of the Interdisciplinary Research Process That Feature Techniques Demonstrated to Achieve Integration

The case for integration is strengthened by the development of new models of the IRP that feature techniques demonstrated to achieve integration. These models agree on four key points:

- Integration should be the *goal* of the interdisciplinary research enterprise.

- Integration is a *process.*

- Integration is *achievable* if one pays careful attention to process.

- Integration is something that we must *create* using techniques demonstrated to achieve the goal of a more comprehensive understanding (Repko, 2007, p. 13).

The model presented in this book understands interdisciplinary integration to be a cognitive process that uses techniques and strategies (STEPS) demonstrated to create common ground, integrate theories, and construct a more comprehensive understanding of the problem. While generalists may view integration as an ideal that is almost impossible to realize, it is, in practice, very achievable.

The Publication of Groundbreaking
Integrative Work on a Wide Range of Complex Problems

BOX 8.1

The integrationist case is supported by the growing number of publications that view integration as the primary methodology of interdisciplinarity. These include Rick Szostak's (2004) *Classifying Science: Phenomena, Data, Theory, Method, Practice*; the National Academies' (2005) *Facilitating Interdisciplinary Research*; Sharon J. Derry, Christian D. Schunn, and Morton Ann Gernsbacher's (2005) *Interdisciplinary Collaboration: An Emerging Cognitive Science*; John Atkinson and Malcolm Crowe's (2006) *Interdisciplinary Research: Diverse Approaches in Science, Technology, Health and Society*; David McDonald, Gabriele Bammer, and Peter Deane's (2009) "Research Integration Using Dialogue Methods"; Rick Szostak's (2009) *The Causes of Economic Growth: Interdisciplinary Perspectives*; Matthias Bergmann et al. (2012), *Methods for Transdisciplinary Research: A Primer for Practice*; and Allen F. Repko, William H. Newell, and Rick Szostak's (2012) *Case Studies in Interdisciplinary Research*. These books provide a number of examples, several rather detailed, of successful interdisciplinary integration. They and a large number of journal articles offer a variety of insights into interdisciplinary research, share a common conception of interdisciplinarity, and appreciate that integration is the best way to address complex problems that confront us. See also Szostak's (2019) *Manifesto of Interdisciplinarity*, which urges the centrality of integration. The examples they showcase do not fully address the difficulties of interdisciplinary integration, but they stand as convincing counter examples to claims that barriers of fragmentation, epistemology, ideology, and perspective make integration unachievable.

The best evidence that integration is not impossible is that it has so often been achieved across a wide range of topics.

The Insistence on the Centrality of Integration by
Leading Interdisciplinary and Transdisciplinary Organizations

---

BOX 8.2

The *Handbook of Transdisciplinary Research* calls integration "the core methodology underpinning the transdisciplinary research process" (Pohl et al., 2008, p. 421). In the field of landscape research, for example, the issue is not whether to integrate but "how to apply new and specific interdisciplinary and/or transdisciplinary techniques" (Tress, Tres, Fry, & Opdam, 2006, p. i).

The Swiss Academies of Arts and Sciences Network for Transdisciplinary Research (td-net) devoted an international conference to the topic in 2009. The Australian-based Integration and Implementation Sciences network provides an academic base for synthesizing pertinent knowledge, concepts, and methods to address complex problems (Bammer, 2005). And the Association for Interdisciplinary Studies (AIS) has long promoted integration as the distinctive characteristic of interdisciplinarity and the goal of interdisciplinary studies.

The Availability of Ways to Adjudicate Between Competing Explanations and Theories

Generalists prefer to present a menu of competing disciplinary explanations, theories, and meanings rather than one integrative explanation, theory, or meaning to avoid adding more explanations that will result in confusion. Integrationists make two counterarguments. First, competition for the best explanations is inherently unequal if the integrative explanation is omitted. Integrationists in the field of criminology, for example, see several distinct advantages in integrating existing discipline-based theories. Integration, they argue,

- facilitates the development of central concepts that are common to several theories,

- provides coherence to a bewildering array of fragmented theories and thereby reduces their number,

- achieves comprehensiveness and completeness, and thereby enhances their explanatory power,

- advances scientific progress and theory development, and

- integrates ideas about crime causation and social control policy (Henry & Bracy, 2012, p. 261).

What is needed, say integrationists, is not more granular and conflicting explanations but the integration of the best elements of these explanations.

Second, generalists' concern over "theory competition" is overblown, say integrationists, because it presumes that there is no way to adjudicate between competing theories. As shown elsewhere in this book, there are, in fact, multiple ways to adjudicate between competing theories. One test is to ask, "Which understanding is most consistent both with the contributing concepts and theories *and* with the available empirical evidence?"

If more than one comprehensive understanding passes the consistency test, then the second test can readily distinguish between/among them: "Which one produces the more desirable solution when applied to the complex problem under study?" The problem, however, is not an abundance of integrative explanations, but a scarcity of them. *We need to move beyond debate between disciplinary theories to discussion and critical evaluation of more comprehensive understandings produced through interdisciplinary integration (especially when they form the basis for public policy).* (See Box 8.3.)

BOX 8.3

Philosophy of Integration

While there is considerable consensus that integration is critical for interdisciplinarity, there is nevertheless some ambiguity about what precisely is meant by the term. O'Rourke, Crowley, and Gonnerman (2016) examine the different meanings that have been attached to the concept. One distinction they draw is between the sort of cognitive integration of ideas that we have stressed and a sort of social integration that needs to occur early in interdisciplinary research teams by which team members come to understand each other and what they might contribute. We appreciate the importance of this quite different form of integration: If students perform group work, they should be conscious of the need for group understanding. They need to achieve a common ground to facilitate communication before they can achieve the sort of cognitive integration we stress. O'Rourke et al. (2016) also distinguish between those who believe that there are clear strategies for achieving integration and those who feel that only broad guidance can be given. We pursue a middle course in this book: We highlight strategies that have proven useful to past researchers but appreciate that integration is a creative act and that researchers may develop yet more useful strategies in future. O'Rourke et al. (2016) further distinguish an emphasis on integrating conflicting insights from a more general treatment of integration. We will often stress conflict in what follows—for this is both common and challenging to address—but also often recognize that sometimes integration involves adding together complementary insights.

We raised the distinction between full and partial interdisciplinarity above. Our discussion there suggests that full integration is always feasible (with the partial exception of the humanities where there may always be ambiguity in our understanding of works of art), and thus should be the goal of interdisciplinary inquiry. Yet since integration is a creative act, we should applaud a researcher that achieves only partial integration. Their work may set the stage for others to achieve a fuller integration. In a research team, different members may contribute a partial integration that feeds into a full integration.

The Importance of Perspective Taking

We emphasize the importance of perspective taking: This, as noted in Chapter 2, means viewing some problem, object, or phenomenon from a standpoint other than one's own. In interdisciplinary work, perspective taking involves appreciating alternative disciplinary perspectives. This does not mean abandoning one's own disciplinary beliefs to the views of another discipline. It does mean being aware of more than one way to account for natural and social phenomena and processes. For example, in *Catching the Light: The Entwined History of Light and Mind*, physicist Zajonc (1993) describes how philosophical and psychological arguments influenced the understanding of the nature of light waves:

> Light . . . has been treated scientifically by physicists, symbolically by religious thinkers, and practically by artists and technicians. Each gives voice to a part of our experience of light. When heard together, all speak of one thing whose nature and meaning has been the object of human attention for millennia. During the last three centuries, the artistic and religious dimensions of light have been kept severely apart from its scientific study. . . . The time has come to welcome them back, and to craft a fuller image of light than any one discipline can offer. (p. 8)

Importantly, interdisciplinary openness to other perspectives is of a special nature. It involves exploring the roots of conflict between disciplinary insights, which is the focus of STEP 7 (Chapter 9). The point to be stressed here is that the interdisciplinary mind tends to go beyond mere *appreciation* of other disciplinary perspectives (i.e., multidisciplinarity); it *evaluates* their capacity to address a problem and assesses their relevance.

The Importance of Cultivating Certain Qualities of Mind

Successfully performing integration and engaging in perspective taking require that you consciously cultivate these qualities of mind:

- Thinking inclusively and integratively, not exclusively

- Being responsive to each perspective but dominated by none (i.e., not allowing one's strength in a particular discipline to influence one's treatment of other relevant disciplines with which one is less familiar)

- Maintaining intellectual flexibility

- Thinking both inductively and deductively

- Thinking about the whole while simultaneously working with parts[4]

INTEGRATION IN THE BROAD MODEL

This section addresses three key questions concerning integration in the interdisciplinary research process pursued in this book: (1) What is integrated? (2) How is it integrated? (3) What does the result of integration look like?

What the Model Integrates

The broad model integrates *insights* concerning a particular complex problem. These are produced by disciplines, subdisciplines, interdisciplines, schools of thought, professional fields, and one's basic research. Some of these insights may be from disciplines that are epistemologically distant. Insights are not limited to those produced inside the academy and may include expertise from stakeholders outside the academy. This expertise may be from government agencies, private industry, or not-for-profit organizations. Such expertise may partly reflect the disciplinary background of the stakeholder and/or the disciplinary orientation of the organization that the stakeholder represents. This expertise is expressed in the form of insights into the problem. Theories and the insights they produce along with their concepts and assumptions constitute the "raw material" used in integration. Depending on the combination of disciplines that produced them, these theories and insights do not usually fit together naturally or easily (with exceptions in the natural sciences), though they may be somewhat complementary.

The flexibility of the broad model enables it to work with disciplinary theories and their key concepts in the natural and social sciences. The model is also able to work with specific objects such as text, photos, film, or other objects that are the focus of the humanities and applied arts, and use tools from a specific set of collaborating disciplines to analyze them. Moreover, the model can integrate any disciplinary material in the fabric of time, culture, and personal experience and address timeless questions concerning the human condition.

A Caveat

Perspectives of disciplines are *not* what are being integrated. The interdisciplinary research process involves integrating disciplinary *insights*—not their perspectives—concerning a particular problem. As we have seen in previous chapters, we employ disciplinary perspective in evaluating disciplinary insights before integrating insights. Three examples illustrate our point:

- Earth science views planet Earth as a large-scale and highly complex system involving the four subsystems of geosphere, hydrosphere, atmosphere, and biosphere. When this perspective is applied to a particular problem, say damming a river system such as the Columbia, the insight that Earth science may generate (in the form of a scholarly monograph, journal article, or report to a public agency) is that building the system of dams is feasible given the geological characteristics of the Columbia Basin.

- Sociology views the world as a social reality that includes the range and scope of relationships that exist between people in any given society. When this perspective is applied to a particular problem, say repeated spousal battery, the insight into the problem that sociologists may generate is that it is caused by male unemployment or the desire for patriarchal control.

- Art history views art in all its forms as representing a culture at a given point in time and therefore providing a window into that culture. When this perspective is applied to a specific work of art, say a post–Civil War Currier & Ives lithograph, the insight that art historians may generate concerning the meaning of the work is that it expresses the optimism of a culture that has embraced the concept of progress by conquering nature.

How the Model Integrates

The interdisciplinary research process uses a process approach to achieve integration. It proceeds in a steplike fashion, but also involves reflecting on and possibly revisiting earlier STEPS. It has three components: (1) identifying conflicts in insights and locating their sources (STEP 7, the subject of Chapter 9), (2) creating common ground between conflicting disciplinary concepts or assumptions (STEP 8a, the subject of Chapter 10) or theories (STEP 8b, the subject of Chapter 11), and (3) constructing a more comprehensive understanding of the problem (STEP 9, the subject of Chapter 12). Though these STEPS can usefully be understood separately, each one is closely connected to those that immediately precede and follow it. (*Note:* While we focus on integrating across conflicting insights in the next chapters, many of the strategies outlined there can be applied when disciplinary insights are complementary.)[5]

Summary of How the Interdisciplinary Research Process Integrates

The key characteristics of interdisciplinary integration are as follow:

- It makes the process of integration explicit and transparent by breaking it down into discrete STEPS that require reflecting on earlier STEPS.

- It maps complex problems to reveal their complexity and causal links.

- It connects disciplines that are epistemologically distant or close.

- It uses all disciplinary methodological and philosophical tools that are relevant to the problem.

- It draws upon all relevant disciplinary theories and insights, including those derived from one's own basic research and those of stakeholders outside the academy, and critically evaluates these to locate sources of conflict.

- It uses various techniques (see Chapters 10 and 11) to modify insights and theories in order to create common ground.

- It melds insights concerning a particular phenomenon until the contribution of each becomes inseparable.

- The cognitive process involved in performing integration is both compelling and elusive. It is compelling because the idea of integration is heavily supported by theory; it is elusive because other than the broad model, none of the specialized approaches delineates explicit actions, operations, or STEPS that make it possible (but not inevitable in every case) to achieve integration in a wide range of contexts.

What the Result of Integration Looks Like

Interdisciplinarians have not always been clear about how to achieve integration; they have been even less clear about the result of integration. The definition of integration stated earlier specifies that the integrative process should produce a result possessing certain characteristics. Newell (1990) relates how he and other practitioners used to think of integration as analogous to completing a jigsaw puzzle (p. 74). For a time, this analogy was commonly used. However, comparing the result of integration to a completed jigsaw puzzle is problematic in at least three critical respects. Discussing these will deepen our understanding of integration both as a process and as a result.

The Result of Integration Accommodates Epistemological Differences

The first characteristic of the resulting integration is that it *accommodates* (but does not fully resolve) epistemological differences. The pieces of a jigsaw puzzle are finely milled to fit together as tightly as possible. This is not so with disciplinary insights, which pertain to a particular problem or object and thus "fit" to some degree on that limited basis. The problem of fit is minimized in the natural sciences where fit is achieved merely by reconciling alternative conceptualizations of the same phenomenon. For example, when addressing the energy crisis, physics may focus on how a power plant works (i.e., the alternative formulations of thermodynamics) and chemistry on how energy is released from chemical bonds in coal. The insights may be fully complementary but require using the technique of redefinition (see Chapter 10), perhaps to address differences in scale (Newell & Green, 1982, pp. 25–26). However, when an issue connects the natural sciences to the humanities or social sciences as in the embryonic stem cell research controversy, insights become less complementary, so the problem of fit becomes more challenging (Kelly, 1996, p. 95).

The Result of Integration Is Something New and More Comprehensive

The second characteristic of the resulting integration is that it is new and more comprehensive. The jigsaw puzzle pieces, when assembled, form a picture that is not new because it existed before the pieces were assembled (i.e., the picture on the puzzle box). With disciplinary insights, however, there is no predetermined pattern that one can consult to see if one is "getting it right." In fact, working with disciplinary theories and insights is comparable to working with puzzle pieces from different puzzles mixed together with no picture for guidance. The absence of pattern is particularly challenging when working in the social sciences and the humanities where theories and insights from different disciplines tend to conflict more sharply. For example, in examining the meaning of the Temple Mount in Jerusalem to Judaism, Christianity, and Islam, religious studies scholars may focus on the sanctity of the Mount based on the sacred writings of these faith traditions, whereas political scientists may focus on the political implications of physical and administrative control of the site. These differences make the task of creating common ground difficult, though not impossible. Once common ground is created and integration is performed, you can see whether the new understanding is consistent with or can accommodate the conflicting disciplinary insights.

When disciplinary insights are integrated, they generally form something that is truly new—a new understanding, a new meaning, or an extended theoretical explanation.[6] When a disciplinary expert produces an insight into a problem and the insight appears in a book aimed at a disciplinary audience or is an article published in a peer-reviewed journal, the insight is considered new, complete, and authoritative in the view of the discipline. It is not perceived as missing some piece from another discipline. From an interdisciplinary standpoint, however, the same insight is only partial and only one of many possible explanations of the problem or interpretation of the object, especially if other disciplines have produced insights into it.

The resulting integration or more comprehensive understanding is "new" in four ways:

- It is the result of an explicit, reflexive, and iterative integrative process.

- It is inclusive of the relevant disciplinary insights but not dominated by any one of them.

- It did not exist prior to the integration of the separate disciplinary theories and/ or insights.

- It is not merely a juxtaposition of separate disciplinary insights in a multidisciplinary fashion.

Though new, the integrated result should be viewed as only provisional until it is tested according to criteria presented in Chapter 13. Not every new idea is a good idea.

The Result of Integration Is "Larger" Than the Sum of Its Parts

The third characteristic of the resulting integration is that it is "larger" than the sum of its constituent parts, not in spatial terms but in *cognitive* terms. The puzzle picture before it was cut up encompassed a predetermined area expressed in square inches or centimeters. When the pieces are fitted together, the completed puzzle is not larger (or smaller) than the sum of its pieces. By contrast, the new understanding created by the activity of integration constitutes what Boix Mansilla (2005) calls "a cognitive advancement," meaning that it would have been unlikely through single disciplinary means (p. 16). The new whole or more comprehensive understanding is larger than its constituent parts in another way: It cannot be reduced to the separate disciplinary insights from which it emerged. The act of integration adds new understanding (Newell, 1990, p. 74). Consequently, the understanding is cognitively "larger" compared to what could be achieved by merely gathering up individual specialty insights and using them to view the problem from a series of disciplinary perspectives the way multidisciplinarity does. The understanding is likewise "larger" in that nothing relevant has been excluded. Disciplinarity tends to exclude, whereas interdisciplinarity strives for inclusion.

The Defining Characteristics of the Result of Integration Summarized

The defining characteristics of the result of integration are summarized here:

- It is motivated by the instrumentalist goal of using the resulting integrative understanding to solve a real-world problem, better understand the human condition (past and present), and open new pathways of research.

- It is formed from disciplinary and stakeholder insights that do not fit together naturally or easily, though they have the potential to be somewhat complementary.

- It is created without benefit of a preexisting integrative pattern (though a recognized approach to achieve integration is used).

- It has characteristics that differ from any of those of the contributing insights.

- It is inclusive of relevant information and is thus often messy with all sorts of caveats.

- It may be partial or full, depending on the range of disciplines the researcher was able to embrace.

- It may assume a wide variety of forms (see Chapter 13 for examples of these forms).

QUESTIONS RAISED BY THIS DISCUSSION OF INTEGRATION

What Does Integration Change?

Does integration change only the contribution of each discipline, or are the disciplines themselves somehow changed? The answer to the first half of the question is that the contributions of each discipline are changed because one cannot integrate insights before they have been modified to create common ground. Most practitioners agree that disciplinary contributions—that is, assumptions, concepts, and theories—must change for integration to succeed. Precisely how this occurs will be demonstrated in Chapters 10 and 11. Interdisciplinary and disciplinary research are symbiotic: Interdisciplinary insights resulting from integration can feed back into disciplinary research; encourage new questions; change concepts, theories, and methods; and encourage interdisciplinary researchers to pay attention to a wider range of phenomena. The answer to the second half of the question is that the feedback effect on the contributing disciplines themselves is neither inevitable nor essential though it can happen.

Must Integration Result in a Clear-Cut Solution to a Problem for a Study to Be "Successful" and Truly Interdisciplinary?

The answer to this question is "Not necessarily." Those in the humanities and in the fine and performing arts resist the impulse to provide a single best integration (or best meaning), preferring instead to respect the ambiguity inherent in the art objects they critically examine and *lay out the range of possibilities for integration.* Interdisciplinarians in the natural and social sciences, by contrast, seek to integrate on behalf of others, presenting their new, more comprehensive understanding as a finished product or "clear-cut solution." To conclude that the integrative effort has "failed" because of the absence of a feasible solution, even though the effort revealed a new understanding or new areas to be investigated, would be a mistake. Integration is a process that may proceed toward ever fuller understanding. As long as the integration adds *something* "new" to the understanding of a complex problem, even an increased appreciation of its complexity, or the range of possibilities for integration, or of new avenues for research, the work should be viewed as successful. Testing the quality of interdisciplinary work is the focus of Chapter 13.

Will Integration Always Resolve All Conflict?

This question overlaps with the previous one. Seipel (2002) cautions students and teachers alike that we should not expect that the process of integration will always result in a neat, tidy solution in which all contradictions between the alternative disciplinary insights are resolved. "Interdisciplinary study," he writes, "may indeed be 'messy'" because the problems it studies are messy (p. 3). Differing insights into a problem and accompanying tensions between disciplines may not only provide further understanding, he says, but should be seen as a "healthy symptom" of interdisciplinarity. The richest interdisciplinary work is that resulting from a research process that works through these tensions and contradictions between disciplinary systems of knowledge with the goal of integration, the creation of new knowledge (p. 3). One such example of integrative work is William Dietrich's (1995) *Northwest Passage: The Great Columbia River,* which was introduced earlier and will be referenced in later chapters to illustrate certain STEPS of the integrative process.[7]

Chapter Summary

This chapter examines both the process and the product of integration. As a defining characteristic of interdisciplinarity, integration is what differentiates interdisciplinarity from multidisciplinarity. This chapter discusses the controversy between generalists and integrationists over the role that integration should play in interdisciplinary work. It emphasizes the critical importance of perspective taking and of cultivating certain qualities of mind. It explains how the broad model integrates. The chapter concludes by answering three questions raised by this discussion of integration.

Moving forward, Chapters 9 to 12 explain STEPS 7 to 9 of the IRP. STEP 7 calls for identifying conflicts between insights and locating their sources; STEP 8 calls for modifying insights and/or theories through the creation of common ground between their concepts and/or assumptions; and STEP 9 calls for constructing a more comprehensive understanding of the problem.

Notes

1. But Newell (personal communication, January 8, 2011) asserts that interdisciplinarians "must eventually come to terms with linkages between insights."

2. Some interdisciplinarians continue to object to the notion of process with respect to integration. For counterarguments, see Szostak (2012), "The Interdisciplinary Research Process," in Repko, Newell, and Szostak, *Case Studies in Interdisciplinary Research,* 2012, pp. 3–19.

3. An illustration of this is Raymond C. Miller's (2nd Ed. 2018) study of international political economy where he outlines three competing theories but refuses to integrate them.

4. Piso, O'Rourke, and Weathers (2016) identify a long list of capacities that are developed at each step of the interdisciplinary research process.

5. Repko, Szostak, and Buchberger (2020) discuss how the interdisciplinary research process advocated in this book achieves the goals, while transcending the limitations of other integrative strategies such as contextualization, conceptualization, and problem-solving.

6. Recent interdisciplinary work has shown that the more comprehensive understanding is not *inevitably* new or novel. For example, the more comprehensive understanding that Marilyn Tayler (2012) achieves is that one of the positions was correct all along, only now we can appreciate (in a way even that its advocates could not) *why* it is correct. Her understanding is "new" in the limited sense that it casts an old position in a new light.

7. The result of integration is valued not as an end in itself, but for the more comprehensive understanding, cognitive advancement, or new product it makes possible. Marsha Bundy Seabury (2002) writes of her hope that students will *move toward integration,* and thus reach a more comprehensive understanding. The metaphor of "moving toward integration" does *not* mean "a graph-like progression whereby students gradually move from lower forms of thinking on up to more holistic, abstract thinking, ending in the upper-right quadrant of the page" (Seabury, 2002, p. 47). Sometimes the "'goal' may be *not a position but a motion,* meaning that students should be able to move among levels of abstraction and generalization [italics added]," which is part of the integrative process (p. 47).

Exercises

Defining Integration

8.1 Compare the definition of integration to the definition of interdisciplinary studies presented in Chapter 1. What insight(s) does the definition of integration add to your understanding of interdisciplinarity and to the interdisciplinary research process?

Generalists

8.2 What is the strongest argument that generalists make for understanding interdisciplinarity "loosely"?

8.3 Explain how a *generalist* interdisciplinarian would go about developing a policy recommendation for combating homegrown terrorism.

Integrationists

8.4 What is the strongest argument that integrationists make for insisting that full integration be the goal of interdisciplinarity?

8.5 Explain how an *integrationist* interdisciplinarian would go about developing a policy recommendation for combating homegrown terrorism.

(Continued)

(Continued)

Supportive Theories

8.6   How do Clark's and Bromme's theories complement Nikitina's, and how do all three theories strengthen the theoretical basis of interdisciplinarity?

Perspective Taking

8.7   What does the discussion on perspective taking add to our understanding of perspective as it is presented in Chapter 2?

To Contextualize or to Conceptualize

8.8   How might the broad model of integration described in this book approach each of the following situations?

   a.   Smog attributable, in part, to diesel exhaust emissions from trucks/lorries

   b.   Application to rezone 100 acres of farmland and woodland that would permit the building of residential housing (single and multifamily) and commercial buildings that would increase employment in an economically depressed area

   c.   Proposed legislation that would ban religious garb that obscures the identity of the person wearing it

Image by Gerd Altmann from Pixabay.

©iStockphoto.com/andrewgenn

9

IDENTIFYING CONFLICTS AMONG INSIGHTS AND THEIR SOURCES

LEARNING OUTCOMES

By the end of this chapter, you will be able to

- Identify conflicting insights
- Locate sources of conflict among insights
- Communicate your research to the appropriate audience

GUIDING QUESTIONS

How and why do you identify conflicts among disciplinary insights?

How and why do you identify the sources of conflicts among disciplinary insights?

How do you write about conflicts among insights?

CHAPTER OBJECTIVES

Conflict among the insights produced by scholars from different disciplines reflects the different perspectives that they apply. The existence of conflict is not just an inconvenience that somehow keeps popping up when reading the literature on a problem; rather, it is endemic, inevitable, and central to the interdisciplinary enterprise. Conflict is what one typically discovers when viewing a complex problem from the perspective of several disciplines, but not always; we will address the case where insights are different but complementary from time to time below.

Students are sometimes horrified when they first encounter disagreements among scholars. It shakes their confidence in the entire scholarly enterprise. Yet conflicts are central to the process of scholarly discovery. Scholars disagree, and then they marshal evidence and argument that support competing hypotheses. Over time, scholarly communities generally achieve some consensus, often around a view that combines elements of competing hypotheses. Disagreements within disciplines motivate disciplinary research. Disagreements across disciplines are often ignored or unappreciated. The interdisciplinary researcher thus serves a valuable role in identifying and seeking to transcend cross-disciplinary differences in insights.

This chapter explains the importance of identifying conflicts among disciplinary insights and locating their sources. It concludes by suggesting how to communicate this information in a research paper. Identifying conflicts and locating their sources (STEP 7) is preparatory to creating common ground (STEP 8, Chapters 10 and 11) and constructing a more comprehensive understanding (STEP 9, Chapter 12).

B. INTEGRATING DISCIPLINARY INSIGHTS

7. **Identify conflicts among insights and their sources.**
   - ○ **Identify conflicting insights.**
   - ○ **Locate sources of conflict among insights.**
   - ○ **Communicate your research to the appropriate audience.**
8. Create common ground among insights.
9. Construct a more comprehensive understanding.
10. Reflect on, test, and communicate the understanding.

IDENTIFY CONFLICTING INSIGHTS

The immediate challenge for the interdisciplinarian in this STEP is to identify conflicts among disciplinary insights. This is necessary because these conflicts stand in the way of creating common ground and, thus, of achieving integration. The need for integration generally arises out of conflicts, controversies, and differences (whether large or small).

Conflicts among insights are generally discovered when conducting the full-scale literature search. Such conflicts may occur *within* disciplines as well as *across* disciplines.

Conflicts Within a Discipline

Conflicting insights into a problem may be produced by authors within a discipline. For example, in the threaded problem of suicide terrorism, the literature search

TABLE 9.1 ● Conflicting Insights Into the Causes of Suicide Terrorism Within the *Same* Discipline	
Subdiscipline	**Insight of Author**
Cognitive Psychology	Individuals are driven to commit acts of violence that "they are psychologically compelled to commit" (Post, 1998, p. 25).
	"Self-sanctions can be disengaged by cognitively restructuring the moral value of killing so that the killing can be done free of self-censuring restraints" (Bandura, 1998, p. 164).
	"In most cases, the perpetrators sacrificed themselves in the name of a nationalistic rather than a religious idea" (Merari, 1998, p. 205).

revealed three insights from cognitive psychology, a subdiscipline of psychology. (The relative importance of these insights is determined, in part, by how often other writers cite these insights and by how much scholarly debate they generate.) Though these insights are from the same discipline, the insights of each author clearly conflict, as Table 9.1 shows.

Some students may be tempted to gloss over the differences in these insights, but in doing so, they risk overlooking important clues to how each author's insight differs from the others. Oftentimes these differences are nuanced rather than stark because they are within a discipline and reflect the discipline's perspective. Only by closely reading each author's phrasing (reproduced in this table) is it possible to detect these differences.

Conflicts Across Disciplines

When facing conflicts among insights produced by authors from different disciplines, you first have to determine if the authors are talking about the same thing. The way to decide if they are is to map each author's insight. They may simply be talking about different phenomena or relationships among phenomena; in such a case, you

TABLE 9.2 ● Conflicting Insights Into the Causes of Suicide Terrorism Within a Discipline and Across Disciplines	
Discipline or Subdiscipline	**Insight of Author**
Cognitive Psychology	Individuals are driven to commit acts of violence that "they are psychologically compelled to commit" (Post, 1998, p. 25).
	"Self-sanctions can be disengaged by cognitively restructuring the moral value of killing so that the killing can be done free of self-censuring restraints" (Bandura, 1998, p. 164).
	"In most cases, the perpetrators sacrificed themselves in the name of a nationalistic rather than a religious idea" (Merari, 1998, p. 205).
Political Science	"Terrorism can be understood as an expression of political strategy" (Crenshaw, 1998, p. 7).
	"The principal difference in means between sacred terror and secular terror derives from the special justifications and precedents each uses" (Rapoport, 1998, p. 107).
Cultural Anthropology	Suicide terrorists act out of a universal heartfelt human sentiment of self-sacrifice for the welfare of the group/culture (Atran, 2003b, p. 2).

Note: Cultural anthropology is a subdiscipline of anthropology.

may find that the different insights are complementary and can be combined into a more comprehensive understanding (with the map itself serving as a common ground connecting these different insights).

Often, though, you will find that authors disagree about the same phenomenon or causal relationship. An example is Bal's (1999) study of the enigmatic graffito introduced in Chapter 3. History's focus on the graffito's "pastness" conflicts with art history's view that the writing is subservient to the image itself and the wall on which the graffito was painted. Similarly, linguistics' focus on the graffito's exclamation "Note" (or "pay atten-tion") conflicts with rhetorical analysis's broader focus of helping us "read" not just the words of the poem, but the wall as well in terms of its red color and brick composition (Bal, 1999, pp. 8–9).

In general, there is opportunity for far greater conflicts to arise between insights across disci-plines because their perspectives and assumptions differ. This is shown in the study of the causes of suicide terrorism in Table 9.2.

LOCATE SOURCES OF CONFLICT AMONG INSIGHTS

While the discovery of conflicting insights may at first be disconcerting, the student should be confident that they can carefully proceed to identify the sources of these conflicts. Not surprisingly, the sources of conflict will generally be found in elements of disciplinary perspective. *The three most common sources of conflict among insights are concepts, theories, and the assumptions underlying them.* Concepts are technical terms that represent a phenomenon or an idea and are basic components of insights. Assumptions refer to the suppositions an author makes about the problem and typically reflect the discipline's philosophical assumptions (noted in Chapter 2). Theories, which increasingly dominate the scholarly discourse within the disciplines, heavily influence the questions asked, the phenomena investigated, and the insights produced. Given the importance of theories to interdisciplinary work, a more extended discussion of them follows below. Advanced students and professionals often find themselves having to deal with insights produced by one or more theories and need to determine why these theory-based insights conflict.

We encourage you to organize information as you read as shown in the tables in this chapter. Table 9.2 is an example of relevant insights organized by discipline. The close juxtaposition of the insights makes it easier to identify points of conflict that might otherwise escape your notice.

NOTE TO READERS

Disciplines also disagree about which phenomena to study. When disciplinary insights differ because they are focused upon different phenomena, the best strategy (which we will call "organization" in Chapter 10) is to map the different insights. As noted above, it may then be possible to combine the insights, stressing how the phenomena emphasized in different disciplines interact. A sociologist may stress the social influences on a certain criminal behavior, while a psychologist stresses the personal influences. The interdisciplinary researcher can combine these insights, stressing how social forces influence individuals. Disciplinary insights may also disagree because they are grounded in research that employed different methods and data sources. Graduate students should be particularly alive to this possibility (and can refer to our discussion of methods in Chapter 7 in investigating it). Differences in methods will generally be associated with differences in theories—since disciplines choose compatible sets of theories and methods—and thus will be identified when the student investigates theoretical sources of conflicts below.

Concepts as Sources of Conflict Among Insights

Conflicts in insights often involve embedded terminology or concepts. (However, those in the fine and performing arts are quick to point out that insights are not exclusively expressed in language, but may also be expressed in form, movement, and sound.) Interdisciplinarians commonly encounter two problems concerning concepts.

One arises when *the same concept masks different contextual meanings in the relevant disciplinary insights* (Bromme, 2000, p. 127). When the same concept is used by two disciplines to describe some aspect of the problem, the researcher needs to look closely for differences in meaning. Concerning the example of acid rain in Chapter 7, the alert interdisciplinarian will discover that the concept of "efficiency" has related but different meanings for biologists and physicists (energy out/energy in), economists (dollars out/dollars in), and political scientists (influence exerted/political capital expended) (Newell, 2001, p. 19). Creating common ground when the same concept masks different contextual meanings is relatively easy because the integrative concept (in this case, "efficiency") already exists and merely awaits discovery.

Another problem arises *when different concepts are used to describe similar ideas.* In the threaded example of suicide terrorism, Table 9.3 shows that the subdiscipline of cognitive psychology uses the concept of "special logic," whereas political science uses the concept of "strategic logic." Importantly, both concepts share the same assumption: The agents (i.e., terrorists) are rational. In other words, the basis for common ground already exists

TABLE 9.3 ● Differing Concepts Used by Disciplinary Theories to Address the Causes of Suicide Terrorism		
Discipline or Subdiscipline	**Theory**	**Concept**
Cognitive Psychology	Terrorist Psycho-logic	Special logic
	Self-Sanction	Displacement
	Suicidal Terrorism	Indoctrination
Political Science	Collective Rational Strategic Choice	Strategic choice
	"Sacred" Terror	Strategic logic
Cultural Anthropology	Kin Altruism	Religious communion

between these two concepts, and thus between the insights in which they are embedded and between the theories that produced them. This is an example of "integrating as we go." However, this partial integration should be viewed as *provisional* because at this point, we do not know if the assumption of rationality applies equally to the other concepts, their insights, or their theories. This will not become clear until STEP 8 is completed (see Chapters 10 and 11).

Students should be aware that conceptual sources of conflict are often the easiest to address. This is because the conflicts are more *apparent* than *real*: Scholars seem to be disagreeing because they are using terminology in different ways. Once the differences in terminology are identified, the student can use the strategy of "redefinition" to create terminology that erases or reduces the conflict among insights (see Chapters 10 and 11).

Assumptions as Sources of Conflict Among Insights

Interdisciplinarians work with assumptions when they find that they are the source of conflict among insights. Chapter 2 establishes that every discipline makes a number of assumptions. These assumptions include what constitutes truth, what counts as evidence or proof, how problems should be formulated, and what the general ideals of the discipline are (Wolfe & Haynes, 2003, p. 154). Undergraduates majoring in a discipline absorb these assumptions by taking advanced coursework in the discipline and becoming acculturated in the discipline. However, interdisciplinary students who have not had much or any coursework in a particular discipline need to locate these assumptions to achieve integration.

Assumptions can be of three types:

a. Ontological (regarding the nature of "reality"): For example, each social science makes an ontological assumption about the rationality of individuals, whether they are rational or irrational. If, for example, the student is trying to develop an interdisciplinary understanding of those persons carrying out suicide bombings, the student will likely encounter a variety of scholarly assumptions about the bomber's state of mind, ranging from rational to irrational. Other ontological assumptions by the social sciences concern whether persons act autonomously or as a product of their culture.

b. Epistemological (regarding the nature of knowledge of that "reality"): Epistemology basically answers the question "How do and can we know what we know?" In essence, epistemology is a way of testing any belief or assertion of truth. Each discipline tests for truth in different ways. In biology, for example, the assumption that all plants need sunlight may seem obvious. However, to establish that this is true, the biologist must conduct experiments to demonstrate consistently and conclusively that plants die without sunlight. Experiments often lead to new knowledge, such as that forms of artificial light can cause plants to grow or that some plants are not very dependent on light. In the humanities, by contrast, epistemology becomes much more subjective. A poem may acquire validity not because it is liked by a large number of people, but because its meaning has stood the test of time.

c. Value laden: The social sciences make value assumptions about diversity, justice, and truth; the humanities often deal quite directly with questions of value; and the natural sciences make value judgments about which problems are worth studying and what knowledge is worth developing (Newell, 2007a, p. 256).

Examining Bal's (1999) analysis of the graffito that draws upon many disciplines illustrates how ontological, epistemological, and value-laden assumptions can easily conflict. Concerning ontological assumptions, just as disciplinary communities differ over the state of mind of a person who carries out a suicide bombing, so, too, will scholars differ over the state of mind of the person who wrote the graffito. This person's state of mind, like that of the suicide bomber, can range from rational to irrational. Another ontological assumption that may be disputed by scholars from the relevant disciplines concerns whether the author was acting autonomously or merely as a product of the culture.

The epistemological assumptions brought to bear by the relevant disciplines are no less problematic. As applied to the graffito, one cannot establish that it has a singular meaning by conducting an experiment. Epistemology is much more subjective in the humanities than in the social sciences or the natural sciences.

Making value-laden assumptions about the graffito is also problematic because these, too, are likely to generate conflict. As Bal (1999) says, "an exposition is always also an argument" (p. 5). What, then, are researchers to do in such circumstances? They should recognize that conflicting assumptions are a natural feature of the interdisciplinary landscape.

Students at all levels should be mindful of disciplinary assumptions to more easily identify the sources of conflict among disciplinary insights. One effective way to probe the assumptions of a discipline is to step back from the disciplinary insights into the problem and ask the simple question, "How does this writer see the problem?" By stepping back from the insights of economists on occupational sex discrimination (OSD) as shown later in Tables 9.8, 9.11, and 9.13, for example, one can see clearly that economics views OSD as an economic problem and the result of rational decision making.

Students should be aware that there are strategies for addressing differences in assumptions. In Chapter 10, for example, we will discuss the strategy of "transformation." The student can recognize that there is a continuum between opposing assumptions such as rationality versus nonrationality. By deciding where on this continuum a particular decision-maker may lie, the student can draw appropriately on theories that assume rationality versus those that assume nonrationality.

Theories as Sources of Conflict Among Insights

Theories play an important role in the integrative phase of the IRP. In this section, we discuss how to identify conflicts between theories. In later sections, we discuss how theories themselves, and the assumptions on which they rest, can be important sources of conflict among insights, both within and across disciplines. Each of these discussions includes a table that organizes information as it is gathered (see Box 9.1)

BOX 9.1

The approach used here to organize information about theories is the same as the approach used earlier for insights: develop a taxonomy of the most relevant theories whether they are from a single discipline or from several disciplines. This taxonomy should include each theory's name, its insight into the problem stated in the author's own words (so as to avoid skewing its meaning), its key concepts that are used in expressing the theory, and its core assumption(s) about the problem. The systematic juxtaposing of each theory's defining elements in a taxonomy such as Table 9.5 makes it easier to detect latent commonalities as well as sources of conflict needed to perform subsequent STEPS of the IRP. Locating these elements in each book and article requires paying close attention to individual words, syntax, and the order in which sentences and ideas unfold. It is better to include too much information than too little for the simple reason that it is impossible to know in advance what information will ultimately prove critical to creating common ground and performing integration.

It is worth pointing out that Table 9.6, for example, consists largely of quotations rather than paraphrases of each writer's words. Paraphrasing an author's carefully crafted statement runs the risk of unintentionally distorting the meaning or skewing definitions of complex and unfamiliar concepts. But sometimes the author's explanation of a theory is so expansive (and never summarized succinctly) that students have no choice but to paraphrase.

Theories and Their Insights

Theories are themselves sources of insights. Theories bring together concepts and make a claim about their relationship.

Focus on Each Theory's Concepts. Picking up the threaded example of suicide terrorism, Table 9.4 shows the relevant theories and their insights into the causes of suicide terrorism. These theories and their insights clearly conflict, but it is difficult to locate the sources of conflict without probing more deeply by focusing on each theory's concepts as shown in Table 9.3.

TABLE 9.4 ● Theories and Their Insights Concerning the Causes of Suicide Terrorism	
Theory	**Insight of Theory**
Terrorist Psychologic	Individuals are driven to commit acts of violence that "they are psychologically compelled to commit" (Post, 1998, p. 25).
Self-Sanction	"Self-sanctions can be disengaged by cognitively restructuring the moral value of killing so that the killing can be done free of self-censuring restraints" (Bandura, 1998, p. 164).
Suicidal Terrorism	"In most cases, the perpetrators sacrificed themselves in the name of a nationalistic rather than a religious idea" (Merari, 1998, p. 205).
Collective Rational Strategic Choice	"Terrorism can be understood as an expression of political strategy" (Crenshaw, 1998, p. 7).
"Sacred" Terror	"The principal difference in means between sacred terror and secular terror derives from the special justifications and precedents each uses" (Rapoport, 1998, p. 107).
Kin Altruism	Suicide terrorists act out of a universal heartfelt human sentiment of self-sacrifice for the welfare of the group/culture (Atran, 2003b, p. 2).

Focus on Each Theory's Assumptions. Scholars make assumptions as they develop a theory about the problem. These typically reflect the more general assumption(s) of the discipline that produced the theory. Picking up the threaded example of suicide terrorism, the right-hand column in Table 9.5 shows how each theory's assumption tends to reflect the more general assumption of the discipline that produced the theory.

Table 9.6 is a composite of Tables 9.3 and 9.5.

Theories From the Same Discipline and Their Insights

Theories from the *same* discipline are likely to have *far less* conflict because they typically reflect the discipline's perspective (in a general sense). Nevertheless, these theories *may* conflict in important ways, depending on the discipline's internal coherence. For example, economics is characterized by tight unification of theory

TABLE 9.5 ●	The Assumptions of the Relevant Theories on the Causes of Suicide Terrorism Reflecting the More General Assumption of the Discipline That Produced It		
Discipline or Subdiscipline	**General Assumption of Discipline or Subdiscipline**	**Theory**	**Assumption(s) of Theory**
Cognitive Psychology	Group behavior can be reduced to individuals and their interactions, and humans organize their mental life through psychological constructs (Leary, 2004, p. 9).	*Terrorist Psycho-logic* *Self-Sanction* *Martyrdom*	Humans organize their mental life through psychological constructs.
Political Science	Individual and group behavior is motivated primarily by a desire for or the exercise of power. "[H]uman beings, while they are undeniably subject to certain causal forces, are . . . in part intentional actors, capable of cognition and acting on the basis of it" (Goodin & Klingerman, 1996, pp. 9–10).	*Collective Rational Strategic Choice* *Identity*[a]	"Terrorism may follow logical processes that can be discovered and explained" (Crenshaw, 1998, p. 7). There is no separation between religion and politics. Religious identity explains "political" behavior.
Cultural Anthropology	Cultural relativism (the notion that people's ideas about what is good and beautiful are shaped by their culture) assumes that systems of knowledge possessed by different cultures are "incommensurable" (i.e., not comparable and not transferable) (Whitaker, 1996, p. 480).	*Fictive kin*	Personal relationships shape people's ideas about what is good.

Source: Repko, A. F., Newell, W. H., & Szostak, R. (Eds.). (2012). *Case studies in interdisciplinary research.* Thousand Oaks, CA: Sage.

a. Identity theory is already an interdisciplinary theory, though narrowly so.

TABLE 9.6 ● Taxonomy of Relevant Theories on the Problem of Suicide Terrorism			
Theory	**Insight of Theory Stated in General Terms**	**Concept**	**Assumption**
Terrorist Psycho-logic	"Political violence is not instrumental but an end in itself. The cause becomes the rationale for acts of terrorism the terrorist is compelled to commit" (Post, 1998, p. 35).	Special logic (Post, 1998, p. 25)	Terrorists are rational actors who organize their mental life through psychological constructs.
Self-Sanction	"Self-sanctions can be disengaged by reconstruing conduct as serving moral purposes, by obscuring personal agency in detrimental activities, by disregarding or misrepresenting the injurious consequences of one's victims, or by blaming and dehumanizing the victims" (Bandura, 1998, p. 161).	Moral cognitive restructuring (Bandura, 1998, p. 164)	
Martyrdom	"Terrorist suicide is basically an individual rather than a group phenomenon; it is done by people who wish to die for personal reasons. . . . Personality factors seem to play a critical role in suicidal terrorism. . . . It seems that a broken family background is an important constituent" (Merari, 1998, pp. 206–207).	Indoctrination (Merari, 1998, p. 199)	
Collective Rational Strategic Choice	"This approach permits the construction of a standard that can measure degrees of rationality, the degree to which strategic reasoning is modified by psychology and other constraints, and explain how reality is interpreted" (Crenshaw, 1998, pp. 9–10).	Collective rationality (Crenshaw, 1998, pp. 8–9)	"Terrorism may follow logical processes that can be discovered and explained" (Crenshaw, 1998, p. 7).
"Sacred" Terror	"Holy" or "sacred" terror is "terrorist activities to support religious purposes or terror justified in theological terms" (Rapoport, 1998, p. 103).	"Holy" or "sacred" terror (Rapoport, 1998, p. 103)	
Identity	"Religious identity sets and determines the range of options open to the fundamentalist. It extends into all areas of life and respects no separation between the private and the political" (Monroe & Kreidie, 1997, p. 41).	Identity	Religious identity explains "political" behavior.
Fictive Kin	Loyalty to an intimate cohort of peers who are emotionally bonded to the same religious and political sentiments (Atran, 2003a, pp. 1534, 1537).	"Religious communion" (Atran, 2003a, p. 1537)	Personal relationships shape people's ideas about what is "good."
Modernization	Explains the process of historical and cultural change and why some cultures "modernize" or transform themselves politically, economically, and technologically following the Western model while others do not (B. Lewis, 2002, p. 59).	Modernization	Terrorists behave rationally in response to these exogenous factors.

(nevertheless we have seen competing theories within economics of OSD), whereas English literature is characterized by fragmentation and has many theoretical approaches. The following examples of interdisciplinary work from the natural sciences, the social sciences, and the humanities each show theories as sources of conflict concerning a particular problem.

From the Natural Sciences. Smolinski[*] (2005), *Freshwater Scarcity in Texas.* Table 9.7 illustrates the conflicting insights from three geological theories relevant to the problem of water scarcity in global terms. These theories are then applied to the local problem of freshwater scarcity in Texas. In the absence of prior discussion of these theoretical approaches, the insight of each theory is conveyed in general terms before being applied to the problem.

From the Social Sciences. Fischer (1988), "On the Need for Integrating Occupational Sex Discrimination Theory on the Basis of Causal Variables." Table 9.8 identifies four major economics theories that provide insights into the causes of OSD. For each theory, conflict among the insights is noticeably sharpened when it is applied to the particular problem of OSD.

TABLE 9.7 ● Conflicting Theories and Their Insights Concerning the Problem of Freshwater Scarcity in Texas From the Same Discipline			
Discipline	**Theory**	**Insight of the Theory Stated in General Terms**	**Insight of the Theory Applied to the Problem**
Earth Science	Global Warming	The production of greenhouse gases has caused, and will continue to cause, the planet's temperature to rise, and this negatively impacts freshwater availability.	Global warming will significantly increase the rates of evaporation, exacerbating an already critical freshwater shortage in Texas.
	Overexploitation	Too much water is taken from aquifers (water-bearing rock formations), and this practice is seriously degrading the quality of the remaining volume of water in these aquifers.	Increased rates of water removal from the Ogallala Aquifer under Texas coupled with decreased rates of recharge have led to an overall drawdown of the water table (the uppermost limit of the water contained in the Ogallala Aquifer).
	Infiltration	There is more than enough water in the nation's groundwater systems to satisfy demand, but saline water is increasingly contaminating it.	In Texas, saline and brackish water is infiltrating and contaminating the remaining groundwater.

Discipline	Theory	Insight of the Theory Stated in General Terms	Insight of the Theory as Applied to the Problem
Economics	Monopsony Exploitation	"The profit-maximizing monopsonist . . . hires labor up to the point of where marginal labor costs equal marginal revenue (or marginal value) . . . [and thus] pays workers a wage less than their value contribution to the firm" (p. 27).	Focuses on the demand side of OSD, explaining it as "the result of collusive behavior of male employers in discriminating against females" (p. 31)
	Human Capital	"Each worker [human capital] is viewed as a combination of native abilities and raw labor power plus specific skills acquired through education and training" (p. 28).	"Focuses on the supply side of OSD, characteristics of females— particularly their education and training levels" (p. 31)
	Statistical Discrimination	"Statistical discrimination exists when an individual is evaluated on the basis of the average characteristics of the group to which he or she belongs, rather than on his or her personal characteristics" (p. 29).	"Emphasizes group (rather than individual) characteristics of female job applicants. It is female group characteristics that make women a poor choice for risk adverse employers" (p. 31)
	Prejudice	This model is "based on the notion that some employers indulge their own sexual prejudices (or the prejudices, real or perceived, of their employees and customers) in making hiring and other personnel decisions" (p. 30).	Says that "the male employer doesn't hire women because of his own tastes or preferences for female discrimination" (p. 31)

TABLE 9.8 ● Conflicting Theories and Their Insights Concerning the Problem of Occupational Sex Discrimination (OSD) From the Same Discipline

Source: Fisher, C. C. (1988). On the need for integrating occupational sex discrimination theory on the basis of causal variables. *Issues in Integrative Studies, 6,* 21–50.

From the Humanities. Fry (1999), *Samuel Taylor Coleridge: The Rime of the Ancient Mariner.* Coleridge's *Rime* illustrates the importance of identifying conflicting theories and their insights concerning a text or work of art. In literature, critical essays on a text typically begin work using different theoretical approaches (p. v). These are of great interest to the interdisciplinarian because they are informed by a set of coherent assumptions that can be articulated, compared, and modified as the process of integration proceeds. Table 9.9 shows five theoretical approaches and their insights concerning the text.

Examining a text from the stance of two or more theoretical approaches, while valuable, is not, by itself, interdisciplinary; nor is juxtaposing them and leaving it up to the reader to somehow perform the integration and create new meaning. What is needed for the analysis of the text to be fully interdisciplinary is the identification of a theory, concept, or assumption that can be modified and used as the basis for integrating these conflicting interpretations.

Theories From the Same Discipline and Their Basic Assumptions

Theories from the *same* discipline are likely to have far less conflict because they typically share the discipline's basic assumptions. Each theory is undergirded by one or more assumptions that typically reflect its disciplinary origin. Yet there are also assumptions that are specific to a particular theory. These more specific assumptions are often, though not always, made explicit by the author.

From the Natural Sciences. Smolinski[*] (2005), *Freshwater Scarcity in Texas.* Even though theories from the same discipline tend to share its underlying assumptions, Table 9.10 illustrates that when these theories focus on a particular problem, their assumptions about the problem can still differ significantly. Such is the case with the theories explaining the reasons for freshwater scarcity in Texas. This situation makes more challenging the student's task of creating common ground among them.

TABLE 9.9 ● Conflicting Theoretical Approaches and Their Insights Concerning the Meaning of *The Rime of the Ancient Mariner* From the Same Discipline			
Discipline	**Theoretical Approach**	**Insight of the Approach**	**Insight of the Approach as Applied to the Text**
Literature	Reader-Response	The meaning of a work is not inherent in its internal form, but rather is cooperatively produced by the readers (what they bring to the text) and the text (Murfin, 1999c, p. 169, 1999e, p. 108).	The major issue of the poem is not its implied moral values, but the process of arriving at moral values, and that process is about reading (Ferguson, 1999, p. 123; Murfin, 1999e, p. 108).
	Marxist Criticism	Literature is a material medium that reflects prevailing social and cultural ideologies and also "transcends or sees through the limitations of ideology" (Murfin, 1999b, p. 144).	*The Rime* is susceptible to an historical approach that reveals an early instance of ecological concern, an emerging interest in hypnotism, and that raises Protestant issues regarding free will, choice, election, and damnation, as well as broader, philosophical questions regarding "epistemological consensus" (Simpson, 1999, pp. 152, 158).
	The New Historicism	"Literature is not a sphere apart or distinct from the history that is relevant to it" (Murfin, 1999c, p. 171).	Reconstructing the historical context of a literary text like *The Rime* is, by itself, treacherously difficult to achieve, and all the more so because we have been conditioned by our own place and time (Murfin, 1999c, p. 171).

Discipline	Theoretical Approach	Insight of the Approach	Insight of the Approach as Applied to the Text
	Psychoanalytic Criticism	"A work of literature is a fantasy or a dream, and psychoanalysis can help explain the mind that produced it" (Murfin, 1999d, p. 225).	*The Rime* is a constructed world unified by language and symbols and is, in fact, "a reaction against the horrifying loss of a boundariless, pre-oedipal world in which the infant, mother and natural world are one" (Murfin, 1999d, p. 232).
	Deconstruction	"A text is not a unique, hermetically sealed space but is perpetually open to being seen in the light of new contexts and has the potential to be different each time it is read" (Murfin, 1999a, p. 268).	*The Rime* should be seen in terms of its "linguistic strangeness, as a 'series of dislocations—translations, displacements, metonymies'—that 'dares its audience to make sense of it'" (Eilenberg, 1999, p. 283).

From the Social Sciences. Fischer (1988), "On the Need for Integrating Occupational Sex Discrimination Theory on the Basis of Causal Variables." Fischer's study is an example of how a professional interdisciplinarian may see the need to make explicit the conflicting assumptions of theories advanced by a particular discipline, in this case, economics. After identifying the four major mainstream economic theories that address OSD, Fischer states the assumptions of each one. He says, for instance, that Monopsony Exploitation Theory

TABLE 9.10 ● Conflicting Theories and Their Assumptions and Insights Concerning the Problem of Freshwater Scarcity in Texas From the Same Discipline

Discipline	Theory	Assumption of the Theory	Insight of the Theory as Applied to the Problem
Earth Science	Global Warming	Global warming is the result of rapid agricultural and technological advances.	Global warming will significantly increase the rates of evaporation, exacerbating an already critical freshwater shortage.
	Overexploitation	Strict conservation legislation aimed primarily at agriculture will solve the problem.	Increased rates of water removal from the Ogallala Aquifer coupled with decreased rates of recharge have led to an overall drawdown of the water table (the uppermost limit of the water contained in the Ogallala Aquifer).
	Infiltration	Freshwater scarcity already exists, as a result of either global warming or overexploitation.	In Texas, saline and brackish water is infiltrating and contaminating the remaining groundwater.

TABLE 9.11 ⬤ Conflicting Theories and Their Assumptions and Insights Concerning the Problem of Occupational Sex Discrimination (OSD) From the Same Discipline			
Discipline	**Theory**	**Assumption of the Theory**	**Insight of the Theory as Applied to the Problem**
Economics	Monopsony Exploitation	OSD is an economic demand problem.	Focuses on the demand side of OSD, explaining it as "the result of collusive behavior of male employers in discriminating against females" (p. 31)
	Human Capital	"OSD is an economic supply problem: Women make rational choices regarding home responsibilities and career commitment" (p. 29).	"Focuses on the supply side of OSD, characteristics of females—particularly their education and training levels" (p. 31)
	Statistical Discrimination	OSD is an economic demand problem: "Discrimination exists because . . . its benefits to the employer overweigh its costs" (p. 30).	"Emphasizes group (rather than individual) characteristics of female job applicants. It is female group characteristics that make women a poor choice for risk adverse employers" (p. 31).
	Prejudice	OSD is an economic demand problem: "Sex role socialization helps form [employer] tastes . . . For example, . . . employers tend to believe that women can not and should not do hard physical work . . . [and] 'can't handle responsibility'" (p. 31).	Says that "the male employer doesn't hire women because of his own tastes or preferences for female discrimination" (p. 31)

Source: Fischer (1988). On the need for integrating occupational sex discrimination theory on the basis of causal variables. *Issues in Integrative Studies, 6,* 21–50.

"assumes that men, in their role as husbands, employers, workers, consumers, and legislators, have power over female occupational choices [italics added]" (p. 26). These economic theories, their assumptions, and corresponding insights into the problem of OSD are summarized in Table 9.11.

Since these theories are from the same discipline, they share the discipline's overall perspective that the world is a rational marketplace and its fundamental assumption that humans are motivated by rational self-interest. However, when these theories address the causes of OSD, it is clear that they conflict more than they overlap, even though they are from the same discipline. It is appropriate for researchers to explain, as Fischer does, why there is need to go beyond these particular theories: Each of them, he says, offers an "incomplete explanation of the causes of OSD," requiring that the analysis be broadened to include the "relevant work in related disciplines if an interdisciplinary understanding is to be achieved" (Fischer, 1988, p. 31).

From the Humanities. Fry (1999), *Samuel Taylor Coleridge: The Rime of the Ancient Mariner.* The theories of reader-response, Marxist criticism, New Historicism, psychoanalytic criticism, and deconstruction shown in Table 9.12 are prevalent in literary criticism (Fry, 1999, pp. v–vi). They are not monolithic schools of thought but, rather, umbrella terms, each of which covers a variety of approaches to textual criticism. The insights associated with each theory, therefore, are not definitive, but expressive of each critic's way of applying the theory to the text.

TABLE 9.12 ● Conflicting Theories and Their Assumptions and Insights Concerning the Meaning of *The Rime of the Ancient Mariner* Within the Same Discipline				
Discipline	**Theory**	**Assumption of the Theory**	**Insight of the Theory as Applied to the Text**
Literature	Reader-Response	*The Rime* is about its author and the deluded reader.	The major issue of *The Rime* is not its implied moral values but the process of arriving at moral values and that process is about reading (Ferguson 1999, p. 123; Murfin, 1999e, p. 108).
	Marxist Criticism	"*The Rime* is driven by an 'agenda' of remystifying the world and thus undermines the reader's confidence in rationality and rationalist theories" (Simpson, 1999, p. 158).	*The Rime* is susceptible to a historical approach that reveals an early instance of ecological concern, an emerging interest in hypnotism, and that raises Protestant issues regarding free will, choice, election, and damnation, as well as broader, philosophical questions regarding "epistemological consensus" (Simpson, 1999, pp. 152, 158).
	The New Historicism	"*The Rime* reflects and challenges any number of ideologies or value systems, ranging from Christianity to the radical political standpoints on the French Revolution to the slave trade" (Modiano, 1999, p. 215).	"Reconstructing the historical context of a literary text like *The Rime* is, by itself, treacherously difficult to achieve, and all the more so because we have been conditioned by our own place and time" (Murfin, 1999c, p. 171).
	Psychoanalytic Criticism	"The 'horror' at the heart of *The Rime* is a symptom of the not yet self's casting off the maternal and semiotic (loosely associational) in favor of the paternal and symbolic" (Murfin, 1999d, p. 231).	*The Rime* is a constructed world unified by language and symbols and is, in fact, "a reaction against the horrifying loss of a boundariless, pre-oedipal world in which the infant, mother and natural world are one" (Murfin, 1999d, p. 232).
	Deconstruction	"All texts, including *The Rime*, are ultimately unreadable (if reading means reducing a text to a single homogenous meaning)" (Murfin, 1999a, p. 269).	*The Rime* should be seen in terms of its "linguistic strangeness, as a 'series of dislocations—translations, displacements, metonymies'—that 'dares its audience to make sense of it'" (Eilenberg, 1999, p. 283).

Theories From Different Disciplines and Their Basic Assumptions

Insights of theories from *different* disciplines, though focused on the same problem, generally conflict because they reflect the differing assumptions of their respective disciplines. Be certain that these authors are indeed talking about the same thing. As we saw in our earlier discussion of concepts, it is possible that authors from different disciplines may use similar-sounding terminology to discuss quite different problems or different aspects of a particular problem.

Locating the sources of conflict between theories from all relevant disciplines is critical to performing STEP 8. The following examples of threaded work illustrate how theories from *different* disciplines conflict because their basic assumptions about the problem conflict.

From the Social Sciences. Fischer (1988), "On the Need for Integrating Occupational Sex Discrimination Theory on the Basis of Causal Variables." Table 9.13 is an extension of Table 9.8 and includes theories from the other relevant disciplines on the causes of OSD. The conflict between economic theories and theories from the other relevant disciplines becomes stark when they are juxtaposed in this manner.

TABLE 9.13 ● Conflicting Theory-Based Insights and Their Corresponding Assumptions on the Problem of OSD Within and Between Disciplines			
Discipline or School of Thought	**Theory**	**Assumption**	**Insight of the Theory as Applied to the Problem**
Economics	Monopsony Exploitation	OSD is an economic (supply and demand) problem.	Focuses on the demand side of OSD, explaining it as "the result of collusive behavior of male employers in discriminating against females" (Fischer, 1988, p. 31)
	Human Capital		"Focuses on the supply side of OSD, characteristics of females—particularly their education and training levels" (p. 31)
	Statistical Discrimination		"Emphasizes group (rather than individual) characteristics of female job applicants. It is female group characteristics that make women a poor choice for risk adverse employers" (p. 31)
	Prejudice		Says that "the male employer doesn't hire women because of his own tastes or preferences for female discrimination" (p. 31)
History	Institutional Development	OSD is a historical problem.	"Sex segregation at entry to firms is perpetuated over time, and done so without the need for further overt sex discrimination" (p. 34).

Discipline or School of Thought	Theory	Assumption	Insight of the Theory as Applied to the Problem
Sociology	Sex Role Orientation	OSD is a social problem.	"Female socialization encourages the acceptance of responsibility for domestic work, and a nurturant and helping orientation for child care. . . . Female socialization, on the other hand, discourages authoritativeness or aggressiveness, physical prowess, and quantitative or mechanical aptitude. It is argued that sex role orientation produces different traits in females, and employers use their knowledge of these traits to decide what jobs should be 'female jobs'" (p. 34).
Psychology	Male Dominance	OSD is a male problem.	"Men have socio-economic incentives to continue monopolizing their privileged status in the labor market, and that they can best do this by maintaining the traditional male–female division of household production" (p. 36).
Marxism	Class Conflict	OSD is an ideological problem.	"Some workers—women in particular—are channeled into less desirable jobs and segregated from other workers to keep workers in general from developing a class consciousness and acting collusively to overthrow capitalism. Here, OSD is seen as a necessary act in preserving the institutions of capitalism" (p. 32).

Source: Fischer (1988). On the need for integrating occupational sex discrimination theory on the basis of causal variables. *Issues in Integrative Studies, 6,* 21–50.

Having identified all relevant theory-based insights into the problem of OSD, Fischer is ready to proceed with STEP 8b, modifying theories, the subject of Chapter 11.

From the Humanities. Bal (1999), "Introduction," *The Practice of Cultural Analysis: Exposing Interdisciplinary Interpretation.* At several points, Bal's discussion of the graffito reveals conflicts between the relevant theoretical approaches used by different disciplines. For one thing, by stressing the methodological explicitness of cultural analysis, Bal is implicitly challenging postmodernism, with its disdain for explicitness and method (p. 4). Also, cultural analysis's emphasis on self-reflexivity places this theory at odds with the modernist emphasis on the object and/or text as opposed to the viewer and/or reader (p. 6). For Bal, what counts as evidence is the discovery in the text of layered meanings using the technique of close reading. One example of layered meaning is the word *Note* at the beginning of the graffito, which, on one level, is a direct address to the reader in the present, but on another level, is an address to a past someone (pp. 7–8).

We have spent more time addressing theories than the other sources of conflicts because they are a bit more challenging to address than differences in concepts or assumptions. Once these differences in theory have been identified, the student can determine what the best strategy to deal with these is. In addition to the strategies of "organization," "redefinition," and "transformation" introduced briefly above, we will introduce a strategy of "theory extension" in Chapter 11 in which the student adds elements of one theory to another.

COMMUNICATE YOUR RESEARCH TO THE APPROPRIATE AUDIENCE

Once you have identified all of the relevant insights and theories and have located their sources of conflict, you should communicate this information to the appropriate audience. For undergraduates, a paragraph that concisely summarizes the results of STEP 7 is often sufficient. For graduate students, members of research teams, and practitioners, a more detailed accounting may be called for. In all cases, you want to be clear about the precise nature of the conflicts you have identified, and their sources; you may also identify how you identified these conflicts. There are multiple ways to do this, as evidenced in the touchstone examples by Fischer (1988) and Bal (1999). These examples offer two different types of interdisciplinary literature, each of which is aimed at a somewhat different audience.

From the Social Sciences. Fischer (1988), "On the Need for Integrating Occupational Sex Discrimination Theory on the Basis of Causal Variables." Fischer's peer-reviewed journal article on OSD illustrates the interdisciplinary research process to a professional audience. Theory is central to his purpose, and his discussion of it dominates the essay, but in clearly defined ways. First, Fischer is concerned to describe concisely each theory and its underlying assumption. Consider, for example, his description of human capital theory:

> The human capital theory of OSD focuses on the relatively high mobility and
> intermittent nature of employment women tend to experience. . . . Because
> of domestic responsibilities, women tend to be in and out of the labor force
> more frequently than men and thus acquire less on-the-job training (OJT)
> than their male counterparts. This, it is argued, adversely affects female
> occupational opportunities in two ways. First, fewer women than men acquire
> sufficient human capital for jobs which require substantial previous experience.
> Second, while women are out of the labor force, their job skills depreciate. It
> is thus rational for women who anticipate intermittent employment to choose
> occupations which require relatively little time to acquire the necessary job
> skills and which require job skills that do not depreciate rapidly from nonuse.
> The combined impact of reduced job experience and the incentive to minimize

depreciation of job skills results in women being concentrated in service, sales, clerical and labor jobs and underrepresented among operators, managers, and professionals. (pp. 28–29)

Second, Fischer contrasts each theory with the theory that precedes it. Contrasting differences precedes comparing differences. In what follows, Fischer contrasts monopsony theory (the first economic theory) with human capital theory (the second economic theory):

> While monopsony theory focuses on the demand side of OSD, human capital theory offers a supply-side explanation of OSD. Each worker is viewed as a combination of native abilities and raw labor power plus specific skills acquired through education and training. The latter is commonly referred to as human capital. (p. 28)

Third, Fischer applies the theory specifically and clearly to the problem of OSD: "In summary, the human capital theory of OSD holds that economic incentives lead women to segregate themselves into female occupations. It is economically rational for women to continue to pursue traditional female jobs" (p. 29).

Finally, Fischer briefly evaluates each theory. Concerning human capital theory as applied to the problem of OSD, he says that "it is bound to be controversial since it implies that OSD is largely the result of choices that women make regarding home responsibilities and career commitment; that is, it tends to 'blame the victim' for OSD" (p. 29).

From the Humanities. Bal (1999), "Introduction," *The Practice of Cultural Analysis: Exposing Interdisciplinary Interpretation.* Bal's interdisciplinary essay differs from Fischer's work in two ways. First, unlike Fischer, who assumes that the reader has little prior knowledge of the topic or theories relevant to it, Bal assumes that the reader is already familiar with the theory and concepts of cultural analysis. Second, Bal, in contrast to Fischer, privileges the theory of cultural analysis over other approaches. Fischer tries not to privilege one discipline over the others, but he nonetheless ends up organizing them in a framework that is essentially economic and that uses economic concepts such as demand, supply, and labor market. Bal, however, privileges cultural analysis because "it is an interdisciplinary practice" (though imperfectly so) (p. 1). She uses the graffito to demonstrate what cultural analysis can and should be, and to answer critics who fault cultural analysis for its lack of "methodological explicitness" (p. 4).

She contrasts other approaches with cultural analysis, beginning with history:

> Cultural analysis as a critical practice is different from what is commonly understood as "history." It is based on a keen awareness of the critic's situatedness

in the present, the social and cultural present from which we look, and look back, at the objects that are always already of the past, objects that we take to define our present culture. (p. 1)

Once she explains this basic contrast, Bal is able to expand the contrast, emphasizing three differences. The first is that cultural analysis probes "history's silent assumptions in order to come to an understanding of the past that is different" (p. 1). The object, of course, is not merely to have an understanding that is just different but that is different because it is integrative and, thus, interdisciplinary. Her second point is that cultural analysis does not attempt to project on the past, and thus on the object in question, what she calls "an objectivitist 'reconstruction'" (p. 1). She means that cultural analysis accepts that there will remain, even after one has undertaken the most comprehensive examination of the moment in time and of the object in it that is possible, an element of ambiguity and mystery. Bal's third point of contrast is that cultural analysis, in contrast to much history writing, does not seek to impose on the past "an evolutionist line," meaning that the object was an inevitable product of knowable historical "developments."

These examples show how interdisciplinarians, writing for different audiences, approaching different problems, and using different methods, go about describing conflicting theories. Fischer's approach can profitably be applied to theories relating to almost any problem.

Chapter Summary

STEP 7 calls for identifying conflicts between insights and theories and locating sources of conflict. Experts in the *same* discipline often study the same problem but produce different insights concerning it. Art critics, for example, study the same painting but arrive at very different understandings of its meaning. This tendency is even more pronounced when experts in *different* disciplines approach the same problem. These insights typically conflict. However, it is not enough to say that insights differ or even conflict. Interdisciplinarians must probe more deeply and discover *why* they conflict. Our discussion of STEP 7 emphasizes three possible sources of conflicts between disciplinary insights: their embedded concepts, their underlying assumptions, and their theories.

One important insight of this chapter is the great diversity of interdisciplinary work, as is evident in the touchstone examples by Fischer (1988), Bal (1999), and Repko (2012). Different audiences, different problems, and different purposes for writing are inevitably reflected in different approaches to the interdisciplinary task. Using a combination of narrative and tables is helpful when dealing with many variables. Identifying conflicts and their sources is foundational to STEP 8: creating common ground among insights and theories by modifying their concepts and assumptions.

Exercises

Concepts Embedded in Insights

9.1 Identify two insights from academic journals or books, each from a different disciplinary perspective, that focus on the concepts *sustainable* or *sustainability*. Compare and contrast how each author defines this term and the context in which it is used. Decide which of the following (if any) problematic situations apply here:

  a. Though the concepts are the same, they mask different contextual meanings.

  b. The concepts are used to describe similar (but not exactly the same) ideas.

  c. The concepts, as used, are so different in meaning that differences cannot be overcome without modifying their meaning.

Assumptions

9.2 Determine from each author's insight into sustainability (see above) the kind of assumption(s) each makes.

  a. Ontological

  b. Epistemological

  c. Value laden

Theories

9.3 Identify two theories from the *same* discipline on the topic of sustainability and answer the following questions:

  a. What concepts does each author use?

  b. What assumption(s) does each author make?

  c. What insight or argument does each theory advance?

9.4 Repeat the exercise of 9.3 using two theories from two *different* disciplines and answer the same questions.

Image by Niek Verlaan from Pixabay

10

CREATING COMMON GROUND AMONG INSIGHTS

Concepts and/or Assumptions

LEARNING OUTCOMES

By the end of this chapter, you will be able to

- Identify the six core ideas about creating common ground
- Explain how to create common ground among conflicting concepts or assumptions
- Identify four techniques used to modify concepts and assumptions
- Explain how to create common ground when ethical positions conflict

GUIDING QUESTIONS

What is common ground?

How do you create common ground among conflicting concepts or assumptions?

How do you create common ground when there is ethical disagreement?

Why do we speak of "creating" common ground?

CHAPTER OBJECTIVES

The basis for collaborative communication across disciplines and integration of conflicting insights is the creation of common ground. Creating common ground is undoubtedly the most challenging task that one faces in the interdisciplinary research process (IRP). It requires a combination of original thought, close reading, analytical reasoning, creativity, and intuition.

This chapter is divided into two sections: the first section defines interdisciplinary common ground and presents six core ideas that form the basis for creating it; the second section explains how to create common ground. While there is no guarantee that common ground can be achieved in every case, this chapter and the following chapter are guided by the idea that creating common ground among conflicting disciplinary insights is possible if one uses an appropriate technique.[1]

STEP 8 of the IRP calls for creating common ground among conflicting insights. Though creating common ground is the focus of a single STEP, we split the discussion of it into two sub-STEPS: the first addresses the creation of common ground among concepts and/or assumptions (the subject of this chapter), and the second deals with the creation of common ground among theories (the subject of Chapter 11).

B. INTEGRATING DISCIPLINARY INSIGHTS

7. Identify conflicts among insights and their sources.

8. **Create common ground among insights.**

    o **Identify the six core ideas about creating common ground.**

    o **Create common ground among conflicting concepts and/or assumptions.**

    o **Create common ground among conflicting ethical positions.**

    o **Create common ground among insights that have a theoretical base.**

9. Construct a more comprehensive understanding.

10. Reflect on, test, and communicate the understanding.

ABOUT INTERDISCIPLINARY COMMON GROUND

Interdisciplinary common ground involves modifying one or more concepts or theories and their underlying assumptions. Assumptions undergird concepts and theories, which in turn undergird insights. Common ground is not the same as integration, but it is preparatory to and essential for integration.[2]

Creating common ground is like building a bridge to span a chasm. The near side is the place of conflicting insights and the lack of a common language (STEP 7, see Chapter 9); the opposite side is the product of the process of integration: the more comprehensive understanding (STEP 9, see Chapter 12). Unless the interdisciplinarian first builds the bridge of common ground to connect the two sides, the integrative enterprise cannot succeed.

The bridge metaphor is useful because it shows how, in operational terms, the three STEPS in the integrative process (identifying conflicts, creating common ground to reconcile them, and producing a more comprehensive understanding) are connected yet discrete.

This definition and description of common ground rests on six core ideas that form the basis for creating common ground: Thinking integratively, common ground is created regularly, it is necessary for collaborative communication, it plays out differently in contexts of narrow versus wide interdisciplinarity, it is essential to integration, and it requires using intuition.

Thinking Integratively

Psychologists tell us that the human brain is designed to process information integratively. For example, the surprisingly multifaceted task of deciding what, when, where, how, and how much to plant in a vegetable garden requires integrative thinking to deal with considerations of nutrition, taste preferences, availability of land, and cost. We do not consciously reflect on all the components of such decisions because of our natural capacity to process information integratively.

But a person's *natural* thinking process stands in sharp contrast to a person's *learned* thinking process. Much of modern education teaches students to think in three non-integrative ways:

- *Disciplinary categories.* From kindergarten onward, students are taught to think in disciplinary categories. They are told that knowledge is found in clearly marked "boxes" or disciplines called math, social studies, English, and art (though this is slowly changing; see Lenoir and Klein, 2010). Learning, students discover, occurs through a process of knowledge fragmentation, compartmentalization, and reductionism.

- *Right or wrong answers.* Students are trained to think in terms of answers to questions as being either right or wrong. Standardized tests that promote a focus on facts over reasoning lead to education that, for example, asks students to read a novel like *Tom Sawyer* and choose whether Huck Finn is a good or bad influence on Tom, even though most of us recognize that humans are a mixture of good and bad, that mixture can change with the circumstances, and that the very concepts of "good" and "bad" are contested.

- *For or against something.* Though debates on controversial topics are effective ways to engage students and teach debating techniques, they reinforce the idea that the point of it all is to win and that the objective in confronting alternative perspectives is to choose one and reject the rest.

The interdisciplinary enterprise, however, is not like prosecuting a case, defending a client, or just adding another pro or con opinion to the many pro or con opinions already available on the issue. Rather, the interdisciplinary enterprise is about building bridges that join together rather than building walls that divide. It is about creating commonalities rather than sharpening differences. It is about inclusion rather than exclusion. It is about producing understandings and meanings that are new and more comprehensive rather than those using a single disciplinary approach. And this requires a different kind of thinking and mode of analysis, one that draws (critically but sympathetically) on most, if not all, available disciplinary perspectives and their insights. So, instead of asking if Huck was a good or bad influence on Tom, the interdisciplinarian would ask about the whole package of Huck's influences (positive, negative, and mixed) on Tom, and probably about Tom's influences on Huck as well.

Common Ground Is Created Regularly

Common ground is something we create regularly in our daily lives. In everyday communication, we encounter people who have different perspectives than we have on a wide range of issues. Since our—and their—everyday perceptions of facts and events depend on the different ways we view a certain situation, we are only able to comprehend each other by regularly achieving common ground.

Cognitive psychology explains successful communication between individuals having different perspectives by exploring the way our brain constructs perceiving, seeing, and acting. Common ground theory says that "every act of communication presumes a common cognitive frame of reference between the partners of interaction called the common ground" (Bromme, 2000, p. 119). The phrase "common cognitive frame of reference between the partners of interaction" simply refers to everyday social interaction where two individuals enter into each other's frame of reference, attempt to discuss a problem, try to identify sources of disagreement concerning it, and arrive jointly at a resolution of it. Common ground theory postulates further that "all contributions to the process of mutual understanding serve to establish or ascertain and continually maintain this common ground" (p. 119). This theory applies to both oral and written communication.

The theory assumes that any successful verbal encounter represents an act of cooperation by both parties. When we communicate, we do so to attain a certain goal or to respond to a certain question, whether verbalized or unspoken. Though common ground theory was developed to explain everyday interactions, cognitive psychology is now applying it to communication across academic disciplines, especially the natural sciences.

Common Ground Is Necessary
for Collaborative Communication

Collaborative communication relies on common ground. One occasion when creating common ground is required is when people (or disciplines) use different concepts or terms to describe the same thing. For example, the terms *city* and *suburb* have different meanings across disciplines. The second occasion is when people take opposing positions on a particular issue stemming from conflicting assumptions or values. If these are the sources of difference or even outright conflict, then the solution must be to create common ground. Common ground is *not* needed, however, where authors *seem to be disagreeing* but may not be because they are talking about different things—altogether different phenomena or variables. Mapping their arguments will reveal if this is so. Alternatively, we could think of the map as the common ground that connects these insights. This is what the technique of organization (discussed below) involves.

Other occasions when creating common ground is required include when (1) research scientists trained in different disciplines need to develop a collaborative language, (2) planners need to develop a comprehensive approach to community development, (3) businesses need to produce a comprehensive strategy to produce and market a new product, (4) counselors need to reconcile partners who are contemplating divorce, and (5) policy makers need to develop broad support for a particular legislative action.

But what about issues that involve diametrically opposed ethical values, deeply held religious beliefs, or sharply conflicting political ideals? Such issues are often emotionally charged, supposedly reducing notions of common ground to mere wishful thinking. But even when dealing with such issues, it is often possible to create common ground (see the discussion of conflicting ethical positions below).

Creating Common Ground Plays Out Differently in Contexts of Narrow Versus Wide Interdisciplinarity

Narrow interdisciplinarity draws on disciplines that are epistemologically close (e.g., physics and chemistry). **Wide interdisciplinarity** draws on disciplines that are epistemologically much further apart (e.g., art history and mathematics). The epistemological presuppositions of the natural sciences promote a focus on facts—on what something is and how it works. Scientific presuppositions do not allow us to assign value (in a moral or ethical sense) to facts as do the epistemological presuppositions of the humanities that give us access to the moral dimension of reality. This means that it should be easier to discover commonalities among insights produced by the natural sciences concerning a problem (given the similarity in their epistemological focus) than it is to find commonalities between insights produced by disciplines spanning the natural sciences and the humanities, whose epistemologies differ widely. In general, *the greater the epistemological distance between disciplines, the more challenging it is to create common ground between their insights.*

Creating Common Ground Is Essential to Integration

The connection between creating common ground and performing integration is established by a study conducted by the Interdisciplinary Studies Project (Project Zero), Harvard Graduate School of Education. It examines exemplary practices of interdisciplinary work at the precollegiate, collegiate, and professional levels. The study finds that interdisciplinary thinking occurs as the mind performs a *complicated chain of cognitive operations* in which it integrates disciplinary ideas. At the juncture of disciplines, the mind is involved in two cognitive activities: (1) overcoming internal monodisciplinarity (i.e., the preference for a single and simplistic disciplinary perspective), and (2) *attaining integration*. A key finding of the study is the possibility that there exists "a central cognitive process," expressive of the *dialogical tendency of the human mind,* that manifests itself in interdisciplinary thinking (Nikitina, 2005, p. 414).

Creating Common Ground Requires Using Intuition

Many interdisciplinarians believe creating common ground requires using intuition. **Intuition** is the natural ability to understand or perceive something immediately without consciously using reason, analysis, or inference (Welch, 2007, p. 3). This definition may be satisfying to students working in the humanities where creativity and spontaneity are prized, but it may be disconcerting to students working in the natural sciences and hard social sciences where rational or logical methods are highly valued. In the sciences, intuition is typically seen as a form of common sense. In fact, science advances not so much by incremental expansion of knowledge but more often by discontinuous leaps of creative or intuitive thought (Csikszentmihalyi & Sawyer, 1995, p. 242; Kuhn, 1996). Historians of science generally agree that both logic and intuition are involved in scientific discovery. Scientists work hard on a problem and gather relevant information. But the insight often comes while they are taking a break from work, allowing their subconscious mind to do its work while they are walking in the park or taking a bath.

Intuitive understanding is experienced in a number of modes. It may occur as

- a moment of instant assessment of a complex situation or object;

- a "gut instinct"—visceral, emotional, and empathetic—in a social situation (D. G. Myers, 2002, pp. 33–38);

- the result of mulling over a problem that is "on the back burner" of our mind, "where insight into a persistent conundrum that had been dismissed from consciousness comes to light 'out of the blue'" (Welch, 2007, p. 3);

- a "common sense" insight into "the complexity of ordinary decision-making, which is embedded in collective cultural standards" (Gerber, 2001, p. 72);

- creativity, which is intertwined with the process of inspiration, imagination, artistic expression, and symbolic understanding (D. G. Myers, 2002, pp. 59–61; Darbellay, Moody, Sedooka, & Steffen, 2014); or

- insight, which is understanding the underlying structure of a problem and attaining a synthesis of the relationships among its disparate elements (Csikszentmihalyi & Sawyer, 1995, p. 329; Dominowski & Ballob, 1995, p. 38; Myers, 2002, p. 28).

Intuition has been the subject of extensive research in cognitive psychology. According to James Welch (2007), subconscious thought processes confront reality in ways that are superior to conscious processes. Our subconscious thoughts are not merely cataloguing discreet packets of experiential information, but accumulating an adaptive multidimensional matrix of associations (p. 6).

Intuitive insights do not just happen. They are, says Arthur Miller (1996), the result of a process of mental gestation whereby a highly trained mind has been purposefully focusing on a particular problem for a length of time (p. 419). This finding has important implications for interdisciplinary research, which features an "iterative process that works toward solution through an interweaving of generative and cognitive processes, not a big bang that comes all at once, then not again" (Sill, 1996, p. 144). Maya Lin, the designer of the Vietnam War Memorial, speaks of thinking and reading about the war for a long time before "laying an egg" with the idea of scars, leading to a monument that scrapes the skin of the Earth.

Interdisciplinarians are conflicted concerning the role of intuition in interdisciplinary research. Some argue against an uncritical acceptance of intuitive insight, arguing that any valid understanding of reality must be based upon logic and empirical methods. Others assert that interdisciplinarity, seen from the perspective of postmodernism, "must be highly individual, unspecifiable, and institutionally anarchical" (Welch, 2007, p. 5). Szostak (2002), however, views this dichotomy between structure and intuition as essentially false. In good interdisciplinary fashion, he argues that *both structure and intuition* are necessary for developing new integrative approaches (pp. 131–137). Welch (2007) agrees and argues for what he calls *equilibrium between intellect and intuition*. This equilibrium is expressed in what he calls "**integrative wisdom.**" This he defines as the synthetic interaction between "inspiration, intellect, and intuition" (p. 149).

> Wisdom is the synthesis of all avenues of insight—rational, experiential, intuitive, physical, cultural, and emotional. [It] breaks down all boundaries between categories of knowledge and returns them [to] holistic understanding. Wisdom creates equilibrium among these faculties, minimizing their individual weaknesses and achieving synergy. (Welch, 2007, pp. 149–150)

An Example of How Intuition Helps to Achieve Common Ground

The story of Helen Keller provides an example, though imperfect, of how intuition helps achieve common ground between two persons who have been unable to communicate with each other. Helen, after an illness in infancy, was left unable to see or hear, and thus unable to speak or communicate with anyone. Though everyone had given up on Helen, her young teacher, Anne Sullivan, did not, believing that she could find a way to communicate with Helen. For some time, Anne's best efforts proved fruitless, and Helen grew more and more incorrigible—until one day when they were at the well outside the cabin where they were staying, and Helen knocked over the bucket of drinking water that Anne had just drawn. In that moment, water became more than water. In an intuitive flash, Anne realized that she could use the spilled water to make the sign for "water" in the palm of Helen's wet hand. It worked. Helen understood. Anne had achieved common ground with Helen. Water became the key that ended Helen's terrible isolation and enabled her to comprehend and communicate with her world. The result was a new and an amazingly productive life. (*Note:* This example overlooks the fact that one usually has to redefine disciplinary concepts to create or even discover common ground. And without using redefinition or some other integrative technique discussed in this chapter, common ground, and thus integration, can seldom be achieved.)

This book reflects Szostak's stress on balance and Welch's call for equilibrium between intellect and intuition. When it comes to intuition, says Welch (2007), interdisciplinary studies, "with its emphasis on practical problem solving, cannot afford to dismiss such a potentially powerful [cognitive] faculty for integrative understanding" (p. 148).

Also, adherence to a steplike research process cannot automatically resolve all problems, including the challenging problem of creating common ground. That is why students are well advised to leave room, in some cases a great deal of room, for an "intuitive leap" or a "eureka moment" when, after a period of struggle, reflection, and analysis, they suddenly discover how to create common ground.

We should emphasize the implications of this discussion for student or scholarly work patterns. It is very easy not to work hard enough: Unless you have consciously gathered enough relevant information, you will not be able to subconsciously develop common ground. It is also possible to work too hard and not give your brain the free time it needs to process subconscious thoughts and feed these into consciousness. Recall that scholarly inspiration most often comes while the conscious mind is at rest: walking in a park or taking a bath. Note also that if you put off researching until just before your paper is due you will not have the time to relax and let inspiration flow.

CREATE COMMON GROUND AMONG CONFLICTING CONCEPTS AND/OR ASSUMPTIONS

The task of creating common ground among conflicting concepts or assumptions begins with deciding how to proceed. This is followed by the researcher choosing which technique(s) to use to modify the concepts or assumptions. Added considerations apply to creating common ground when ethical positions conflict.

Decide How to Proceed

The researcher has in earlier STEPS decided how comprehensive their study will be: how many insights and disciplines it will encompass. They may find it necessary to revisit that decision in this STEP, for it will be harder to create common ground across a large number of insights and disciplines. Graduate students and scholars should try to cast their gaze widely. Even they should recognize a tradeoff between creativity and manageability: Common ground across diverse insights may generate a more comprehensive understanding that is very novel and useful, but may prove to be challenging.

We have recommended in earlier STEPS that the researcher map the phenomena and interactions among these relevant to their research question. If the insights that have been identified each address different parts of this map, then the map itself can serve as the common ground (see the strategy of "organization" below) that integrates these insights (which still should have been evaluated) into a coherent understanding. The authors of these insights may *appear* to disagree but are in fact talking about different things. More often, insights will overlap, addressing the same phenomenon or (more likely) relationship between two phenomena. The researcher then needs to decide how to create common ground among overlapping insights. They should move from the easiest to hardest ways of creating common ground, looking first at concepts, then if necessary at assumptions, and only then if necessary at theories.

Decide How Comprehensive the Study Will Be

Researchers need to decide how *comprehensive* their study will be. Most undergraduates and even some graduate students will find themselves limited in the number of disciplines on which they do draw. But these students can and should seek to *integrate fully the insights* of the disciplines from which they draw. For members of interdisciplinary research teams and solo practitioners, their studies will be comprehensive: They will draw on all relevant disciplines and all relevant insights and seek full integration.

Why should you focus on concepts before addressing assumptions? There are two reasons. First, concepts are easy to identify, whereas assumptions can be difficult to identify (because authors use concepts overtly and explicitly but often fail to acknowledge most, if not all their assumptions). Second, and less obviously, probing for the source of conflict between concepts may well uncover conflicting assumptions.

Begin Looking for Concepts. Return to the data table created earlier (Chapter 6, Table 6.1) to see which insights make explicit reference to one or more concepts. You may discover that only a few authors reference one or more concepts, while others do not reference any. *To create common ground using concepts, all the authors must use one or more concepts that reference the problem in some way, even though these concepts may have different (apparent or real) meanings.* For a concept to be used as the basis for creating common ground, it must be applicable to all, not just a few, of the insights. Examples of how to modify concepts for this purpose appear below.

Work With Assumptions When This Seems More Promising. Consult the data table you created earlier to identify the assumptions of each author's insight. If this information was not collected earlier, it will have to be now. One strategy for locating assumptions is to identify the discipline that has produced each of the insights and refer to the tables on disciplinary assumptions in Chapter 2. This often requires close *rereading* of the insight to see how it reflects the overall assumptions of the discipline that produced it. For example, Martha Crenshaw (1998), a political scientist, writes that her approach to understanding the cause of suicide terrorism "permits the construction of a standard which can measure degrees of rationality, the degree to which strategic reasoning is modified by psychology and other constraints, and explain how reality is interpreted" (pp. 9–10). As a political scientist, Crenshaw is likely to share one of the discipline's major assumptions that are noted in Table 2.12. Her statement about the importance of measuring "degrees of rationality" (as well as other statements that she makes in her essay) appears to reflect the modernist and secular assumption that "human beings are . . . in part intentional actors, [and] capable of cognition and acting on the basis of it" (Goodin & Klingerman, 1996, pp. 9–10). Thus, from the text of Crenshaw's insight and from the assumption statements concerning political science in Table 2.12, it is relatively easy to construct an assumption statement as follows: "Suicide terrorists follow logical processes that can be discovered and explained."

Work With Theoretical Explanations Only When All Authors Use Them to Explain the Cause of the Behavior in Question. Authors in the social sciences typically base their insights on theoretical or causal explanations. How to integrate different theoretical explanations of the same problem is the subject of the following chapter.

Another situation commonly encountered is working with insights of authors from applied fields (such as education or business) and/or interdisciplinary fields (such as bioethics or women's studies). The defining elements of these fields are not included in Chapter 2. Consequently, the researcher will have to carefully examine each insight and look for statements that reveal the author's assumptions (i.e., what the author believes to be true about the problem).

A Best Practice When Working With Concepts and Assumptions

A best practice when working with concepts and assumptions is to follow the principle of least action. This means making sure that the changes made in them are the smallest possible to still create sufficient common ground on which to construct the more comprehensive understanding. The rationale for using this principle is essentially grounded in the conservative laws of thermodynamics: Nature finds the path that requires the least expenditure of energy. It will also make it easier to communicate the results of your research to the appropriate audiences.

Techniques Used to Modify Concepts and Assumptions

Four main techniques have been identified for creating common ground. Researchers will often find that they need to combine techniques. Though each technique has wide applicability, researchers can start with these guidelines:

- When insights conflict because of differences in terminology, the technique of "redefinition," which strives for shared understandings of terminology, is recommended. (*Note:* This is the most common technique.)

- When insights (appear to) conflict because authors are talking about different parts of a complex set of interactions, the techniques of "organization" (discussed above) is recommended.

- When insights differ because different authors and disciplines stress the importance of the phenomena they study, the technique of "extension," which adds the variables or ideas pursued in one theory or explanation or assumption to another theory or explanation or assumption, will often prove to be useful.

- When insights conflict because different authors make opposing assumptions (such as rationality versus nonrationality), the technique of "transformation," which identifies a continuum between opposing assumptions, is recommended.

Examples of each technique are drawn from problem-based course projects, published literature, and student papers that are explicitly interdisciplinary. As with the touchstone examples written by professional interdisciplinarians, the problem-based course project and student papers illustrate many, but not all, of the possible features of an interdisciplinary research paper. The categorization of these examples refers to the area of the researcher's training and the orientation of the topic rather than to the disciplines from which insights were drawn.

The Technique of Redefinition

The technique of **redefinition** involves modifying or redefining *concepts* in different texts and contexts to bring out a common meaning. In interdisciplinary work, the technique of redefinition is sometimes referred to as "textual integration" (Brown, 1989, cited in Henry & Bracy, 2012, p. 264). As noted earlier, each discipline has developed its own technical vocabulary to describe the phenomena it prefers to study. Since every discipline has its own vocabulary expressed as concepts, it is necessary for the interdisciplinarian to create a common vocabulary to facilitate communication among disciplines—that is, to "get them on the same page." Since most disciplinary concepts as well as assumptions are couched in discipline-specific language, the technique of redefinition is used in most efforts to create common ground (Newell, 2007a, p. 258). The trick is to modify terms as little as possible while still creating common ground. *Redefining a concept might also involve some modification of the assumption(s) underlying the concept.*

Example of the Admission, Review, and Dismissal Meeting. The importance of finding common ground when trying to achieve coherent understanding of a complex problem is illustrated in the example of an admission, review, and dismissal (ARD) meeting in special education. The purpose of this meeting is to develop a comprehensive approach to providing individualized instruction for a student with learning disabilities. Those attending the meeting include administrators, various specialists, the student, the student's parents, and the facilitator whose job it is to move the discussion toward an integrated plan for the student's educational needs for the coming year. The facilitator asks each person—the speech pathologist, the social studies teacher, the neurologist, an assistant principal—to propose a solution designed to meet the student's ongoing educational needs for the coming year. The specialists commonly use highly technical concepts or language to describe the student's disability. Perhaps sensing that the parents do not understand what the specialists are saying, the facilitator asks the specialists to "translate" the technical jargon into language that the parents, and indeed all in attendance, can understand. The facilitator attempts to find common ground among the various proposals (typically grounded in theory) offered by the specialists and the parents. Then, building on that common ground, the facilitator proposes an integrative solution.

The facilitator's role in the ARD meeting is similar to that of the interdisciplinary student who is attempting to produce an integrative understanding of, say, the causes of the high rate of obesity among adolescents. Each discipline interested in the problem brings its perspective to the table. And experts in each of these disciplines attribute the causes of the problem to various factors. The task of the student working on this problem, like that of the ARD facilitator, is to encourage each viewpoint to be expressed, identify conflicts and their sources, and then nurture the emergence of one or more points of agreement. This latter activity is creating or finding common ground.

Two lessons can be drawn from the above narrative. The first is the role of technical language (i.e., concepts) in establishing common ground. The second is the importance of recognizing that underneath the technical language used by the various specialists are disciplinary perspectives (i.e., theories, concepts, and assumptions) on how to treat a child with learning disabilities. In interdisciplinary work, one must take into account not only disciplinary terminology, but also disciplinary perspectives.

Working With Concepts. Concerning concepts, then, researchers should do two things. First, pay close attention to how the same concept may have different meanings when used by different disciplines or theories that are focusing on the same problem.

Second, be alert to how experts from different disciplines use *different concepts* in their discussion of the same problem and where different concepts have *overlapping meanings*. Both are common occurrences. From these, it is often possible to identify one concept that can be modified by using the technique of redefinition. In cases where different researchers use the same concept in quite different ways, it may be best to identify the distinct meanings of each rather than seek a common redefinition (Bergmann et al., 2012).

When redefining a concept, avoid using terminology that tacitly favors one disciplinary approach at the expense of another. Using the technique of redefinition can reveal commonalities in concepts that may be obscured by discipline-specific language. Once this language is stripped away, the concept can be redefined, enabling it to become the basis for creating common ground between the conflicting insights. Sometimes this occurs in conjunction with other integrative techniques, as shown in these threaded examples.

From the Humanities. Silver* (2005), *Composing Race and Gender: The Appropriation of Social Identity in Fiction.* Creative writing, says Silver, like all other disciplines, sees the world through its own "peephole" or perspective. "I love this peephole deeply," she confides, "but I also want to see the entire truth [because] truth is fundamental to fiction" (p. 75). For Silver, seeing the "entire truth" as a writer of fiction involves crossing disciplinary boundaries. One way fiction writers do this is by appropriating (i.e., assuming) social identities, which are reflected in their characters. Silver uses the modification technique of redefinition to resolve an ethical dilemma that exists when fiction writers, actors, and filmmakers regularly and uncritically appropriate a person's identity. That dilemma is how to engage in this practice in an ethical, by which Silver means truthful or authentic, way. The disciplines that Silver finds most relevant to the topic are sociology, psychology, cultural studies, and creative writing. The challenge for Silver was identifying a concept that these disciplinary insights shared concerning this common practice. This concept, she concluded, was "implicature," which denotes either (a) the act of meaning, implying, or suggesting one thing by saying something else, or (b) the object of that act (*Stanford Encyclopedia of Philosophy,* 2010). By redefining "implicature" to mean "the ultimate level of empathy that one person can have with another," Silver makes it possible to practice appropriation in a way that is ethical rather than hypocritical.

From the Natural Sciences and the Social Sciences. Delph* (2005), *An Integrative Approach to the Elimination of the "Perfect Crime."* Delph questions whether advances in criminal investigatory techniques are able to eliminate the possibility of the "perfect crime." She defines a "perfect crime" as one that goes unnoticed and/or as one for which the criminal will never be caught (p. 2). Of the several disciplines and subdisciplines that are relevant to crime investigation, the three that Delph finds most relevant are criminal justice, forensic science, and forensic psychology. Delph identifies the current theories of these rapidly evolving subdisciplines and finds that the source of conflict among them is their preference for two different investigatory methods and reliance on two kinds of evidence. Forensic science analyzes physical evidence, whereas forensic psychology analyzes behavioral evidence. Each approach constructs a "profile" of the criminal, with forensic science using physical evidence and forensic psychology using a combination of intuition informed by years of experience and information collected from interviews and other sources.

Delph creates common ground among the conflicting approaches by redefining the concept of profiling to include both forensic science, with its emphasis on physical evidence, and forensic psychology, with its emphasis on "intuition" born of extensive experience and insights derived from crime scene analysis. This redefinition of criminal profiling enables her to bridge the physical (i.e., forensic science) and behavioral sciences (i.e., forensic psychology and criminal investigation). Forensic scientists do not need to use profiling as long as they have adequate evidence to analyze. But in the absence of such evidence, profiling can move the investigation forward by using a combination of "intuition" born of extensive experience and insights derived from crime scene analysis (p. 29). In this way, the redefined concept of profiling serves as common ground among the specialized knowledge that criminal investigation, forensic science, and forensic psychology offer.

From the Social Sciences. Schoenfeld* (2005), *Customer Service: The Ultimate Return Policy.* Schoenfeld draws from the disciplines of psychology, sociology, and management to address an all-too-often overlooked and underappreciated aspect of consumerism, customer service. She defines customer service as "anything we do for a patron that embraces their experience" (p. ii). The goal of her study is to probe "the deeper levels of providing customer service," which is another way of saying "to develop a holistic approach to the customer experience" (pp. 3–4). Schoenfeld distinguishes between the concept of customer service (any steps that are taken to satisfy and retain customers' loyalty while they are in the store) and the concept of customer relationship management, or CRM (any steps taken to satisfy and retain customers when they are not in the store), and seeks to create common ground between them (p. 6). Her approach is to identify theories generated by psychology, sociology, and anthropology—including social exchange theory, expectancy theory, reasoned action, role theory, and attribution theory—that explain customers' expectations, behaviors, and habits. Schoenfeld observes that these theories describe the concept of customer service in two different ways: from the perspective of the customer and from the perspective of the merchant or store owner instead of from just one or the other. To create common ground, she redefines the concept of customer service so that it includes both perspectives. The focus of the concept is unaltered, or altered only slightly.

NOTE TO READERS

At times, the best strategy may *not* be to seek one shared definition of a term, but rather to clarify the differences between different uses of the same term (Bergmann et al., 2012). For example, Szostak (2016) argues that the word *globalization* can mean many things such as "expansion in trade," "spread of democracy," or "watching foreign movies." One might best address differing insights by clarifying what different authors mean. Piso (2016) warns us, though, to be careful of ignoring what may be common elements of definition, such as a shared feeling that globalization is troublesome.

The Technique of Extension

Extension refers to increasing the scope of the "something" that we are talking about. Whereas the focus of redefinition is linguistic, the focus of extension is conceptual. It involves addressing differences or oppositions in disciplinary concepts and/or assumptions by extending their meaning beyond the domain of the discipline that originated them into the domain(s) of the other relevant discipline(s) (Newell, 2007a, p. 258). Extending into the domain of a different discipline almost inevitably involves some sort of modification.

Example of Extending Concepts. The following is an example of a concept that was birthed in one disciplinary domain and later extended into other disciplinary domains. In the threaded example of the graffito, Bal extends the concept of exposure so that it includes three different perspectives on a message to a lover written on a brick wall (i.e., graffito).

From the Humanities. Bal (1999), "Introduction," *The Practice of Cultural Analysis: Exposing Interdisciplinary Interpretation.* The challenge Bal faces with the graffito is how to expose its fullest meaning while not privileging any single disciplinary perspective. Her strategy is to analyze it from three perspectives (but not disciplinary ones) simultaneously: from the perspective of its author, from the perspective of the subject (i.e., the author's beloved), and from the perspective of one who is reading the graffito and pondering its meaning. Bal uses the technique of extension to create a common vocabulary centered on the verb *to expose*, to which she connects three nouns: *exposition, exposé,* and *exposure.* These are the three meanings or insights that this close reading of the graffito brings together. The verb refers to making a public presentation or to "publicly demonstrate." "It can be combined with a noun meaning opinions or judgments and refer to the public presentation of someone's views; and it can refer to the performing of those deeds that deserve to be made public" (pp. 4–5). The graffito, as an exposition, brings out into the public domain the deepest held views and beliefs of the author. Exposition, says Bal, "is also always an argument. Therefore, in publicizing these views the author objectifies or exposes himself as much as the subject. This makes the graffito an exposure of the self. Such exposure is an act of producing meaning, a performance" (p. 2).

Example of Extending Assumptions. Assumptions as well as concepts can and often are extended (e.g., the concept of sustainability has been extended in recent decades from economic development to include the ecology, culture, and political system of a country). The example of extending assumptions that follows addresses the assumption of rationality, which some authors make explicit and others leave implicit.

From the Social Sciences and the Humanities. Repko (2012), "Integrating Theory-Based Insights on the Causes of Suicide Terrorism." In previous STEPS, Repko identifies the relevant insights (noting that all of them explicitly espouse a particular theoretical explanation) and the sources of conflict among them. Given the great diversity of concepts used and their conflicting meanings, he concludes that common ground cannot be created by redefining any one of these concepts. So, unable to work with concepts, he decides instead to work with assumptions. He begins by reflecting on the taxonomy of theory-based insights in Table 9.6 (Chapter 9).

Repko observes that insights from the same discipline usually share the same assumption. For example, the political science theories of collective rational strategic choice and sacred terror share the assumption that terrorists are rational actors who follow logical processes that can be discovered and explained.

He also observes that conflicts exist between the assumptions of different disciplines:

> The assumption underlying self-sanction theory (from cognitive psychology) is that understanding the behavior and motivation of suicide terrorists requires studying *primarily* the mental life and the psychological constructs of *individual* terrorists. By contrast, the assumption of identity theory (from political science) is that understanding the behavior and motivation of suicide terrorists requires studying their cultural as well as their religious identity, but not at the expense of taking into account personality traits (inherent and acquired). (p. 145)

A deeper probing of the assumptions of these theories reveals a commonality that both share, namely the *goals* of suicide terrorists. These are understood not in terms of self-interest as rational choice advocates would have it, but rather as "moral imperatives" or "sacred duties." It so happens that this deeper—and extended—assumption is also shared by the theories of fictive kin, strategic rational choice, "sacred terror," martyrdom, terrorist psycho-logic, and modernization. He concludes that the common ground assumption shared by all of the theory-based insights to varying degrees is this: *The goals of suicide terrorists are "moral" and "sacred"—and, thus, rational—as defined by their beliefs* (p. 145).

NOTE TO READERS

We will see many examples of theory extension in the next chapter.

The Technique of Transformation

The technique of **transformation** is used to modify concepts or assumptions that are not merely different (e.g., love, fear, selfishness) but opposite (e.g., rational, irrational) into continuous variables (Newell, 2007a, p. 259). For example, Amitai Etzioni (1988) in *The Moral Dimension: Towards a New Economics* addressed the problem of how to overcome diametrically opposed concepts and assumptions about the rationality (economics) or irrationality (sociology) of humans. His solution was to transform them by placing them on opposite ends of a continuous variable called "the degree of rationality." By studying the factors that influence rationality, he found that it is possible to determine in principle the degree of rationality exercised in any given situation. One can then draw upon theories that assume either rationality or irrationality, rather than having to reject one or the other. Similarly, Etzioni treated "degree of trust" and "degree of governmental intervention" as continuous variables, making it possible to explore and estimate determinative influences in any particular context rather than as dichotomous assumptions to accept or reject.

The value of transformation in creating common ground is this: Rather than force us to accept or reject dichotomous concepts and assumptions, continuous variables allow us to integrate opposing insights. The effect of this strategy is not only to resolve a philosophical dispute, but also to extend the range of a theory (Newell, 2007a, p. 260). Transforming opposing assumptions into variables allows the interdisciplinarian to move toward resolving almost *any* dichotomy or duality, as illustrated in these examples.

From the Social Sciences. Englehart* (2005), *Organized Environmentalism: Towards a Shift in the Political and Social Roles and Tactics of Environmental Advocacy Groups.* Englehart was concerned that anti-environmentalism was becoming institutionalized in American politics during the G. W. Bush administration. To ensure that environmental responsibility becomes an integral part of our society, she proposes that environmental advocacy groups integrate their social and political agendas. These groups assume various active roles in society: They challenge and pressure the government with an environmental ethic; they are actors in the political arena who influence policy making by lobbying and campaigning in election cycles; and they are what sociologists call "social movement organizers" who mobilize the public to take action on pressing environmental issues.

To better understand the roles and tactics of environmental groups, Englehart examines them in light of relevant theories, including social movement theory (in its several variations), rational choice theory, collective identity theory, and structural network theory. By comparing these theories and the insights they have generated, she finds that for environmental groups to grow and recapture the political initiative, they must change their approach to what they do and how they do it.

Creating common ground among the various theories and insights requires that she use the technique of transformation. This involves transforming opposing theoretical assumptions so as to extend the scope of social movement theory. This resulted in what Etzioni (1998) refers to as transforming the "I" of self-interested economics and political advocacy and the "We" of collective identity in social movements into a jointly maximized "I" and "We" for environmental advocacy. Englehart advocates using face-to-face relationships within an environmental organization to shift members along the continuum from "I" to "We," and then to extend the "We" (for the purposes of interorganizational networking) to include those with differing environmental values. Practically, this will cause environmental organizations to concentrate their efforts on educating and politicizing the social arena and creating their own political opportunity structures through innovative mobilization strategies so as to challenge current antienvironmental political action. Integration via transformation of these theoretical and disciplinary insights, Englehart argues, would result in a bottom-up, grassroots, coalition-driven social emphasis that, when combined with the traditional top-down, legislative-driven political pressure, will help environmental advocacy groups recapture the political initiative (pp. 58–63).

From the Natural Sciences and the Humanities. Arms* (2005), *Mathematics and Religion: Processes of Faith and Reason.* Arms compares faith and reason, which are often seen as polar opposites. "People think," she says, "that religion finds its home in the heart and faith, while mathematics belongs in the brain and reason" (p. i). The disciplines of her focus are mathematics, philosophy (i.e., logic), and religion. Logic, she finds, is the fulcrum discipline for mathematics and religion because both rely on it. Religion employs logic, albeit according to its own rules and within its own frame of reference. Logic is also used in determining the provability of mathematical statements, and this requires that one employ deductive reasoning. Gödel's Incompleteness Theorems, says Arms, show that we cannot prove necessary truths in mathematics. But by his Completeness Theorem, we know that first-order logic, sometimes called mathematical logic, is complete, and therefore at least trustworthy (p. 5). She also draws upon sociology and Durkheim's theory of religion and extracts from the latter his definition of religion as a socially constructed belief system, which she employs in her study.

The belief in the existence of a Christian God and the belief in the completeness and consistency of mathematics are not only belief systems, Arms says, but faith-based belief systems, and very different ones at that (pp. 66–67). She uses the concept of faith to continue the idea that mathematics and religion still have the possibility of certainty. Her reasoning runs as follows:

> We take it on faith that reason is a good thing. Since reason is an object of faith, it is reasonable to assume than an object of reason can become an object of faith. Faith is justifiable in keeping belief in the certainty of mathematics. Mathematics has made it clear to us that we cannot depend on it purely through reason. And even if Gödel and his Incompleteness Theorems had never come about, there would still be things in mathematics that are not provable. There are plenty of problems that have never been solved, and many that may never be solved. It took mathematicians over 300 years to solve Fermat's Last Theorem, but they had faith that it was true and that they would find a solution. In mathematics, it is common to prove something using an idea that we do not know is true, but assume it is. (p. 76)

To find common ground between faith and reason, Arms transforms the dichotomies of faith and reason and, by implication, the dichotomies of mathematics and religion. In the end, she confesses that she had been under the impression that her logic could "go any-where"; that "science trumped religion, and [that] logic trumped science." Therefore, logic was obviously stronger than faith. Then she learned that her "dear logic," while complete, could not prove even mathematics. This rude awakening kept faith "afloat" and enabled her to accept "the complementary nature of reason and faith" (p. 80).

From the Social Sciences. Boulding (1981), *A Preface to Grants Economics: The Economy of Love and Fear.* Kenneth Boulding's study of research grants involved him prob-ing the complexities of human behavior that motivates grant bequests. More particularly, Boulding sought a way to transform the debate about whether human nature in general is selfish or altruistic, as described in Newell's (2007a) summation:

> Boulding (1981) recognized that both benevolent behavior (studied by sociologists) and malevolent behavior (studied by political scientists) can be understood as other-regarding behavior (positive and negative, respectively). He then arrayed them along a continuum of other-regarding behavior. The self-interested behavior studied by economists became the midpoint on that continuum because its degree of other-regarding behavior is zero. Thus, he set out a way to transform the debate about whether human nature in general is selfish or altruistic into a choice of where on the continuum of motivations people are likely to fall in the complex problem under study. By combining into a single continuum with self-interest the motivations of love and hate/fear that

support or threaten the integrative mechanisms binding societies and politics together, Boulding used the technique of transformation to integrate the differing conceptions of human nature under economics, sociology, and political science. (p. 259)

The Technique of Organization

The technique of **organization** creates common ground by clarifying how certain phenomena interact and mapping their causal relationships. More specifically, organization (1) identifies a latent commonality in the meaning of different concepts or assumptions (or variables) and redefines them accordingly and (2) then organizes, arranges, arrays, or maps the redefined concepts or assumptions to bring out a relationship between them (Newell, 2007a, p. 259).

Organization focuses on mapping the overall relations among distinct variables or clusters of distinct variables (see Wallis [2014] for examples). For example, "organization" enables us to see cultural values, social norms, institutional policy, historical precedent, religious beliefs, and so forth, as providing a context within which rational decisions are made. These influences from outside the individual may constrain the acceptable means by which personal goals are pursued. That is, individuals may internalize some of those values, norms, beliefs, and so on, and alter their personal goals accordingly. And it is possible that, over time, the combined effects of the goals of many individuals may reshape societal values, norms, policies, and the like. *How much* and *how quickly* societal constraints may constrain or alter individual goals, and individuals alter the values of groups, institutions, and entire societies, may well vary from one issue to another and from one era to another. That is, the speed and extent of these influences of society on individuals, and of individuals on society, can be thought of as empirical questions that need to be researched. In this example, the map of relations between society and individuals serves as the common ground between theories that focus on one or the other.

At this point in the IRP, the map of the problem created earlier may need to be refined so as to show all causal relations (some of which may have been overlooked when the map was first drawn). Sometimes we may already have used organization without realizing it when we first mapped the problem. *What we are doing here that is new is appreciating that each discipline tends to make assumptions that privilege its own phenomena.* So economists stress individual rational decisions, and sociologists stress the influences of other people, but we can see how those interrelate. Thus, we are indeed coping with the core assumptions of these disciplines.

From the Social Sciences. Etzioni (1988), *The Moral Dimension: Towards a New Economics.* We turn again to Etzioni's book, this time to show how the technique of organization can be extended from individual concepts and assumptions to large-scale models, major theoretical approaches, and even entire disciplines. Etzioni's use of organization is summarized here by Newell (2007a):

> Etzioni argued that there are several identifiable large-scale patterns of interrelationships between the "rational/empirical" factors studied by economics and the "normative/affective" factors studied by sociology. One such pattern I call an envelope. Here the rational behavior studied by economics is bounded, limited, or constrained by the normative factors studied by sociologists. Thus, rational economic behavior functions within a normative sociological envelope. Another pattern might be called inter-penetration. Some sociological factors directly influence economic behavior, while some economic factors directly influence social behavior. Thus, social relationships can have an effect on how economic information is gathered and processed, what inferences are drawn, and what options are considered. And a third pattern can be referred to as facilitation. Etzioni points out that the "free individuals [studied by economics] are found only within communities [studied by sociologists], which anchor emotions and morals" (xi). Thus, sociological factors such as communities can actually facilitate individual economic behavior. (p. 259)

In this example, the technique of organization can make macro-level applications to bring out the relationship among commonalities of meaning within contrasting disciplinary concepts or assumptions.

It may be useful at this point to work through a hypothetical example (drawn from Repko, Szostak, & Buchberger, 2020, Ch. 8). Imagine you are researching the causes of burglary and read an analysis of crime by an economist. The economist calculates the potential burglar's costs and benefits. The benefits are the money the burglar can receive by selling what he steals. The costs are the probability of being caught multiplied by the penalty imposed for burglary. The economist calculates the penalty that needs to be imposed in order for the costs to exceed the benefits and thus crime to be deterred. In critically evaluating this work in terms of disciplinary perspective, you would appreciate that the economist is assuming the burglar is acting rationally and only cares about financial costs and benefits.

You then read an article by a sociologist. (You should read multiple papers in each discipline, of course, but we will simplify the process for this example.) The sociologist studies peer pressure and recommends after-school programs for at-risk teens as the best policy for reducing crime. And then you read an article by a psychologist that identifies certain personality traits associated with criminal activity, and proposes a counseling strategy to reduce crime. (You can usefully reflect on how you would critically analyze these other works.)

You must avoid becoming frustrated at this point. Three experts study the causes of crime and reach three distinct conclusions. What are you to think? If you were to map these three arguments, you could see that you have three potentially complementary arguments: peer pressure, personality, and economic calculations might all influence the likelihood of burglary. Each discipline has assumed that the things it studies are most important, but you as an interdisciplinarian can see value (but also limitations) in each.

There is, though, a more interesting—and more likely—outcome than that the causal factors identified by each discipline operate independently. Perhaps economic calculations are influenced by peer pressure? Perhaps only people with certain personality characteristics perform such calculations? Indeed, perhaps personality only has an indirect influence on crime by affecting how people respond to peer pressure and calculate costs and benefits? By exploring possibilities such as these, you can "organize" the insights drawn from each discipline into a coherent explanation. Your diagram of the links among the phenomena engaged by different authors serves as common ground. (There are, of course, still other influences on crime that could be included in a more comprehensive map. However, recognizing that interdisciplinary analysis is an ongoing process, you should not insist that your maps include every possible variable.)

The Value of These Techniques

The value of these techniques is that they enable us to create common ground when working with concepts (and their underlying assumptions). They replace the either/or thinking characteristic of the disciplines with both/and thinking characteristic of interdisciplinary integration. Inclusion, insofar as this is possible, is substituted for conflict. Creating common ground *does not* remove the tension between the concepts and the insights they produce; it *does* reduce this tension, making integration possible (but not guaranteed).

Create Common Ground Among Conflicting Ethical Positions

Students commonly work with issues that involve conflicting values and rights. Examples include the value of an unborn's life versus the value that the mother assigns to her freedom of choice, the value of equality for women versus the value of a cultural tradition that denies such equality, the right of the terminally ill to end life with dignity versus the right of society to sustain life, and the value of using fertilizers to increase crop production versus the value of using organic (i.e., sustainable) farming techniques.

Values and rights involve ethics (a subdiscipline of philosophy). Ethics (a type of philosophical theory) is concerned with how the world *should* work rather than how it *does* work (which is the function of scientific theory). Though we tend to focus in this book on insights about how the world does work, researchers often have to grapple also with insights about how it should work.

Ethical evaluations are made at two points in the IRP. The first occurs during the literature search. Here students should strive *not* to allow their personal views to skew the *selection* of insights concerning the issue. As noted earlier, such skewing is a common practice in disciplinary research, not to mention partisan politics and debate, but has no place in quality interdisciplinary work where all relevant viewpoints should be accorded equal voice. The second is made when analyzing the problem (STEP 6, the subject of Chapter 7). Here students should strive *not* to allow their personal views to skew their *evaluation* of insights with which they may disagree. Note that in both cases our views about how the world *should* work can bias our understanding of how it *does* work.

A proven way to work with insights that conflict because their ethical positions conflict is to use Szostak's (2004) classification of five broad types of ethical analysis and decision making:

1. *Consequentialism:* where an act is judged in terms of whether its consequences are good or not

2. *Deontology:* where an individual's act conforms to certain rules such as the Golden Rule (i.e., "Do unto others as you would have others do unto you."), the Kantian categorical imperative (i.e., "An act is ethical if and only if it is in accord with general principles that everyone would want to live by."), and arguments from "rights"

3. *Virtue:* where individuals are urged to live in accord with one or more virtues, such as honesty

4. *Intuition/experience:* where unique insight into an act, person, relationship, group, or decision is based on "knowing" from experience and reliance on one's subconscious (An act is thus judged to be good if it feels right.)

5. *Tradition:* where a behavior, relationship, group, or decision is judged based on its conformity to a historical, cultural, or societal practice (pp. 124, 194)

These five types of decision-making processes, says Szostak (2004), are mutually exclusive and exhaustive. They "are the five ways in which *any* person might make *any* decision. In terms of ethics," he says, "these five processes describe the five ways in which any person might evaluate *any* act or outcome" (p. 195).

Arguments for the Validity of These Types

Szostak (2004) advances three arguments for the validity of these five broad types of ethical analysis.

1. They each start from valid premises. It makes sense to judge acts by consequences. But it also makes sense to judge acts in terms of virtues or some predetermined rules. And it makes sense to respect (albeit critically) a society's traditions, for there is good reason to suspect these have been (imperfectly) selected to serve the society. And it makes sense to not do things that make us feel guilty.

2. Each of us uses these five types of decision making all the time: We rationally evaluate big decisions (going to university), follow certain rules (be nice to strangers, say), identify ourselves as kind or courageous or honest, do what others do (say, when buying clothes), and act on intuition (when dating). Indeed, we usually do and should use more than one type when making particular decisions: A little rational evaluation of dating choices is a good idea, for example.

3. Each approach draws upon the other approaches for justification. Rule utilitarians, for example, justify following rules by arguing that we don't have the time or cognitive capacity to rationally evaluate each decision, and thus should follow rules that generally lead to good outcomes. In turn, utilitarians, when asked why we should focus on happiness as the consequence we care about, argue that our intuition tells us that humans want to be happy (Hooker, 2000). Indeed, our desire to avoid guilty feelings provides a powerful motive for ethical behavior in general.

Ethical disputes often degenerate into mutual incomprehension as each participant imagines that his or her own view is obviously correct. An important first step for the interdisciplinary researcher is to simply recognize that quite different conclusions can be reached from equally valid ethical premises. As when trying to integrate insights about how the world works, interdisciplinary researchers should strive to see the good even in ethical insights with which they disagree. (*Note:* It is also important to appreciate, though, that many key values can be justified in terms of each of the five types of ethical analysis, including honesty and responsibility; it is thus a mistake to think that respect for diversity must yield an "anything goes" ethical outlook [Szostak 2004].) We can then usefully investigate why certain authors make certain ethical arguments.

Determine If Insights Conflict Over Ethics

Though authors sometimes make their ethical views explicit, they often do not. They may be driven to reach a particular conclusion because of ethical reasons but will couch their entire argument in scientific terms. *We can usefully use the techniques above to achieve common ground among the scientific arguments employed by different authors. But unless we also strive to address the different ethical positions that motivate them, we will likely have limited success in changing their minds.*

Ethical conflicts are most likely when insights involve policy proposals. These by their nature tend to combine elements of policy goals (that is, how we think the world should work) with ideas about how to achieve those goals (which are hopefully grounded in an understanding of how the world does work). If insights disagree about goals, then ethical analysis will generally be necessary. If insights conflict only about the means to achieve certain shared goals, then ethical analysis may be unnecessary.

Students can determine if insights (regarding either goals or means) conflict because of ethical disagreements (rather than because of conflicting theories, concepts, or assumptions) by asking certain questions of each insight. Note that these questions correspond to the five broad types of ethical analysis and decision making. The following questions are designed to help you identify which type of analysis—consequences, rights, and so on—is at issue:

- Do the insights differ on the desired *consequences* of individual actions?

- Do the insights differ on the *rights* that should be accorded an individual or group?

- Do the insights differ over the choice of which *virtues* should be appreciated or established (i.e., either at the group level, at the individual level, or by adherence to some universal code)?

- Do the insights differ in their beliefs about human *intuition* (e.g., "If one believes that human intuition is grounded in genetics or 'the gift of god' one will expect universal intuitive behavior; if it is thought that intuition is grounded in experience, one will not" [Szostak, 2004, p. 195])?

- Do they differ over the role that *tradition* (e.g., the test of time) should play when evaluating an act or attitude?

Create Common Ground

The first task is to determine whether different authors are applying the same or different types of ethical analysis. Once ethical statements have been identified in a work, it is generally straightforward to determine which of the five types of ethical analysis is being employed (though it may at times be difficult to know whether a particular value or rule is being invoked). The easiest case is when the same type of ethical analysis is being employed. We can then usually develop a continuum as described in the discussion of transformation and illustrated in the work of Englehart and Etzioni above. Recall from their works that they created common ground (though Etzioni did not use this term) between conflicting ethical views. For example, you could develop a continuum between one's right to act and one's right not to be hurt by the actions of others.

However, when different types of ethical analysis are employed, a different strategy must be employed. One possibility is organization: This is a useful strategy when different ethical perspectives focus on different parts of a problem. One common example is when consequentialism identifies outcomes but virtue ethics focuses on how to achieve these. It is then possible to outline how the results of one type of ethical analysis feed into application of another type.

Redefinition may also be useful (both within and across types of ethical analysis). The psychologist George Lakoff has noted that policy advocates "frame" political issues in a way that appeals to ethical sentiments. That is, they carefully choose their words to impart ethical meaning. The most obvious cases are the "right to life" and "right to choose" labels employed in the abortion debate, but debates regarding welfare can contrast appeals to social responsibility and personal responsibility, and policy toward unions can contrast a right to work with an appeal to solidarity. Though careful redefinition on its own is not likely to lead to political consensus on any of these issues, it can at least uncover hidden assumptions and meanings that might set the stage for other common ground strategies.

Extension may also be useful. Consequential analysis tends to focus on a small set of outcomes: Happiness or incomes are the most common. But potentially any consequence could be embraced, including the values and rules and traditions and even gut feelings urged by other types of ethical analysis. An intervention that raises incomes but leaves people feeling guilty may thus fail a consequentialist test. The four other types of ethical analysis also lend themselves to extension: Other values or rules or traditions or emotions might be invoked that lessen conflict with other types of ethical analysis.

An important example here is the ethical matrix developed by Mepham (2000) for rational ethical analysis. The components of the matrix are based on the work of Beauchamp and Childress (2001) on bioethics that has gained wide support in medicine and medical ethics. They introduced the "four principles approach" through which decision makers were guided to consider four core values: nonmaleficence (doing no harm), beneficence, autonomy, and justice. Note that the first two of these address consequences; the third and fourth express important values; the appeal to justice might also be understood as an appeal to rights; and valuing autonomy means valuing what feels good or appropriate to individuals. As an analytical tool, the matrix

> has the three principles (wellbeing, autonomy and justice) on the horizontal axis. On the vertical axis one lists the interest groups—that is, the people, organizations, communities, and so on—who stand to be affected by the

decisions being made. The task then is to identify and document the ethical impacts of the matter under consideration in each cell of the matrix. While this task can be undertaken through desk-based research, it is also a dialogue tool when undertaken through group discussion. (McDonald, Bammer, & Deane, 2009, p. 110)

The matrix is particularly useful in situations where, for example, decision makers are concerned that their diverse or conflicting values have the potential to influence their decision on a particularly sensitive matter such as the ethical impacts of introducing a proposed technology (Mepham, 2000, p. 168). "Once the cells of the matrix are filled in," explain McDonald et al. (2009),

> its users weigh the relative importance of the issues identified. Different people might give different weights to a given potential ethical impact on a particular interest group. Through discussion, the users of the matrix reach agreement about how the options under consideration, if implemented, will affect different interest groups with respect to their wellbeing, autonomy and entitlement to justice. (p. 110)

McDonald et al. (2009) identify four primary strengths of the matrix as an approach to prepare for integration:

- It is able to anticipate what the values of different stakeholders might be and how they will be differently impacted by the options available for implementing the initiative.

- It affirms the salience of values and conflicts.

- It is grounded in people's own ways of seeing values.

- It is conceptually straightforward. (p. 113)

However, they point to this critical limitation: The matrix fails to provide any clear guidelines on how to move toward a consensus on (i.e., integrate) the values it identifies (p. 113). Yet by making explicit both the nature of different ethical arguments and the motivations behind these, it may set the stage for the various strategies described earlier in this chapter to achieve a greater degree of consensus.

In the end, the researcher should evaluate ethical statements just as one would any other argument. Assuredly, it is doubtful that common ground can be created in all ethical disputes. There are, though, strategies that can at least soften the conflicts involved.

BOX 10.1

The Psychology of Ethical Disagreement

Welch (2017) discusses several psychological mechanisms through which (especially) ethical preferences influence how we see the world:

- *Cognitive dissonance*: Humans feel discomfort when they encounter new information that disagrees with strongly held beliefs. They may thus unconsciously avoid seeing such information, ignore it, or simply place more value on information that does agree with their prior beliefs. (Recall our discussion of confirmation bias in Box 1.4.)

- *Carnivalization*: This involves people seeing themselves as underdogs and blaming those in power for unethical behaviors. Environmentalists vilify corporate elites, while nonenvironmentalists vilify scientists and protestors. As Welch notes, carnival thinking can lead to doubt about the very possibility of applying objective expertise to problem solving. In its extreme, carnivalization leads to conspiracy thinking, in which we imagine that powerful groups are working against us. Since such groups would work in secret, it is hard to debunk conspiracy theories.

- *Groupthink*: People tend to associate with people they agree with. (Recall that social media makes it easy to do so online, but also provides opportunities to encounter different views.) Conversing with those we agree with serves to reinforce our prior beliefs.

Welch reminds us that humans are inherently capable of achieving common ground but warns us that these other mechanisms can interfere with our ability to do so. The interdisciplinary researcher must be wary that others are reaching biased conclusions because of the mechanisms above. And they must interrogate their own beliefs and mental processes. Welch urges "undecidability" (a particular kind of open-mindedness), compassion, and empathy.

Chapter Summary

This chapter discusses how to create common ground among conflicting insights by modifying one or more concepts. One way to do this is to modify one or more assumptions underlying the concept. Assumptions undergird concepts, which in turn undergird insights. The first section of the chapter defines common ground, explains that creating common ground is natural, occurs regularly, is necessary for collaborative communication, plays out differently in contexts of narrow versus wide interdisciplinarity, is critical to integration, and requires using intuition. The second section details how to modify conflicting concepts and assumptions by using one or more techniques: *redefinition*, *extension*, *transformation*, and *organization*. These are illustrated from student and professional work. The section also explains how to create common ground when there is ethical conflict.

The discussion of STEP 8 in this and in the following chapter is guided by the idea that disciplinary insights are potentially complementary if their concepts and theories and the assumptions underlying their concepts and theories are sufficiently modified. But there is no guarantee that common ground can be achieved in every case, such as when it comes to deeply held religious beliefs. Chapter 11 continues the discussion of STEP 8 by focusing on how to modify disciplinary theories.

Notes

1. This assertion is opposed to the belief that disciplines are different worlds capable of being understood only from the inside. Weingast (1998) agrees that disciplinary insights are potentially complementary: "In the past, interaction among scholars using different perspectives has tended to emphasize their seeming irreconcilability, as if Kuhn's 'competing paradigms' provides the unique program for interaction among different approaches in the social sciences. In recent years an alternative program has emerged, emphasizing the complementarities among different approaches. This new program acknowledges differences not as competing paradigms but as potentially complementary approaches to complex phenomena. This suggests a more fruitful interaction among scholars of different approaches, where not only the tools and techniques of the other become relevant, but also too do the phenomena under study" (p. 183).

2. From the earliest conceptions of interdisciplinarity, researchers have recognized the need for a common or collaborative language. Joseph J. Kockelmans (1979) was the first to use the term *common ground,* seeing it as the basis for collaborative communication—"a common ground"— among research scientists from different disciplines working on large government and industry projects. Common ground, he says, is the fundamental element of all interdisciplinary investigation because without it, "genuine communication between those who participate in the discussion would be impossible" (Kockelmans, 1979, p. 141). Kockelmans was also the first to connect integrating disciplinary insights with developing common ground (pp. 142–143). In explaining how to teach interdisciplinary research practice, Szostak (2007b) stresses the importance of first creating common ground among different disciplinary insights (p. 2).

Exercises

Definition

10.1   How does the definition of common ground complement and extend the definitions of interdisciplinary studies and interdisciplinarity?

(Continued)

(Continued)

Common Ground Theory

10.2　How does common ground theory explain how we are able to comprehend so many different perspectives? What are the implications of the theories of Clark and Bromme for dealing with conflict of all kinds, including values?

Intuition

10.3　How is the story of Helen Keller an example of how intuition helps achieve common ground, and what does this story overlook?

Roles

10.4　How is the role of a marriage counselor similar to the interdisciplinarian's responsibility?

10.5　How is the role that the facilitator plays in an admission, review, and dismissal meeting in special education similar to the role played by the interdisciplinarian in attempting to create common ground among conflicting disciplinary insights?

Best Practice

10.6　How should one proceed with creating common ground in the following situations?

    a.　Some of the authors make explicit reference to one or more concepts, but the authors of other insights do not.

    b.　All of the authors reference the problem using concepts that have different (apparent or real) meanings.

    c.　Some authors use theoretical explanations, but others do not.

Techniques

10.7　What technique(s) might be used in the following situations?

    a.　Some of the authors use concepts, and some use theories.

    b.　All of the authors use the same concept but use different language to define it.

    c.　Authors employ concepts or assumptions and generate insights without formulating theories.

    d.　Authors have conflicting conceptions and assumptions about whether behavior is rational or irrational.

10.8　How is Bal's approach to the graffito an example of extending concepts?

10.9　Why did Repko decide to work with assumptions?

Ethics

10.10　Choose a political issue of your choice. Can you detect appeals to ethics? If so, what types of ethical analysis are implicated? Speculate on how you might go about reducing the degree of ethical conflict.

Image by Niek Verlaan from Pixabay

Image by Gerd Altmann from Pixabay

CREATING COMMON GROUND AMONG INSIGHTS

Theories

LEARNING OUTCOMES

By the end of this chapter, you will be able to

- Define disciplinary theories
- Describe models, variables, concepts, and causal relationships that characterize theories produced by disciplines
- Explain how to create common ground among theories

GUIDING QUESTIONS

What is a theory, and what are the key elements of a theory?

How do you create common ground among theories?

CHAPTER OBJECTIVES

Creating common ground among conflicting insights plays out differently depending on whether you are working with a set of concepts *or* a set of theories. Insights produced by the natural and social sciences generally rely on theories to explain the phenomena they study. Many theories have concepts embedded within them (e.g., consumer theory has the embedded concepts of preferences or tastes, marginal utility or marginal rate of substitution, trade-offs, and personal income). And many social science analyses make use of both theories and concepts (e.g., the economic analysis of international trade makes use not only of the theory of comparative advantage, but also of the concept of a production function). To create common ground among a set of theories, researchers will have to modify them directly through their concepts or indirectly via their underlying assumptions.

The first section of this chapter defines disciplinary theory. The second section describes the models, variables, concepts, and causal relationships that one typically encounters when working with theories produced by the natural and social sciences. The third section applies the four strategies of modification introduced in the previous chapter to create common ground among theories.

B. INTEGRATING DISCIPLINARY INSIGHTS.

7. Identify conflicts among insights and their sources.

8. **Create common ground among insights.**

   ○ **Identify the six core ideas about creating common ground.**

   ○ **Create common ground among conflicting concepts and/or assumptions.**

   ○ **Create common ground among conflicting ethical positions.**

   ○ **Create common ground among insights that have a theoretical base.**

9. Construct a more comprehensive understanding.

10. Reflect on, test, and communicate the understanding.

DEFINING DISCIPLINARY THEORY

A **disciplinary theory** explains a behavior or phenomenon that falls within a discipline's traditional research domain and *may* have a specified range of applicability.[1] Since disciplinary theories focus on only a particular aspect of a complex problem, any one theory will provide incomplete understanding of it. That is, different theories will illuminate different aspects of the question about a complex problem such as the causes of urban crime. The interdisciplinary researcher must then seek a more comprehensive theory that encompasses the phenomena deemed important after evaluating disciplinary theories.

To avoid confusion, we should appreciate here the difference between scientific and philosophical theories. Scientific theories explain why or how the object of study behaves as it does, tells us who is acting, what they do, how decisions are made, how the process unfolds through time, and over what set of circumstances the theory holds. Such theories would focus, for example, on the *causes* or the *effects* of some process such as economic growth. Scientific theories are the more common focus of integration and are the subject of this chapter.

By contrast, philosophical theories deal with how the object of study *should* behave (ethics), how we *know* about the object (epistemological), and the nature or status of *reality* (metaphysical or ontological). Ethical theories would focus on whether economic growth is good rather than how to achieve (or avoid) it. Readers should appreciate that the authors they read will often refer to philosophical theories. Moreover, some theories blend philosophical and scientific elements: They may argue first for how the world should work (ethics) and then how to transform the world in that direction. We discussed in the last chapter how to address ethical disagreements; the researcher may have to do so first before attempting to achieve common ground among the scientific elements of such theories.

MODELS, VARIABLES, CONCEPTS, AND CAUSAL RELATIONSHIPS

When reading insights that have a theoretical base, you will often encounter models, variables, concepts, and causal relationships. You will need to understand these components of theory as you seek to create common ground. We define these terms and explain how they are related to theories using the example of the "broken windows theory of urban crime."

Models

A **model** is a representation that serves to visualize and communicate a theory. It is specific and clear in the same way that a model of a building represents clearly and precisely what an architect plans to build. Models are very helpful in making sense of complex processes and depicting cause-and-effect relationships.

Models may be either graphical (a picture) or mathematical (a set of equations).[2] In this book, graphical models, such as path models, are used to express theory. Figure 11.1 is a path model that depicts "the broken windows theory of crime." It communicates the idea that seemingly trivial acts of disorder (like broken windows) trigger more serious crime.

The model is made up of two elements: variables (the ovals) and the relationship(s) between them (the arrows). The relationship between crime and disorder is represented by the arrow pointing from disorder to crime. The arrow shows the direction of the relationship: Disorder triggers crime.

Models necessarily simplify: Since all phenomena exert influences (directly or indirectly) on almost all other phenomena, every model necessarily abstracts away from the complexity of the real world. The interdisciplinary researcher will always wish to question whether the phenomena and relationships excluded from a model would likely have an effect on the outcomes predicted by the model.

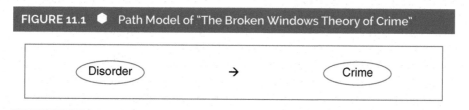

FIGURE 11.1 ● Path Model of "The Broken Windows Theory of Crime"

Disorder → Crime

Source: Remler, D. K., & van Ryzin, G. G. (2011). *Research methods in practice: Strategies for description and causation* (p. 31). Thousand Oaks, CA: Sage.

Variables

A **variable** "is something that can take on different values or assume different attributes—it is something that varies" (Remler & van Ryzin, 2011, p. 31). For example, the broken windows theory is based on empirical evidence demonstrating that the murder rate takes on different values over time and across cities—so murder rate is a variable. The theory attempts to explain this variation between high and low murder rates by way of another variable—disorder. It should be stressed that in saying that one variable influences another, it need not be asserted that only that variable does so. Variables generally though not quite always represent the phenomena studied by disciplines, or some form of data that it is hoped can proxy for phenomena.

Two of the most fundamental yet often confusing terms in research are *independent variable* and *dependent variable*. Figure 11.2 shows that the independent variable is the *cause* and the dependent variable is the *effect*, so the model asserts that crime "depends" on disorder. By convention, the independent variable is symbolized by *X*, while the dependent variable is symbolized by *Y*. Remler and van Ryzin (2011) explain: "Just as cause comes before effect, and *X* comes before *Y* in the alphabet, the independent variable comes before the dependent variable in the causal order of things" (p. 31).

Unfortunately, researchers in different disciplines use different terms to describe independent and dependent variables. For example, in health science research, the independent variable is often called a *treatment* and the dependent variable an *outcome*. Other terms for an independent variable include *explanatory, predictor,* and *regressor*; other terms for a dependent variable include *response, predicted,* and *regressand*. Therefore, the researcher should think carefully about the sequence of the variables and ask which is the presumed "cause" of the problem and which is the "effect" (Remler & van Ryzin, 2011, p. 32). Use of the word *cause* allows for multiple causes. To be a "cause," the independent variable needs only to exert *some* effect on the dependent variable.

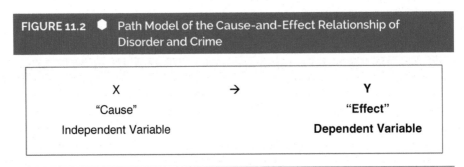

FIGURE 11.2 ● Path Model of the Cause-and-Effect Relationship of Disorder and Crime

X	→	Y
"Cause"		"Effect"
Independent Variable		Dependent Variable

Source: Remler, D. K., & van Ryzin, G. G. (2011). *Research methods in practice: Strategies for description and causation* (p. 31). Thousand Oaks, CA: Sage.

Concepts

In the path model of the broken windows, "disorder" and "crime" are both concepts, and the statement "Disorder is a cause of crime" is a theory. But that theory is stated too vaguely to be researchable: "Disorder" includes a wide array of phenomena, and "crime" includes a wide range of behaviors. To operationalize the theory so that it can be researched, the variable "broken window panes per city block" might be chosen to stand in for the concept "disorder," and the variable "murder rate" might be chosen to stand in for the concept of "crime." These variables have the advantage of being quite specific and readily measurable using available information. Their disadvantage is that they permit researchers to study only one aspect of each concept. (Note that both of these concepts describe phenomena; theories might contain other types of concepts, such as those describing a process such as evolution.)

The broken windows theory is noteworthy, not because it revealed the one true source of crime, but because it hypothesized an oft-overlooked influence among the long list of influences on crime. More generally, the theory that Variable A influences Variable B can be compatible with another theory that Variable C influences Variable B. That is, most theories focus on one explanation of a behavior but do not claim to offer the *only* explanation of that behavior.

Causal Relationships (or Links)

A **causal relationship or link** refers to how change in one variable produces or leads to change in another variable. A causal relationship can be either unidirectional or bidirectional. All positive feedback loops are examples of bidirectional (or mutually reinforcing) causation. The purpose of science, says Szostak (2004), is to identify and understand causal relationships or links.[3] For this reason, "each theory needs to carefully express which causal links it addresses, and under what circumstances the theory holds along those links: that is, its range of applicability" (p. 75).

According to the broken windows theory, cracking down on instances of everyday disorder produces a reduction in more serious crime (p. 31). It is conjectured that this theory is applicable to all cities experiencing social disorder. The social sciences, like their natural science counterparts, typically see theories as assertions about relationships among variables, or (in the language of the humanities) about how some things influence others.

Causal Relationships and Variables

Theories in the natural and social sciences express causal relationships. *How* change in the independent variable causes change in the dependent variable is what researchers call a causal process that underlies this relationship. For example, disorder in a neighborhood evidenced by broken windows and graffiti signals that no one cares and that rules are not enforced. According to the broken windows theory, criminals read these signals and become emboldened to commit crime. This is the causal process theorized in the broken windows theory (Remler & van Ryzin, 2011, p. 32).

Though the causal process noted here is a critical component of the broken windows theory, it does not appear in the model shown in Figure 11.1. So, to provide a full description of the theory, we would need to add to the model a statement of the process that explains *how* change in the independent variable causes change in the dependent variable. We could make this causal process explicit by adding what is called an "intervening variable" to the model (in this case the intervening variable would be attitudes of potential criminals regarding the chances of being punished). An intervening variable is a step(s) in the causal process that leads from the independent to the dependent variable (Remler & van Ryzin, 2011, p. 33). For example, if we identify a positive correlation between school attendance and performance on a test, we might imagine that the intervening variable was some sort of learning. (See Box 11.1 for discussion of variables.)

BOX 11.1

Theories May Contain Macro- and/or Micro-Level Variables Affecting the Construction of a More Comprehensive Theory

The most common distinction in the social sciences is between the micro and macro levels—that is, between the level of individuals and the level of society—though there are plenty of theories that operate at the mezzo (or "in between") level of groups or communities, as well as theories that operate at the international level. In a two-level analysis (as in the following example), *macro* may be used generally to refer to any level larger than individuals. Some theorists choose to use different terms for some levels, and some layer reality differently depending on their focus (e.g., subcultural, cultural, cross-cultural). Since the terminology of levels is not consistent from discipline to discipline, much less from author to author, interdisciplinarians need to exercise care in identifying the level of each variable.

Criminologists Stuart Henry and Nicole Bracy (2012) provide this example: "We might want to look at the range of variables from different theories that have empirically demonstrated to correlate with the propensity of juveniles to join gangs" (p. 263). One set of theories may include variables such as "neighborhood housing density, age, criminality of parents or siblings, degree of neighborhood disorganization, or neighborhood transience rates" (p. 263). (*Note:* These variables operate at the mezzo level [community] or the macro level [society].)

Another set of theories may include variables that operate at the micro level and are internal to the person, such as the biological or psychological development processes of juveniles.

These sets of variables, explain Henry and Bracy (2012), may be interconnected in different ways:

> Theories that see community as shaping the opportunities available for adolescents to make more or less delinquent decisions present a different set of interconnections than those that see such opportunities in the environment only being acted on by juveniles who are predisposed to sensation-seeking behavior through biological or psychological development processes. In this individual-level predisposition case, we might be including variables such as adolescent brain development, domestic abuse, traumatic brain syndrome, high sugar consumption, addictive personality, etc. (p. 263)

Clearly, then, a different theoretical explanation of the problem of why juveniles join gangs would emerge based on whether (a) we integrate variables located at the micro level of analysis (at the level of the individual), or (b) we integrate elements of theory based at the mezzo or macro level of analysis (i.e., at the level of the community and society) since the latter are influenced by the former, and vice versa. Henry and Bracy's (2012) point is this: "*The decision about the nature of the interconnections between variables affects which concepts are integrated from the different discipline-based theories* [emphasis added]" (p. 263). Therefore, a more comprehensive theory (one that has been extended in STEP 9, the subject of Chapter 12) will include a wider array of variables (e.g., often both macro and micro) and provide a much different explanation than do any of the individual disciplinary theories and their corresponding variables (Henry & Bracy, 2012, p. 263).

CREATE COMMON GROUND AMONG THEORIES

Before trying to create common ground among theories, make certain that the different authors are in fact talking about the same thing. Mapping their arguments enables you to distinguish real conflicts (over the same thing) from apparent conflicts (when authors talk about different things). You should also have applied the strategies outlined in earlier chapters for evaluating theories before seeking common ground.

When working with theories that conflict, two situations are commonly encountered: Situation A is that some theories have a broader range of applicability than do others; Situation B is that none of the theories borrow elements from other disciplines. We address the use of the technique of organization when different theories address different parts of a research question in Situation C below.

Situation A: One or More Theories Have a Broader Range of Applicability Than Do Others

At times, one or more theories have a range of applicability that is already broad either because of their disciplinary origins or because they are already somewhat interdisciplinary. (*Note:* The purview of some disciplines is simply broader than that of other disciplines. Geography, history, and anthropology [including physical as well as cultural] are broader in scope than economics, sociology, and political science.) Your task then is to identify the theory that can most easily be extended to encompass all variables engaged by all of the theories you have encountered, and extend it to do so.

Deciding which theory is the most comprehensive can be determined by taking the following substeps: (a) identify all variables or causal factors addressed by any of the theories; (b) reduce these variables to the fewest number possible by categorizing them under a few broad headings (that is, by recognizing variables used by different authors that are broadly similar); (c) determine how many of these categories are included in each theory; (d) if no theory encompasses all categories, determine which theory can most readily be extended to do so; and (e) modify the theory by extending its range of applicability. This extended theory will be used to integrate the other theories in STEP 9 (see Chapter 12). Each of these substeps is discussed below more fully.

Identify All Variables or Causal Factors Addressed by Any of the Theories

All the variables (or at least the phenomena that they proxy for) should have been identified when the problem was mapped in STEP 3 (see Chapter 5). In the threaded example of the problem of suicide terrorism, Repko (2012) identifies eight variables that one or more theories claim are likely causes of suicide terrorism: (a) cognitive constructs that may predispose a person to become a suicide terrorist; (b) influences on a person's mental development; (c) "justification" (an even stronger, more compelling, moral claim that overrides one's natural repugnance to engage in suicide terrorism); (d) "emotion" triggered by traumatic memory, or regret for not exacting vengeance on an enemy; (e) the influence of institutions, both domestic and foreign, that may trigger or exacerbate a collective or an individual sense of political oppression and/or historical loss; (f) a shared sense of place, and identification with a race, an ethnic group, a history, a nation, and/or a religion; (g) emotional bonding to a way of life, traditions, behaviors, values, and symbols; and (h) a fundamentalist faith tradition that relies on sacred writings and charismatic leadership to determine an individual's motivations and actions and a belief that religion must struggle to assert (or reassert) its control over every facet of life.

Reduce These Variables to the Fewest Number Possible by Categorizing Them Under a Few Broad Headings

Categorize the variables under a few broad headings, which can be derived from Szostak's table of phenomena introduced in Chapter 2. Repko (2012) creates four categories: *personality traits, power relations, cultural identity,* and *sacred values. Personality traits* refers to cognitive constructs that may predispose a person to become a suicide terrorist as well as influence a person's mental development. These traits include "justification" (an even stronger, more compelling moral claim that overrides one's natural repugnance to engage in suicide terrorism) and strong "emotion" (triggered by traumatic memory or regret for not exacting vengeance on an enemy) (Reisberg, 2006, pp. 465–470). *Power relations* refers to the influence of domestic and foreign institutions on individuals, on particular groups, and on society as a whole. In the case of Islamic terrorism, such influence may include the policies and activities of Western corporations, UN agencies, foreign military forces, Western support of oppressive regimes, American support of Israel, and the countervailing influence of terrorist organizations such as al-Qaeda and Islamic State of Iraq and the Levant (ISIL). Each and any of these influences may trigger or exacerbate a collective or an individual sense of political oppression and/or historical loss. *Cultural identity* refers to having a shared sense of place and identification with a race, an ethnic group, a history, a nation, and/or a religion. Cultural identity may also include emotional bonding based on a way of life, traditions, behaviors, values, and symbols. *Sacred values* refers to a fundamentalist faith tradition that relies on sacred writings and charismatic leadership to determine an individual's motivations and actions. This factor includes the idea that religion must struggle to assert (or reassert) its control over every facet of life. The perspective of history is subsumed under culture, politics, and religion (Reisberg, 2006, pp. 146–147).

Determine How Many of These Categories Are Included in Each Theory

Regarding the problem of suicide terrorism, Table 11.1 table shows that all of the theories (listed vertically on the left) focus on at least two of the causal factors selected above (listed horizontally across the top). The theories of "sacred" terror, self-sanction, and identity attribute the causes of suicide terrorism to three of the four factors. The value of categorizing the variables and then determining how many of the categories are included in each theory is that it narrows the number of possible candidates for theory modification from the original eight to three: "sacred" terror, self-sanction, and identity theory, all of which may be deemed somewhat interdisciplinary. (The next challenge is to decide which of these three theories should be modified so that it is inclusive of the variables incorporated in other theories.)

TABLE 11.1 ● Theories on the Causes of Suicide Terrorism (ST) Showing Key Causal Factors

Theory of ST	Variables or Causal Factors Broadly Stated			
Theory	Personality traits	Power relations	Cultural identity	Sacred values
Terrorist psycho-logic	Yes	No	No	No
Self-sanction	No	Yes	Yes	Yes
Martyrdom	Yes	No	No	No
Collective rational strategic choice	Yes	Yes	No	No
"Sacred" terror	No	Yes	Yes	Yes
Identity	Yes	Yes	Yes	Yes
Fictive kin	No	No	Yes	Yes
Modernization	No	Yes	Yes	Indirectly

Source: Repko, A. F. (2012). Integrating theory-based insights on the causes of suicide terrorism. In A. F. Repko, W. H. Newell, & R. Szostak (Eds.), *Case studies in interdisciplinary research* (p. 147). Thousand Oaks, CA: Sage.

If No Theory Encompasses All Categories, Determine Which Theory Can Most Readily Be Extended to Do So

You must continue the process of narrowing the number of candidates for theory modification until the theory that is the most comprehensive is revealed. *The most comprehensive theory is one that requires the least possible modification so that it will include all variables.*

What follows is Repko's (2012) description of the process he uses to determine which of the three theories he identifies as interdisciplinary meets this standard:

- Rapoport's (1998) theory of "holy" or "sacred" terror: Though the theory explains suicide terrorism as a political problem caused by terrorist political aspirations, it takes into account how culture and religious values may motivate a person to become a suicide terrorist. But, because the theory does not borrow from psychology, it is unable to explain how a person's mind may be predisposed [to], susceptible to, or shaped by these complex influences and motivated to commit such a horrific act. So rather than force the theory to explain what it was not designed to explain—that is, the shaping of a person's personality and cognitive development—it is better to consider the possibilities offered by the other two.

- Bandura's (1998) self-sanction theory: The theory explains how terrorist organizations convert "socialized people" into dedicated combatants by "cognitively restructuring the moral value of killing, so that the killing can be done free from self-censuring restraints" (p. 164). The process of moral cognitive restructuring involves using religion, politics, and psychology to construe suicide attacks narrowly. This involves using (a) religion to justify such acts by invoking "situational imperatives," (b) the political argument of self-defense to show how the group is "fighting ruthless oppressors" who are threatening the community's "cherished values and way of life," and (c) the psychological device of dehumanization to justify killing "the enemy." Though the theory does not borrow from other disciplines, it explains how cultural and political factors are integrated into the mental process of construal and inform individual decision making. One limitation of the theory is its silence concerning individual personality factors that may influence the would-be terrorist's decision-making process. But since the theory is a psychological theory, this limitation can be overcome by borrowing from other psychology theories so that it would be able to include the influence of personality traits and dispositions.

- Monroe and Kreidie's (1997) identity theory: Identity theory is already interdisciplinary (though narrowly so) because it borrows from other disciplines. Identity theory already addresses three of the four factors—cultural identity, sacred beliefs (i.e., religion), and power seeking (i.e., politics). The theory draws from religious studies by showing that the Islamic fundamentalist conception of religion is, in fact, an all-encompassing ideology that erases all lines between public and private. This theology-based ideology is grounded in religious writings and commentaries on these writings that are viewed by its most devoted followers as sacred, inviolable, nonnegotiable, and worth dying for. As a "sacred ideology" (Marxism never achieved this lofty status) Islamic fundamentalism redefines politics and power. As Monroe and Kreidie show, politics for Islamic fundamentalists is subsumed under an all-encompassing religious faith and is sought and exercised for the purpose of extending the faith. Achieving, maintaining, and extending power is a sacred duty that has priority over all other obligations, including family. Because the theory holds that the self is culturally situated, it is able to explain how culture influences identity formation (Repko, 2012, pp. 147–148).

From his analysis of the three theories, Repko (2012) concludes that Monroe and Kreidie's (1997) identity theory, which is already narrowly interdisciplinary, will require the smallest changes.

Modify the Theory by Extending Its Range

After identifying the theory that is the most comprehensive, you need to extend its range by using the technique of extension. Extending a disciplinary theory involves taking well-known facts normally treated as exogenous (external) to it (such as organisms altering their environment) and making them endogenous (internal) to and mutually interactive with it. This commonly used technique enables a theory to encompass variables emphasized in other theories. When extending a theory, the principle of least action should be followed: making sure that the changes are the smallest possible.

In the present example, it is appropriate for Repko (2012) to use **theory extension** because identity theory already has parts that come from other disciplines. Here, he explains how he extends identity theory to make it fully interdisciplinary:

> For the theory to include the fourth causal factor of personality traits would require the least amount of "stretching and pulling" (i.e., extending). This is because identity theory, as already noted, is based on the psychological concepts of cognition and perspective. These concepts address personality traits in a way that is inclusive of the psychology theories already examined. Monroe and Kreidie [1997] use the concept of cognition in a developmental way, meaning that they are concerned to show how persons are influenced by factors external to themselves—namely culture and societal norms—rather than focusing on individual cognitive abnormalities as Post [1998] does to argue that some persons are psychologically predisposed to commit acts of suicide terrorism. Only slight "stretching" or extending is necessary to have identity theory include personality factors intrinsic to the individual suicide terrorist. Monroe and Kreidie's application of "perspective" to explain suicide terrorist behavior is also helpful in this regard because it effectively delineates the options that terrorists perceive as being available to them. . . . The act of committing a suicide attack in the service of a fundamentalist conception of *jihad* "emanates primarily from the person accepting their identity which means that they have to abide by the tenets of their religion." . . . By borrowing these concepts from psychology, identity theory offers an understanding of human behavior that is based on the interplay of mental constructs in tandem with the exogenous variables of politics, culture, and religion. (Repko, 2012, pp. 148–149)

Once extended, identity theory is of sufficient generality to include the causal factors from psychology. Ideally, an extended theory that is fully interdisciplinary will incorporate all relevant causal factors from the other theories encountered.

Situation B: None of the Relevant Theories Borrow Enough Elements From Other Disciplines

At other times, none of the relevant theories have a sufficiently broad range of applicability to facilitate theory extension. In some cases (see Situation C below), each theory addresses a distinct subset of the variables that are relevant to your research question; in such a case, you will pursue the strategy of organization. In other cases, theories overlap in their coverage of variables, but disagree about how certain variables or relationships behave. In this event, you will have to decide what *parts* of the theories in the set to modify in order to create common ground: their concepts or their assumptions (there is thus an overlap with our treatment of these in Chapter 10). We will employ the techniques of redefinition, transformation (briefly), and extension in what follows.

Earlier we said that a disciplinary theory includes variables and describes relationships among them. A variable may be a phenomenon or, more often, a way of operationalizing a phenomenon. Henry and Bracy (2012) explain how the social disorganization theory of gang formation is built up from phenomena:

> High levels of immigration into cities combined with profit-seeking landlords leads to poor quality, low rent multi-family inner-city housing with high resident turnover; the resultant neighborhood instability fragments communities, resulting in a breakdown of informal networks of social control. Fearful of being victimized, youth band together for self-protection forming subcultures that can form into territorial gangs that protect their own members by exerting fear on non-gang or other gang members and maintain their autonomy by engaging in a variety of delinquency such as vandalism and drug dealing (social disorganization theory of gang formation). (p. 264)

Posited relationships among variables may reflect assumptions such as those that concern human nature or the physical environment. For example, economics assumes individuals make autonomous decisions; sociology assumes that individual decisions are primarily influenced by others (norms, customs, tradition, group membership, advertising, etc.). Economics usually assumes individuals are rational; sociology assumes people's decisions are largely determined by nonrational factors. Economics assumes people make choices to benefit themselves and their household; sociology assumes people are concerned with the well-being of others (their ethnic group or race, their community, their political party, their church, etc.) not just of themselves and their immediate family.

Preparing Theories for Integration Using Concepts

The objective of concept modification is to make selected thoughts, notions, or ideas from different disciplines more compatible, and to express the modified concepts in terms that bring out the commonalty resulting from that modification. Concept modification usually involves the strategy of redefinition and often the strategy of extension as well. Nagy's (2005) study of anthropogenic forces degrading tropical ecosystems in a Latin American country is an example of how to modify concepts that appear to be similar but have different meanings in different disciplinary theories.

Concepts That Have Different Meanings in Different Disciplinary Theories. Nagy (2005) draws on theories from biology, anthropology, and economics to explain the ecological and environmental problems facing Costa Rica. She finds a dichotomy between theories dealing with the environment and those dealing with economics. Nagy explains that the region's economic and environmental problems are linked in a mutual feedback loop: Growing population requires increased economic development, which causes environmental degradation of tropical ecosystems, which worsens living conditions for the poor and widens the income gap between the rich and the poor. Purely disciplinary approaches (of which there are many) suggest that the choice is between economics and environmental science. That is, Costa Rica (or a similar nation) must choose between increasing economic development to help raise living standards (and thereby worsen environmental degradation) and restricting development to protect fragile tropical ecosystems (and thereby reduce living standards).

The challenge for Nagy is to reduce the conflict between these irreconcilable disciplinary stances. While reflecting on the possible reasons for this conflict, she realized that the basic values of the two perspectives are in conflict. These conflicting values express themselves in how each discipline defines "wealth." For environmental scientists, wealth refers to the health of an ecosystem (excluding humans) and to the diversity of species within it. For economists, wealth is accumulated assets derived from development.

Redefine or Extend the Concept so That It Has the Same Meaning Across All the Theories. One way to modify a concept in preparation for modifying the theories that use it is to redefine it. (*Note:* The technique of redefinition is discussed in Chapter 10.) Another way is to extend its meaning. Nagy (2005) does both. She redefines the concept of "wealth" so that it includes both economic development and ecosystem health. She also extends the redefined meaning of wealth from assets valued solely by the marketplace (the result of development) to include assets valued by society as a whole (a healthy and diverse environment). By freeing herself from the marketplace, she can also extend "wealth" from a short-term (economic) to a long-term (environmental) concept (Nagy 2005, pp. 104–108).

Pay Attention to Disciplinary Perspective When Modifying Concepts. Henry and Bracy (2012) urge us to focus on how concepts reflect different disciplinary perspectives.

Concept modification, then, is not the simple task of focusing on similarities and ignoring differences. Rather, it entails figuring out how to utilize those similarities in a way that retains the integrity of the original concept (Henry & Bracy, 2012, p. 264).

These authors take the position that, much like the integration of insights, each redefined concept should be responsive to the perspective of each contributing discipline, but dominated by none of them. They illustrate their point with this example:

> Akers' (1994) "conceptual absorption" approach takes concepts from social learning and social control theory (among others) and merges them together. The control theory concept of "belief," which refers to a person's moral conviction for or against delinquency, is equated to learning theory's "definitions favorable or unfavorable to crime" (differential association). Interestingly, there are parallels here to the theory of human cognitive practice known as "conceptual blending" in which humans subconsciously integrate elements and relations from diverse situations to create new concepts, a process seen by some as being at the heart of the creative process. (Henry & Bracy, 2012, p. 264)

Preparing Theories for Integration Using Assumptions

When speaking of assumptions, we mean assumptions regarding how a particular influence or causal relationship operates. One has to get *behind* the conflict itself to bring out the *source(s)* of the conflict, digging into the text of each theory and juxtaposing their insights to discover why their explanations, arguments, concepts, and use of data conflict.

Assumption modification involves (a) identifying the assumptions of each theory, (b) determining which assumptions are shared by the other theories, and then (c) modifying these assumptions until all relevant theories share the same bedrock assumption.

Identify the Assumption(s) of Each Theory. Disciplinary theories typically reflect the assumption(s) of the disciplines that produce them. Often there will be additional assumptions relevant to particular theories consistent with other assumptions commonly made by that discipline. The assumptions of each theory should be available from the mapping of the problem and from the data table that you developed earlier. While performing earlier STEPS, Repko (2012) found that some authors used concepts (in some cases multiple concepts) while others did not use any. He concluded that he could not create common ground by redefining any of those concepts because they would not (indeed could not) apply to all of the theories. So, unable to work with concepts, he decided to work with assumptions. The assumption(s) of each theory are identified in Table 11.2. Readers will note that these assumptions are primary-level assumptions because they are derived directly from the theory itself.

TABLE 11.2 ● Theories and Their Assumptions on the Causes of Suicide Terrorism	
Theory	**Assumption(s) of Theory (Primary Level)**
Terrorist psycho-logic	Suicide terrorists are born, not made. Understanding the behavior and motivation of suicide terrorists requires studying primarily the mental life and the psychological constructs of individual terrorists.
Self-sanction	Suicide terrorists are made, not born. Understanding the behavior and motivation of suicide terrorists requires studying primarily the mental life and the psychological constructs of individual terrorists.
Martyrdom	Suicide terrorists are made, not born. Understanding the behavior and motivation of suicide terrorists requires studying primarily the mental life and the psychological constructs of individual terrorists.
Collective rational strategic choice	Suicide terrorists follow logical processes that can be discovered and explained. The primary focus of research should be on terrorist groups rather than on individuals.
"Sacred" terror	Suicide terrorists follow logical processes that can be discovered and explained. The primary focus of research should be on terrorist groups rather than on individuals.
Identity	Suicide terrorists are made, not born. Religious identity is (at least in this instance) an effective way to explain the "political" phenomenon of suicide terrorism. Terrorist behavior is essentially rational.
Fictive kin	Suicide terrorists are made, not born. Suicide terrorists are largely the product of identity with and loyalty to a culturally cohesive and intimate cohort of peers that recruiting organizations often promote through religious indoctrination.
Modernization	Suicide terrorists are made, not born. Poverty, authoritarianism, and diminishing expectations inevitably breed alienation and violence. Suicide terrorism is the inevitable result of Islam's failure to embrace Western institutions and values.

Determine Which Assumptions Are Shared by the Other Theories. Table 11.3, which Repko (2012) constructed while performing STEP 7 and is reproduced below, shows which assumptions are shared by other theories. It lists the relevant theories concerning the causes of suicide terrorism (left column), their parent disciplines (center column), and the assumption that each group of disciplinary theories shares (right column). These assumptions are secondary-level assumptions that reflect the assumptions of the parent disciplines.

Juxtaposing the secondary level assumptions of each theory reveals that some of them are shared by more than one theory. For example, the cognitive psychology theories of terrorist psycho-logic, self-sanction, and martyrdom share the assumption typical of psychology: Understanding the behavior and motivation of suicide terrorists requires studying the mental life and psychological constructs of individual terrorists.

TABLE 11.3 ● Sources of Conflict Among Relevant Theories in Terms of Their Underlying Assumptions		
Theory	**Discipline**	**Assumption(s) of Theory (Secondary Level)**
Terrorist psycho-logic	Cognitive Psychology	Understanding the behavior and motivation of suicide terrorists requires studying primarily the mental life and the psychological constructs of individual terrorists.
Self-sanction		
Martyrdom		
Collective rational strategic choice	Political Science	Suicide terrorists follow logical processes that can be discovered and explained.
"Sacred" terror		The primary focus of research should be on the behavior of terrorist groups rather than on the behavior of individual terrorists.
Identity		Religious identity is (at least in this instance) an effective way to explain the "political" phenomenon of suicide terrorism. Terrorist behavior is essentially a rational outworking of one's religious identity.
Fictive kin	Cultural Anthropology	Suicide terrorists are largely the product of identity with and loyalty to a culturally cohesive and intimate cohort of peers that recruiting organizations often promote through religious indoctrination.
Modernization	History	Poverty, authoritarianism, and diminishing expectations inevitably breed alienation and violence. Suicide terrorism is the inevitable result of Islam's failure to embrace Western institutions and values.

The theories of collective rational strategic choice, "sacred" terror, and identity share two assumptions of political science that suicide terrorists follow logical processes that can be discovered and explained and that the primary focus of research should be the behavior of terrorist groups rather than the behavior of individual terrorists. Note that only "sacred" terror theory and identity theory assume that a terrorist's religious affiliation is an effective way to explain the "political" phenomenon of suicide terrorism.

Fictive kin theory is grounded in cultural anthropology and therefore reflects its assumption that the determining factor in shaping the development of a suicide terrorist is the terrorist's loyalty to an intimate cohort of peers, all of whom share an intense devotion to religious dogma.

Modify These Assumptions Until All Relevant Theories Share the Same Bedrock Assumption. The goal of STEP 8 is neither complete elimination of conflict among disciplinary insights nor the removal of all conflict among disciplinary assumptions. Rather, it is the creation of an additional assumption that the contributing disciplines can accept with the least discomfort. Table 11.4 demonstrates how to "drill down" from the secondary-level assumption shared by each group of disciplinary theories to a third level of assumptions shared by theories from other disciplines, and ultimately to the additional assumption that all the theories can share. Fictive kin theory (as shown in Table 11.4) shares with identity theory (introduced in Chapter 10 as an interdisciplinary theory grounded in political science and religious studies) the assumption that religion is an important factor in understanding the development of a suicide terrorist. Finally, modernization theory from history rests on the assumption that suicide terrorism is the result of Islam's unwillingness to embrace Western institutions and values.

The bedrock assumption is that suicide terrorists view their behavior as "moral" and "sacred"—and, thus, rational—as defined by Islamic fundamentalism. The redefined meaning of "rational," then, can serve as the basis for integrating these theories and producing the more comprehensive theory in STEP 9 (see Chapter 12).

TABLE 11.4 ● How "Drilling Down" Into Disciplinary Assumptions Reveals a Bedrock Assumption

Theory	Secondary-Level Assumptions	Third-Level Assumptions	Bedrock Assumption
Terrorist psycho-logic	Understanding the behavior and motivation of suicide terrorists requires studying primarily the mental life and the psychological constructs of individual terrorists.	Suicide terrorism is a rational outworking of the mental life and psychological constructs of individual terrorists.	Suicide terrorists are rational, not in the Western understanding of the term, but as defined by their culture and faith tradition.
Self-sanction			
Martyrdom			
Collective rational strategic choice	Suicide terrorists follow logical processes that can be discovered and explained.		
"Sacred" terror			
Identity	Religious identity is (at least in this instance) an effective way to explain the "political" phenomenon of suicide terrorism. Terrorist behavior is essentially rational.	Suicide terrorists are "rational" in the sense that their behavior is consistent with adherence to their cultural and religious values.	
Fictive kin	Suicide terrorists are largely the product of identity with and loyalty to a culturally cohesive and intimate cohort of peers, whom recruiting organizations often promote through religious indoctrination.		
Modernization	Poverty, authoritarianism, and diminishing expectations inevitably breed alienation and violence. Suicide terrorism is the inevitable result of Islam's failure to embrace Western institutions and values.	Suicide terrorism is a rational response to exogenous factors.	

NOTE TO READERS

For undergraduate students who have little experience working with assumptions of theories, there are two practical benefits of comparing the assumptions of these disciplinary theories in this manner. The first is that it reveals which assumptions are shared by two or more theories, thus reducing the number of assumptions that require modification. The second benefit is that it begins the *partial* integration of the theories by working with assumptions from the "bottom up." Graduate students and professionals who are experienced in working with theories and their assumptions are more likely to arrive at a bedrock assumption intuitively.

Modify Concepts and Assumptions

Sometimes it is necessary to modify concepts *and* assumptions. This is required, for instance, when working with theories based on a grand theory such as evolution (macro level) that are divided on their understanding or application of the grand theory (micro level). A case in point is Ria van der Lecq's (2012) study of the origin of language where theories are traditionally divided between those that emphasize genetic evolution and those that emphasize cultural evolution. "Even if all these theories are coherent (i.e., internally consistent)," she says, "they cannot be valid at the same time in the same respect if the conflicting elements remain unresolved" (p. 216). These "conflicting elements" are two concepts that are central to all four theories, evolution and communication, and their underlying assumptions. Her challenge, then, is to prepare these two sets of theories for integration.

Modify Concepts. Van der Lecq (2012) begins with concepts and notes that though the four theories have different assumptions, they nevertheless share a common understanding of the concept "evolution":

> The [theory-based] insights into the problem of the primary function of language have one thing in common: they take the Darwinian theory of evolution as their point of departure. Although some of the details may be open to interpretation, the outlines of this theory are common knowledge and function as a point of common ground for all four theories. Although GG [grooming and gossip theory (social-brain hypothesis)] and RSt [relevance for status theory (political hypothesis)] have different assumptions regarding the question *how* language developed (continuous versus discontinuous), they use the term *evolution* in its biological sense, and agree that natural selection is its cause. (p. 216)

The theories of GG and RSt do not need to be modified because they already share a common understanding of the concept of "evolution."

Not so with theories of niche construction (NCt) and complexity (Ct) that have conflicting notions of evolution. As van der Lecq (2012) explains,

> NCt extends the Darwinian theory of evolution to include cultural developments in so far as they induce changes in the environment over generations. Ct appears to compare the evolution of languages with the evolution of other cultural constructs: Their evolution is not the result of natural selection but of a "natural" tendency to evolve towards more complexity. (pp. 216–217)

To bridge these opposing conceptions of evolution, van der Lecq (2012) turns to medieval philosophy:

> Medieval philosophers used to solve this kind of terminological problem, which they often encountered when they had to reconcile philosophical insights with religious truths, by making a distinction between a strict (or literal) sense and a broad sense of a term. In our case, we could solve the conflict by making a distinction between a strict sense of the term "evolution," meaning evolution-with-modification-by-natural-selection, and a broad sense for the evolution of knowledge, cultures, societies and institutions. The evolution of language towards more complexity would be an example of evolution in the broad sense. This technique of "distinguishing," as it was called in the Middle Ages, is probably best described as a combination of . . . redefinition and extension. (p. 217)

A second source of conflict among the theories was their different understanding of the concept "communication." Van der Lecq (2012) proceeds in a similar way, but in the opposite direction. She redefines communication to cover both its social and its cognitive function: "the cooperative sharing of information by producing knowledge in the minds of hearers." After all, she says, "it is hard to imagine that the exchange of social information succeeds if the partners in the conversation do not understand each other's messages" (p. 217).

Modify Assumptions. With the conflict over terminology resolved, van der Lecq's next challenge is to reconcile the conflicting assumptions underlying the theories: *Man is a social animal* versus *Man is a political animal*. She proceeds by raising two questions: (1) "Is sociality a prerequisite for communication (GG and NCt), or is it just a by-product of communication (RSt)?" (2) "Is man a social (GG and NCt) or a political (RSt) animal?" To answer these questions, van der Lecq (2012) draws on another perspective, that of computer science and the work of Luc Steels:

> Based on the evidence of language game experiments with robots, Steels (2008) argues that one of the factors that make communication successful is a strong social engagement ("joint attention") of speaker and hearer. Another aspect of

sociality is the ability to adopt the perspective of the other. Without this power of perspective reversal no communication system is possible, according to Steels. Thus, if we adopt Steels' conclusions, sociality is a necessary condition for the emergence of language. (p. 217)

Van der Lecq (2012) explains how Steels's work makes it possible to reconcile the conflicting assumptions that "man is a social animal" and that "man is a political animal":

Again, we may find some inspiration in medieval philosophy, this time in the work of Thomas Aquinas (14th century). Entrusted with the task of reconciling Aristotle's political philosophy with Christian values, he silently extended Aristotle's claim that "man is a *political* animal" to "man is a *political* and *social* animal." Aquinas' motive must have been that for him man is not only a citizen with civic duties, but also an individual with Christian duties. Applying this technique to our case, we could argue that humans use linguistic communication in the context of their family and friends mainly for social reasons, but on the level of the larger community they need language to make coalitions. Thus the common ground assumption is that man is a social and political animal. (pp. 217–218)

Though van der Lecq does not use the term *transformation*, we could imagine a continuum of communicative practices ranging from the social to the political: Her consensus assumption can thus be seen as an exercise in transformation.

For van der Lecq (2012), the result of STEP 8 is the creation of two common ground concepts and one common ground assumption. The common ground concept of "evolution" encompasses Darwinian as well as natural evolution. The common ground concept of "communication" refers to the cooperative sharing of information by producing knowledge in the minds of hearers. The common ground assumption is that humans are both social *and* political animals. The challenge of STEP 9, discussed in the next chapter, will be to construct a new theory that distinguishes between them when necessary.

Situation C: When Theories (Mostly) Address Different Phenomena

There are cases when different disciplinary authors have developed theories that complement each other more than they differ. That is, they address different sets of phenomena. We can then focus primarily on how the process identified in one theory sets the stage for the process identified in other theories, which in turn influences others. These cases generally involve multiple interactions and feedbacks. A map showing these interactions and feedbacks represents the common ground (see Wallis 2014 for a discussion of mapping diverse theories). In these cases, we apply the technique of "organization" (see Chapter 10).

From the Social Sciences. Foisy* (2010), *Creating Meaning in Everyday Life: An Interdisciplinary Understanding.* In this threaded example, Michelle Foisy studies the problem of how people create meaning in their everyday lives. She draws on the perspectives of two disciplines, psychology and philosophy (at least indirectly), because they both have produced widely accepted theories on "meaning" in a general sense. She also draws on a nondisciplinary pop psychology perspective popularized in the movie *The Bucket List.* From these, she identified four theories: Flow Theory and Goal-Setting Theory from psychology (more specifically, positive psychology, a recent offshoot of humanistic psychology); Logotherapy from a philosophically based form of psychoanalysis; and "Bucket List" Theory from a nondisciplinary source that was vetted by her instructor.

Foisy compares these theories using Szostak's "five *W*s," and developed Table 11.5 to juxtapose them and discover their similarities and differences. The table reveals minimal conflict among these theories: They all focus on the individual and overlap in terms of the range of decision-making strategies they promote. Foisy writes, "It is likely that when investigating the topic of meaning, all five decision-making strategies could be appropriate, and the *most* appropriate strategy might vary depending on the nature of the problem and the personality of the individual" (p. 16). The theories differ only in terms of *what* the individual is doing (acting, thinking, or reacting), and *where* and *when* the process occurs.

TABLE 11.5 ● A Summary of Flow Theory, Goal-Setting Theory, Logotherapy, and "Bucket List" Theory Using Szostak's "Five *W*s"				
	Flow Theory	**Goal-Setting Theory**	**Logotherapy**	**"Bucket List" Theory**
Who?	Individual	Individual	Individual	Individual
What?	Act	Act Think	*React*	Think
Why?	Emotional/intuitive	Rational Traditional Emotional/intuitive Value based Rule based	Rational Value based	Rational Traditional Emotional/intuitive Value based Rule based
Where?	Universal	Universal	Universal	*Culture specific*
When?	Change in one direction	Change in one direction	Change in one direction	*Cyclical*

Source: Foisy, M. (2010). *Creating meaning in everyday life: An interdisciplinary understanding.* Unpublished paper.

Note: Italics identifies conflicts among theories.

The challenge for Foisy is to create common ground among theories that conflict minimally and that focus on process. Her narrative first addresses how the theories differ:

> Logotherapy is the only theory that focuses on the individual *reacting*. . . .
> "Bucket List" Theory *is culture-specific* and *cyclical*, whereas the other three
> theories are universal theories that promote change in one direction. A third
> conflict that is not identified in Table 11.5 is the issue of semantics. Lastly, there
> are also gaps between all four theories that need to be addressed. (p. 18)

More problematic, she continues, was that these theories "all seem to address *different parts* of my research question": "Bucket List" Theory focuses on explicitly identifying life goals; Flow Theory focuses on the process of becoming engaged in an activity; Logotherapy discusses people's reactions to their current situation; and Goal-Setting Theory focuses on setting clear goals that the individual can reach (p. 18). The solution to the problem of how to create common ground in this case was for Foisy to map the problem as shown in Figure 11.3.

In this case, the map of the causal processes at work depicts the common ground.

FIGURE 11.3 ⬡ Meaning Construction Model: An Interdisciplinary Answer to the Question, "How Can People Create Meaning in Their Lives?"

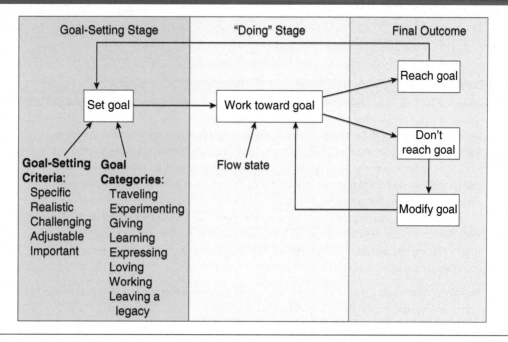

Source: Foisy, M. (2010). *Creating meaning in everyday life: An interdisciplinary understanding.* Unpublished paper.

Chapter Summary

This chapter concludes our discussion of STEP 8 (spanning Chapters 10 and 11) concerning how to create common ground when working with conflicting disciplinary concepts, assumptions, or theories. For concepts, common ground can be created either directly by modifying the concepts themselves or indirectly by modifying their underlying assumptions; for theories, modification involves using various strategies depending on whether (a) one or more theories already have a broader range of applicability than do the others (in which case theory extension can be employed); (b) none of the relevant theories is broad enough to facilitate theory extension but theories disagree about certain phenomena or relationships (in which case redefinition, extension, or transformation might be pursued); or (c) the theories illuminate complementary elements of the research question (in which case organization is used). The modification strategy used in Situation A involves identifying the theory that is already the most comprehensive (narrowly) interdisciplinary theory and then extending its range of applicability. The strategy used in Situation B depends on whether we are working with the theory's concepts or assumptions. Working with concepts involves finding concepts that have similar meanings in different theories, modifying the concepts so they have the same meaning across all the theories, and paying attention to disciplinary perspective when redefining concepts. Working with assumptions involves identifying the assumption(s) of each theory, determining which assumptions are shared by the other theories, and then modifying these assumptions so that all the theories in the set share a bedrock or common assumption. In Situation C, a diagram of the connections among different theories can represent common ground. Once these preparatory measures are complete, the insights and theories are ready for integration in STEP 9.

Notes

1. Concerning theories in the social sciences, William H. Newell (personal communication, February 15, 2011) comments, "Way too many theories in the social sciences ought to be restricted in their range of application but are not. Indeed, one of the contributions of interdisciplinarians critiquing disciplinary theories is to point out appropriate restrictions on their range. That is, disciplinarians—typically thinking that their discipline is most important—have a tendency to overreach when they theorize. Famously, psychologists for decades couched their theories in universalist terms, ignoring cultural (and, earlier, sex or racial) differences. It is by playing one discipline off against another that these needed restrictions in range of applicability become evident."

2. Mathematical modeling is also a method. Such a model can be seen as not only representing, but even clarifying and establishing the consistency of a theory.

3. Szostak (2004) appreciates that a minority of scientific effort is devoted to identifying the nature and internal processes of change in individual phenomena.

Exercises

Models

11.1 Create a model of one of these problems/topics that shows the variables and the relationship(s) among them.

    a. Litter

    b. Labor strike

    c. Child abuse

Variables

11.2 Using the same problem/topic chosen in Exercise 11.1, identify in your model independent and dependent variables.

Macro- and/or Micro-Level Variables

11.3 This chapter states that a set of theories may contain variables from different levels. Construct a model that shows the levels and the variables that may be operative at each level concerning one of the following problems/topics:

    a. Plant/factory/business closing

    b. Rising food prices

    c. Rising anti-immigration sentiment

Causality

11.4 To provide a full description of a theory and model it, add to the model developed in Exercise 11.1 and expanded in Exercise 11.2 a statement of the process that explains how change in the independent variable causes change in the dependent variable.

The Most Comprehensive Interdisciplinary Theory

11.5 Why is identity theory the most comprehensive interdisciplinary theory of the set of theories that Repko (2012) is working with?

Extending a Theory

11.6 How, exactly, does one go about extending a theory?

Situation B

11.7 If none of the theories in the set of theories you are working with has a broader range of applicability than do the others, but they conflict, what should you do?

Situation C

11.8 Why is the diagram the common ground in Situation C?

©iStockphoto.com/erhui1979

12

CONSTRUCTING A MORE COMPREHENSIVE UNDERSTANDING OR THEORY

LEARNING OUTCOMES

By the end of this chapter, you will be able to

- Define *more comprehensive understanding*
- Describe how to enhance your creativity
- Explain how to construct the more comprehensive understanding from modified concepts and/or assumptions
- Explain how to construct a more comprehensive theory from a modified theory

GUIDING QUESTIONS

What is a "more comprehensive understanding"?

How can you enhance your creative capability?

How can you develop a more comprehensive understanding?

CHAPTER OBJECTIVES

We have reached the point at which researchers engage in full integration (defined in Chapter 8). Having modified concepts, assumptions, and/or theories to create common ground, they focus on STEP 9, constructing a more comprehensive understanding or theory of the research question. The chapter defines the term *more comprehensive understanding*. It explores strategies for encouraging the creativity necessary for STEPS 8 and 9. It explains how to construct a more comprehensive understanding from modified concepts and/or assumptions. This discussion is applicable to the humanities, the fine and performing arts, and some applied fields where the focus of integration is directly on concepts and indirectly on their underlying assumptions. The chapter then explains how to construct a more comprehensive theory from a modified theory. This latter discussion is especially useful to those working in the natural sciences and social sciences where theoretical explanations dominate disciplinary discourse concerning a particular complex problem.

B. INTEGRATING DISCIPLINARY INSIGHTS

7. Identify conflicts among insights and their sources.

8. Create common ground among insights.

9. **Construct a more comprehensive understanding.**

   ○ **Be Creative.**

   ○ **Construct the more comprehensive understanding from modified concepts and/or assumptions.**

   ○ **Construct the more comprehensive theory from a modified theory.**

10. Reflect on, test, and communicate the understanding.

DEFINITION OF "MORE COMPREHENSIVE UNDERSTANDING"

A **more comprehensive understanding** is the result of the integration of *insights* and involves a new and more complete and perhaps more nuanced understanding than any of the disciplinary insights could produce.[1] It explains and expands upon the common ground identified in STEP 8 to answer the original research question. *The integration of concepts, assumptions, or theories is a means to the end of integrating insights into a more comprehensive understanding.* Concerning the term *more comprehensive understanding,* authors use a variety of other terms that have similar meanings, such as "complex understanding," "holistic understanding," "interdisciplinary understanding," "integrative understanding," "integrated result," "new meaning," and "interdisciplinary product." What one chooses to call the understanding that results from integration is a matter of preference.

Unpacking this definition deepens our understanding of it and of STEP 9:

- "More comprehensive" refers to the defining characteristic of the understanding or theory: that it combines more elements than does any disciplinary understanding or theory.

- "Understanding" reminds us that the purpose of integration is to better comprehend a particular issue, problem, or question; this will often allow us to better address real-world challenges.

- "Integration" refers to the process used to construct the understanding or theory.

- "Insights" get integrated, not the contributing disciplines themselves or their perspectives.

- "New" references the improbability of any one discipline producing a similar result, and that no one (other than the interdisciplinarian) takes responsibility for studying the complex problem, object, text, or system that transcends the disciplines.

- "More complete" refers to the understanding or theory that includes more aspects, facets, or dimensions than does a disciplinary understanding.

- "Nuanced" refers to the understanding or theory that includes more subtle distinctions than do disciplinary understandings.

To be clear, the more comprehensive understanding or theory is something that one must create from a set of modified concepts, assumptions, or theories. STEP 9 completes the process of integration by explaining how to construct this understanding or theory.

This chapter carries forward the threaded examples of student and professional work used in previous chapters. Each example briefly summarizes the common ground created in STEP 8 before demonstrating how the author constructs the more comprehensive understanding or theory, thus achieving integration. Completing this STEP, or at least attempting to do so, is what constitutes full interdisciplinarity.

BE CREATIVE

We have spoken throughout this book of "creating" common ground and "creating" a more comprehensive understanding. We stress here that the various strategies outlined in this and previous chapters do not lead inexorably to a more comprehensive understanding. When we describe examples of creating a more comprehensive understanding below, we emphasize how researchers built upon preceding STEPS in order to do so. Readers should not be misled into presuming that they will automatically achieve a more comprehensive understanding after performing all STEPS correctly. Instead, researchers need a creative act in which they combine the best elements of disciplinary insights in a novel and useful way. (*Note:* The standard definition of creativity refers to novelty combined with some sort of usefulness or value.)

It is easier in a textbook such as this to guide researchers through the careful conscious STEPS and strategies that we have described in preceding chapters than to guide them through a creative process. It is naturally much easier to guide conscious thoughts than to guide subconscious thoughts. Yet we know that creativity involves subconscious thought processes: We subconsciously make novel connections which then pop into our conscious thoughts. Szostak (2017a) makes some key points regarding subconscious thought:

- Subconscious thought processes take time and energy. We all experience moments in our life when the solution to some challenge we face pops into our conscious thoughts while we are walking in a park or taking a bath. It seems that our subconscious thought processes function best when our conscious mind is at rest. Researchers need to work hard in performing preceding steps, but they need to relax a bit if they hope to achieve creative inspiration.

- Subconscious thought processes may be biased. If a researcher does not want to face the opposition that novel insights often attract, their subconscious thoughts will avoid creativity. If a researcher does not really care about the subject they are researching, their subconscious thoughts may focus elsewhere. If the researcher is hoping for a particular outcome to their research, they may be unable to achieve creative results that point in other directions.

- Perspective taking encourages creativity. If we actively try to see a problem as multiple people would see it, we are far more likely to develop a novel way of seeing.

- Subconscious thoughts are often visual (since they cannot be in words). Buzan (2010) thus recommends a sort of mapmaking effort that goes beyond the sort of maps we recommended in Chapter 4: We should put every relevant concept on a piece of paper, draw connections where appropriate, and stare at the diagram for a while. We may then subconsciously appreciate connections that our conscious thoughts had failed to perceive.

- We often have difficulty describing smells and tastes in words. Likewise, we face challenges in describing how music and art move us. Some authors suggest, then, that certain aromas or kinds of music serve to shut off the conscious mind and trigger subconscious thought processes.

- Creative writing courses often urge potential writers to just write: If you write while limiting conscious guidance, you may find that you are subconsciously guided toward creativity. This technique may prove applicable to nonfiction writing also. (But we hasten to add that students should not turn in a paper that was drafted in this way without first subjecting it to careful conscious revision!)

- We should stress that subconscious creativity is only possible if one has first consciously identified and evaluated relevant insights and thought a bit about common ground. Creativity, then, is a function of both conscious and subconscious thought processes. Thus, the creative person needs to have both "synthetic" skills (to recognize connections that others have missed) and "analytical" skills (to gather and evaluate relevant insights), and (as we shall see) to subject the inspirations produced by the subconscious to careful evaluation.

Perhaps the most important points are these: All humans have creative potential (though there are surely differences across individuals), and this creative potential can be increased by learning and practice. We can teach students strategies such as those above that increase the likelihood of creativity. And we can give them assignments that encourage creative thinking. As with any other human talent, the more one practices creativity the more creative one will be. (One will develop not only creative skills, but a self-confidence that will encourage creative thinking.) We must move past the idea that there are a small number of creative geniuses in the world and that these come up with new ideas without much effort. We can all be creative, but we must prepare our subconscious thoughts by consciously gathering relevant information, and then pursue strategies that encourage creativity. (We will discuss later how to evaluate and clarify our inspirations, and how to persuade others; these are also important parts of the creative process.)

We have mentioned in earlier chapters that there is often a tradeoff between feasibility and creativity. For example, it is easier to connect insights from similar disciplines, but novelty is more likely if you seek insights from quite different disciplines. The tradeoff continues here. Creative strategies take time and effort and are not guaranteed to produce results. You must decide whether you are satisfied with a fairly boring outcome to your research or you seek something more.

CONSTRUCT THE MORE COMPREHENSIVE UNDERSTANDING FROM MODIFIED *CONCEPTS* AND/OR *ASSUMPTIONS*

Constructing the more comprehensive understanding is carried out by using the common ground created in STEP 8 to integrate the disciplinary insights. Students will need to demonstrate how the modified concept or assumption is *inclusive* of the others and fits best with the available evidence. The *form* this more comprehensive understanding takes varies. Narratives are probably more common (along with metaphors and images) in the humanities, whereas theories, models, or simulations are more common in the natural and social sciences.

From the Humanities. In these examples from the humanities, Lisa Silver and Mieke Bal construct their understandings from the insights they modified earlier. It is worth noting that these understandings are framed as practical applications of concepts that were redefined or extended when common ground was created. That common ground is summarized in each example.

Silver* (2005), *Composing Race and Gender: The Appropriation of Social Identity in Fiction.* Silver's purpose is to develop "a cohesive set of guidelines concerning the fictional appropriation of race and gender" (p. ii). Appropriation refers to a person's attempt to take possession of another person's identity for literary, artistic, or entertainment purposes (p. 7).

STEP 8 Summarized: Silver's challenge in STEP 8 was to create common ground among insights from psychology, sociology, and literature concerning the practice of appropriation. These insights defined appropriation in accordance with their particular disciplinary perspective and conflicted substantially. Silver decided that the best way to create common ground among these conflicting meanings of *appropriation* was to redefine it so that its broadened meaning could apply to the professions of acting and filmmaking not normally associated with it.

STEP 9: The broadened meaning of appropriation enabled Silver to integrate "the [literary] 'rules of fiction' with the social and historical implications of appropriation" (p. 11). This, says Silver, makes it possible

> to arrive at a cohesive set of guidelines concerning the fictional appropriation of race and gender. These guidelines require that authors have: (1) an awareness of the implications associated with appropriating a social identity, (2) a valid reason or desire to appropriate despite the implications, and (3) an ability to implicate themselves within the social identity being appropriated. (p. ii)

Silver's more comprehensive understanding of the concept of appropriation is appropriately nuanced. "Implicature," she says, "usually occurs through an awareness of one's own ethnocentric feelings, extended and meaningful contact with the group being represented, and hyper connective thinking" (p. ii). She adds, "Following these guidelines does not guarantee successful appropriation; however, examples of successful and ethical appropriation in fiction tend to meet these guidelines" (p. ii).

Bal (1999), *"Introduction," The Practice of Cultural Analysis: Exposing Interdisciplinary Interpretation.* Bal's challenge is to expose the fullest possible meaning of an enigmatic note or letter painted in yellow on a red brick wall in Amsterdam after World War II:

> Note
>
> I hold you dear
>
> I have not
>
> Thought you up (p. 2)

STEP 8 Summarized: Bal's strategy is to analyze or "expose" the meaning of this note or letter from three perspectives (although not disciplinary ones) simultaneously: from the perspective of the note or letter's author, from the perspective of the subject of the note or letter (i.e., the author's beloved), and from the perspective of one who is viewing the note or letter and pondering its meaning. She does this by extending the meaning of the verb *expose* from a specific and literal definition (i.e., to make a public presentation) so that it is broader, more ambiguous, and metaphorical. She combines the verb's meaning with the triple meaning of the noun *exposé:* "opinion" or "judgment," "the public presentation of someone's view," and/or "the public presentation of those deeds that deserve to be made public" (p. 5).

STEP 9: Bal uses the broadened meaning of the verb/noun *expose/exposé* to construct a more comprehensive understanding of the note or letter that focuses on its three aspects or dimensions. As an exposition, the note or letter

> makes something public, and that event of showing involves articulating in the public domain the most deeply held views and beliefs of a subject. . . . In publicizing these views, the subject of exposing objectifies, exposing himself as much as the object [i.e., his lost love]; this makes the exposition an exposure of the self. (p. 5)

Second, as an exposition, the note or letter is also an argument. The word *Note* has something to say about love and its public discourse, and it argues against oppositions such as public versus private, romantic belief in unique individuality versus the masses, and truth versus fiction.

Third, the note or letter is partly a metaphor of present culture, "a present that carries the past within itself. The present is a museum in which we walk as if it were a city" (p. 5). The note or letter's author brings the past into the present and engages the viewer in both.

From the Social Sciences. Delph* (2005), *An Integrative Approach to the Elimination of the "Perfect Crime."* In this example, Delph decided to work not with theories but with concepts because not all of the most relevant insights came from theories. Consequently, her understanding is conceptual and practical rather than theoretical.

STEP 8 Summarized: Delph's task is to create common ground between two sharply contrasting approaches to criminal investigation and profiling—that used by forensic science (which analyzes physical evidence), and that used by forensic psychology (which analyzes behavioral evidence). She creates common ground between them by redefining the meaning of the concept of "profiling" so that the broadened term includes the specialized kind of knowledge that criminal investigation, forensic science, and forensic psychology all privilege in their approach to profiling.

STEP 9: What is actually integrated, explains Delph, is "the unique knowledge possessed by each of these areas of expertise" and how they apply this knowledge to developing a profile of the perpetrator (p. 30). Criminal profiling, she writes, could achieve its greatest potential if profilers from forensic psychology and forensic science blended their analytical techniques and shared them with local criminal investigators. If achieved, this integrated approach would produce four likely outcomes: (1) quickly reduce the list of possible suspects, (2) predict the prime suspect's future behavior, (3) offer investigative avenues that have been overlooked by police, and (4) empower local law enforcement agencies to use these integrated profiling techniques themselves (p. 32).

CONSTRUCT A MORE COMPREHENSIVE THEORY FROM A MODIFIED THEORY

A theoretical proposition makes a truth claim and generally involves an argument about how one or more variables affect one or more others, though they may refer only to the internal workings of a variable. Here is an example of a proposition: "Children cared for in day care facilities as opposed to their own homes develop social awareness that serves them well upon entering public schools." This proposition also makes a causal argument: Day care facilities cause children to develop social awareness. **Causal arguments** examine the underlying cause for any particular situation or argument, and analyze what causes a trend, an event, or a certain outcome. Thus propositional integration cannot be distinct from causal integration. As noted elsewhere, our use of the word *cause* allows for multiple causes.

Causal or propositional integration refers to combining truth claims from disciplinary theoretical explanations to form an integrated theory—that is, a new proposition that is interdisciplinary and more comprehensive (Lanier & Henry, 2004, p. 343). Propositional integration, explain Paternoster and Bachman (2001),

> is a more formal effort [at integration] because it entails *linking the propositions* and not just the concepts of two or more theorists into a combined theory. . . . Rather than simply usurpation, a propositionally integrated theory must actually meaningfully connect or relate the propositions of different theories into the new theory [emphasis added]. (p. 307)

Critics of causal or propositional integration warn that it can produce an excessive number of variables, making the testing of the resulting integrative theory impractical because of the large sample size required (Shoemaker, 1996, p. 254). Henry and Bracy (2012) counter that proliferation of variables is not inevitable:

Interdisciplinarians may come to realize that some variables from one discipline are being used without much success to explain an aspect of crime that is much better explained by another discipline. Those variables would then be supplanted in the more comprehensive theory by variables from the other discipline or disciplines. (p. 265)

The key challenge in constructing a more comprehensive theory from a modified theory is to determine if the causal arguments or theoretical propositions advanced are logically related to each other.

Strategies to Achieve Causal or Propositional Integration

There are at least five strategies demonstrated to achieve causal or propositional integration and construct a more comprehensive theory:

1. *Sequential or end-to-end*, which implies a sequential causal order

2. *Horizontal or side-by-side*, which implies overlapping influences

3. *Multicausal*, wherein several variables combine to produce an effect

4. *Cross-level or multilevel*, wherein different behaviors occur on different levels

5. *Spatial*, wherein theoretical explanations of the causes or effects of problems are not distributed evenly in space

These strategies are the focus of the following discussion and are illustrated by examples of student and professional work. Noted along the way are the lessons that each strategy holds for interdisciplinary practice.

Sequential or End-to-End Causal Integration

Sequential or end-to-end causal integration links the immediate cause of the problem (in this case, crime) to a more distant cause, and then links that to an even more distant cause (Henry & Bracy, 2012, p. 266). This type integrates *fully complementary explanations of aspects* of the problem into an explanation of the problem as a whole: A causes B, and B causes C, as shown in Figure 12.1.

FIGURE 12.1 ● Sequential or End-to-End Causal Integration

A → B → C = Comprehensive explanation

Source: Henry, S., & Bracy, N. L. (2012). Integrative theory in criminology applied to the complex social problem of school violence. In A. F. Repko, W. H. Newell, and R. Szostak (Eds.), *Case studies in interdisciplinary research* (pp. 259–282). Thousand Oaks, CA: Sage.

Henry and Bracy (2012) describe the causal chain approach to constructing a more comprehensive causal theory of gang violence:

> For example, an arrest for gang violence might be the outcome of the following process of sequential causes over time. Biological defects at birth may lead to low IQ, which leads to learning disabilities in early childhood, which may lead to an inability to follow social norms, which may lead to group and institutional exclusion, which produces reduced self-esteem and [increased] alienation, which generates anger and hostility that results in affiliations with similarly alienated peers, which leads to delinquent peer or gang formation, which leads to law violation, which is reacted to by authorities producing criminal justice intervention and stigmatization and results in an arrest for gang violence. (pp. 266–267)

In this example, an arrest for gang violence is explained by a series of theoretical propositions drawn from several disciplinary theories including labeling theory, subcultural theory, learning theory, cognitive theory, and biological or genetic theory. In sequential integration, no one theory explains the whole sequence. Only when the relevant theories are linked end-to-end in a causal chain may they do so (Henry & Bracy, 2012, p. 267). The theories are complementary, but each captures only a piece of the broader puzzle. It is possible that C can also have some effect on A, making for a feedback loop. In such a case, the technique of organization can be employed to set the stage for the more comprehensive understanding. Care should be taken to ensure that terminology is used consistently throughout the causal process.

Horizontal or Side-by-Side Causal Integration

There are two distinct categories of horizontal or side-by-side causal integration and a spectrum of possibilities between them. One is where the explanations are fully complementary but focus on separate aspects of the complex problem. The explanations can be completely nonoverlapping or some of them may share common variables (but if so, treat these overlapping variables in the same way). Here the challenge of performing horizontal integration is to figure out the relationships among the different causal explanations (or at least some of their variables) using the strategy of organization to construct a more comprehensive theory. Different authors have emphasized different causal relationships, but—as in end-to-end integration—these can be added together. One author says A causes C, and another says B causes C; the interdisciplinary researcher adds the two explanations.

The other category (and the focus of the discussion below) is where the explanations have the same focus but are fully competing as shown in Figure 12.2.

FIGURE 12.2 ● Horizontal or Side-by-Side Causal Integration

Explanation A of Causal Relationship 1 + Explanation B of Causal Relationship 1 + Explanation C of Causal Relationship 1 = More Comprehensive Theory of Causal Relationship 1

Source: Henry, S., & Bracy, N. L. (2012). Integrative theory in criminology applied to the complex social problem of school violence. In A. F. Repko, W. H. Newell, and R. Szostak (Eds.), *Case studies in interdisciplinary research* (pp. 259–282). Thousand Oaks, CA: Sage. Page 268.

Competing explanations from different disciplines usually disagree on the nature of the causal relationship among those variables. They also disagree about which variables are salient. Agreement on the same causal relationship but disagreement on the mechanism by which it operates is found primarily in the natural sciences, or within a single discipline. (*Note:* Some sets of side-by-side theories may fall in between these two categories by being complementary in some ways and competing in others, and by focusing on different aspects to some extent, but also on some overlap between them.)

Horizontal or side-by-side causal integration is illustrated in the following example where competing theories explain the link between individual psychology and different types of sexual assault:

> Some acts of sexual assault may be explained by *self-control theory* which argues that such crimes are the result of a predisposition to sensation seeking and the desire for immediate gratification; but sexual assault may also be explained by *social learning theory* and *low self-esteem [theory]* as a result of the offender being a victim of childhood sexual abuse. Of course the question this raises is whether two acts defined by law as the same crime are actually the same or different behaviors. If different, then each theory would be explaining a different act, even though the law classified them as the same with regard to the harm and consequences for the victim [italics added]. (Henry & Bracy, 2012, p. 267)

An integrated theory, then, would combine these three conflicting causal explanations of sexual assault into a more comprehensive theory. *A general rule in performing horizontal integration is that the common ground must involve some idea of how the different causal explanations are related so that this relationship can form the basis of the more comprehensive theory.*

Multicausal Integration

This and the preceding strategy, though closely related, are treated as separate strategies because multicausal integration sees changes in the subject phenomenon as *the outcome of several different interdependent variables.* This strategy addresses the common case in which a more comprehensive explanation is achieved by combining different propositions or causal variables regarding the same *effect* or behavior. In this case, the explanations cannot simply be added together because the causal variables influence each other.

One cannot know whether different explanations are *complementary* or *conflicting or interdependent* until one has mapped the process under investigation. And then one needs to ask first, "Does Explanation 1 logically preclude Explanation 2?" and second, "Does Explanation 1 inform Explanation 2?" When a discipline assumes that only its variables matter, its conflict with another discipline's insights may be considerable but easy to deal with. *Only if there is some logical conflict—some element of Explanation 1 that precludes the operation of Explanation 2—do we really have a conflict that requires the search for a common ground.* Economists once stressed investment as the primary cause of economic growth, but later came to emphasize technological innovation. The integrative researcher can recognize the value in both explanations, and also that the biggest cause of growth may occur when new technology is embodied in new investment. We thus see the causes as *interdependent* but not really in conflict.

Multiple causation explains the phenomenon or behavior by creating a theory of sufficient generality that it incorporates multiple propositions or causal factors from relevant theories, each of which explains only a part of the phenomenon or behavior: A + B + C + D cause E, which in turn causes Y, as shown in Figure 12.3.

FIGURE 12.3   ●   Multicausal Integration

A (Control Theory = Weak Parental Attachment)
+
B (Social Learning Theory = Underachievement in School)
+
C (Conflict Theory = Conflict With Family)
+
D (Developmental Theory = Alienation From Family)
Cause E (Identification With Peers), Which
Causes Y (Gang-Related Delinquency)

Source: Henry, S., & Bracy, N. L. (2012). Integrative theory in criminology applied to the complex social problem of school violence. In A. F. Repko, W. H. Newell, and R. Szostak (Eds.), *Case studies in interdisciplinary research* (pp. 259–282). Thousand Oaks, CA: Sage. Page 268.

In their study, Henry and Bracy (2012) present theories on the causes of gang-related delinquency that are interdependent:

> Several different theories offer explanations of why adolescents engage in delinquent acts. *Control theory,* for example, has a key concept of parental attachment which is inversely related to delinquency (assuming parents are themselves moral and law-abiding). Combined with other elements such as low commitment to convention and lack of involvement in conventional activities, an adolescent may do poorly in school. *Conflict theory* as well as *developmental theory* argues that family conflict can arise from a variety of internal family dynamics or external societal pressures and can produce alienation of the adolescent from their family. Low commitment to convention can also lead to underachieving at school, which in turn can exacerbate conflict and alienation in the family. Social disorganization also contributes to the alienation of some adolescents from their parents, through a lack of identification, and *social learning theory* shows how alienated and underachieving students can identify more directly with underachieving peers, which in turn can create more alienation and further underachievement as well as lead to deviant and law breaking activity. (p. 268)

Since no one theory explains the causes of gang-related delinquency comprehensively, a more comprehensive theory is needed to identify the way the theories are interdependent. For example, parental attachment (from control theory) may be related to the dynamics of family conflict. An example of arriving at such a theory regarding another complex social phenomenon, suicide terrorism, is shown below.

From the Social Sciences and the Humanities. Repko (2012), "Integrating Theory-Based Insights on the Causes of Suicide Terrorism." In this threaded example, Repko focuses on the causes of suicide terrorism, arguing that this form of terrorism is best understood using the strategy of multiple causal explanation. His study is an example of a problem that has been examined narrowly through single disciplinary frameworks rather than through an integrative paradigm.

Repko's purpose is to construct a comprehensive theory explaining this phenomenon from a set of eight prominent theories on the causes of suicide terrorism. These are rooted in the disciplines of cognitive psychology, political science, cultural anthropology, and history. Of these, only one theory—identity theory advanced by Monroe and Kreidie (1997) and rooted in political science—attempts to cross disciplinary boundaries. Other than Monroe and Kreidie, no other author has recognized the value of taking an explicitly interdisciplinary approach to understand suicide terrorism or has attempted to develop a multiple causation theory that integrates these conflicting explanations.

Consequently, this issue is typically viewed through narrow disciplinary lenses. This produces fragmented and conflicting understandings, which lead to conflicting, unrealistic, and fragmented public policies.

Repko's strategy is two-pronged: to work from the bottom up with assumptions (in STEP 8) while working from the top down with theories (in STEP 9). The goal is a comprehensive theory that combines all salient causal factors addressed by the set of theories and that rests on a commonly held assumption.

This example shows how STEP 9 flows from STEP 8 (see Chapters 10 and 11). *A general rule when using the strategy of multiple causal explanation is that the common ground will have to be some shared understanding of the proximate cause* (e.g., seeing one's goal as sacred). In this case, the common ground could be expressed in the form of either a written narrative or a map depicting causal relationships.

STEP 8 Summarized: The challenge that Repko faced in STEP 8 was twofold. First, he had to identify a theory from the set of eight theories whose range of applicability was already broad but could be extended further. He identified the multiple variables or factors causing suicide terrorism. He then compared each theory-based insight and its theoretical explanation to a list of four causal factors to see which one(s) each theory included. The result of this evaluative process was that of all the theories under review, only three attributed the causes of suicide terrorist behavior to three of the four factors. Next, he narrowed the possible theories from three to one on the basis that Monroe and Kreidie's (1997) *identity theory* would require the least amount of modification or extending since it was already interdisciplinary, though imperfectly so. This decision was guided by the principle of least action, making sure that the changes would be the smallest possible.

The second challenge was to identify an assumption that was common to all of the theories and thereby reduce the conflict among them. Repko identified the assumptions of each theory and compared them to each other. By drilling down more deeply into their assumptions, he was able to reduce the number of conflicts among the theories to only a few. By drilling down even more deeply into these, he discovered a more basic assumption that all eight theories shared: that the terrorists considered their *goals* as "moral imperatives" and "sacred duties," and thus rational, as defined by Islamic fundamentalism (p. 145). This common ground assumption, then, served as the basis for the more comprehensive explanation of suicide terrorism constructed in STEP 9.

STEP 9: To develop a multicausal explanation of suicide terrorism involves taking two actions: broaden identity theory's range of applicability still further by using the technique of extension, and construct a comprehensive multicausal explanation that reflects the bedrock assumption created in STEP 8.

When working with theories, there are two possible approaches. The first approach is to piece together the "good parts" from each of the relevant theories to form a more comprehensive theory. The "good parts" are determined by their compatibility with the available empirical evidence. (*Note:* This applies as well to the selection of parts of theories.) This approach is appropriate when each theory is deeply rooted in the discipline from which it emerges because it avoids the problem of either/or thinking—of having to choose one theory and reject the others.

The second approach (and the one Repko uses) is to add parts from several theories to extend the explanatory range of one of the theories, in this case, identity theory. The integrative technique of extension is used when one of the theories already has parts that come from more than one discipline, as identity theory does. Best practice is to use the former approach and avoid the latter, *except when one of the theories is already interdisciplinary, though imperfectly so, as in the present case.* "Already interdisciplinary" means that parts of the theory are borrowed from other disciplines. What is missing, and what theory extension provides, is one or more additional parts—that is, causal factors—from the other relevant theories that will enable it to account for all the known factors causing the problem.

Identity theory, says Repko, possesses two strengths that make it ideal for extension:

> First, it already addresses three of the four factors—cultural identity, sacred
> values (i.e., religion), and power relations (i.e., politics). Identity theory includes
> religion by showing that the Islamic fundamentalist conception of religion
> is, in fact, an all-encompassing ideology that erases all lines between public
> and private. This theology-based ideology is based on religious writings and
> commentaries on these writings that are viewed by its most devoted followers as
> sacred, inviolable, nonnegotiable—and worth dying for. As a "sacred ideology"
> . . . Islamic fundamentalism redefines Western notions of politics and power.
> As Monroe and Kreidie [1997] show, politics for Islamic fundamentalists is
> subsumed under an all-encompassing religious faith, and is sought and exercised
> for the purpose of extending the faith. Achieving, maintaining, and extending
> power is a sacred duty that has priority over all other obligations, including
> family. Because the theory holds that the self is culturally situated, it is able
> to explain how culture (and religion is a key component of culture) influences
> identity formation. (p. 248)

A second strength compared to the other theories is that it requires the least amount of "stretching and pulling" to extend its explanatory range to include personality factors intrinsic to the suicide terrorist.

> Extending the theory so that it includes the causal factor of personality traits is possible because identity theory . . . is based on the psychological concepts of cognition and perspective. These concepts address personality traits in a way that is inclusive of the psychology theories already examined. Monroe and Kreidie [1997] use the concept of cognition in a developmental way, meaning that they are concerned to show how persons are influenced by factors external to themselves—namely culture and societal norms—rather than focusing on individual cognitive abnormalities as Post [1998] does to argue that some persons are psychologically predisposed to commit acts of suicide terrorism. (p. 248)

Monroe and Kreidie's (1997) application of the concept of "perspective" to explain suicide terrorist behavior is also helpful in this regard because it effectively delineates the options that terrorists perceive as being available to them. As Repko explains,

> The act of committing a suicide attack in the service of a fundamentalist conception of *jihad* "emanates primarily from the person accepting their identity which means that they have to abide by the tenets of their religion" ([Monroe & Kreidie] 1997, pp. 26, 36). By borrowing these concepts from psychology, identity theory offers an understanding of human behavior that is based on the interplay of mental constructs in tandem with the exogenous variables of politics, culture, and religion. (p. 149)

The extended theory, combined with its underlying assumption, enables Repko to construct a comprehensive multicausal explanation of suicide terrorism stated here:

> Suicide terrorism is caused by a complex interaction of variables that are both endogenous and exogenous to the individual. Endogenous variables include psychological predispositions and cognitive constructs developed over time; exogenous variables include the combined influences of culture, politics, and religion, with Islamic fundamentalism providing the perceptual framework that determines an individual's identity, motivation, and behavior. Suicide terrorists manifest varying degrees of rationality in pursuing goals that they consider "moral" and "sacred" according to a theologically based cost/benefit calculus. (p. 153)

We stress here that the more comprehensive understanding involves the interaction among variables identified by different authors.

Lessons for Interdisciplinary Practice. This example demonstrates the utility of first establishing common ground before attempting to integrate the major theoretical causal explanations and construct the multicausal explanation. Students may not need to identify a bedrock assumption underlying all of the theories as Repko does in STEP 8 to arrive at a more comprehensive theory. But more advanced students and practitioners are encouraged to do so for the simple reason that it anchors the comprehensive explanation in a bedrock assumption that is inclusive of all the theories, which in turn enhances the credibility of the explanation itself. This example also demonstrates the utility of theory extension as an effective way to integrate several conflicting causal explanations. Undergraduate students working with only a few theories will find it far easier to perform this integrative task than will graduate students, solo practitioners, and interdisciplinary teams who are working with numerous theories as Repko does.

Cross-Level or Multilevel Causal Integration

Cross-level or multilevel causal integration refers to different levels on which behavior takes place. There is constant interaction among levels, creating a fluid causal process with multiple feedback loops. This strategy differs from any other multiple causation strategy only if there are emergent properties at one level that act causally at another. Otherwise we are just talking about causal links between variables or theories at different levels. An "emergent property" is a characteristic "of a system that cannot be understood by reference to its constituent elements (phenomena and causal links) but only at the level of the system as a whole" (Szostak, 2009, p. 43). For example, consciousness may be an emergent property of human beings that cannot be understood by reference to the details of our biology (Szostak, 2004).[2]

Cross-level or multilevel causal integration involves combining theories from one level with theories from the other levels. Deciding which levels to integrate may require returning to STEP 1 of the IRP and rethinking the kind of complex problem to be studied (see Chapter 3). It may be that the problem, as first framed, is too broad and needs to be narrowed to make the study more manageable. Mapping the problem (if this was not done earlier) should reveal the levels involved. For example, factors contributing to the delinquent behavior of juveniles take place on multiple levels as shown in Figure 12.4.

FIGURE 12.4 ● Cross-Level or Multilevel Causal Integration

Level 5. Culture of materialism with limited moral direction

Level 4. The community and neighborhood

Level 3. Institutionalized educational practices

Level 2. Peer group pecking orders

Level 1. Relations in the family

A family is a primary group that operates at a micro level of society, whereas a school is an institution that operates at a more macro level in society. Relations in the family are part of the relations in school, relations among peers are part of the relations of school and family, and so on. In this situation, the more comprehensive theory of delinquency would be constructed from theories that focus on the different levels on which the behavior takes place and that account for the constant interactions between the family and each of the other levels. In other words, interdisciplinarians are urged to extend their gaze over the widest range of theories that bear on the problem—subject to the concerns regarding feasibility addressed in the preceding paragraph. As we have noted before, researchers are more likely to be creative if they extend their gaze widely.

The following examples of multilevel causal integration presented by Stuart Henry (2009) and Ria van der Lecq (2012) are from the natural and social sciences and demonstrate the use of theories to explain the phenomena they study. Interdisciplinary work in both areas has typically drawn on theories to produce more comprehensive explanations of complex phenomena. In each case, the resulting comprehensive explanation takes the form of a theory that has undergone extension.

From the Social Sciences. Henry (2009), "School Violence Beyond Columbine: A Complex Problem in Need of an Interdisciplinary Analysis." Henry argues that a comprehensive understanding of school violence requires examination of causation across multiple levels of society. Henry works with 12 different theories rooted in a cluster of social and behavioral disciplinary fields: economics, biology, psychology, geography, sociology, political science, Marxist philosophy, feminism, and most recently, postmodernism. His purpose is to combine these theoretical explanations into a single integrated theoretical model that has greater comprehensiveness and explanatory value than any one of its component theories (p. 1). What is crucial, says Henry, are the ways that these causal factors converge over time and culminate in a dramatic eruption of mass violence, or remain bottled up.

His study shows how STEP 8 flows into STEP 9. A *general rule when performing any integrative strategy*, including this one, *is to create common ground among the conflicting theories*. In this case, the goal is to seek a common ground proximate cause, which might be that there is some threshold level of aggravation that triggers school shootings. And then we show how a variety of causes interact to get us there.

STEP 8 Summarized: The challenge facing Henry in STEP 8 is to show that the incidents of violence in schools are not only cumulative between victim and offender, but also operate across multiple levels of society. Of the five levels he examines, only Levels 1 and 2 are presented here.

Level 1 violence includes much student-on-student violence, such as predatory economic crimes in which students use violence and threats to extract material gain from other students. It can also include physical violence, such as fighting between students because of disputes about girlfriends or boyfriends or because of verbal challenges to manhood, reputation, or insults. Much interpersonal violence occurs around proving issues of gender dominance and masculinity. Level 1 violence can also include the relatively rare but dramatic serious rampage homicides, where an individual attacks the whole school or collective elements in it, such as fellow students, teachers, and/or administrators in suicidal-homicidal explosions of hate, rage, or depression. However, the more recent evidence generally shows that when these incidents occur, the "individual" source of violence has previously been the victim of violence over time, and the extent of the extreme violent event is the outcome of the effects of reciprocal victimization at multiple levels rather than at just one. (pp. 11–12)

After the 1999 massacre at Columbine High School, says Henry, critically important Level 2 violence became evident when the role of collective policing of peer group pecking orders through violence, bullying, and exclusion became apparent.

Such peer group policing rejects those who are different from or less accomplished, good-looking, or datable than those at the top of the school social hierarchy. . . . Newman et al. (2004) point out that "among adolescents, whose identities are closely tied to peer relations and positions in the pecking order, bullying and other forms of social exclusion are recipes for marginalization and isolation, which in turn breed extreme levels of desperation and frustration" (pp. 229–230). . . . Indeed, [their] research reveals that four out of five offenders in rampage school shootings had been socially marginalized into outcast cliques (p. 239), and between half and three quarters of shooters (depending on the data source) had been victimized in a variety of ways, including being bullied, threatened with physical violence, persecuted, or assaulted or having their property stolen, for a considerable period of time, in many cases, for years, prior to the decision to commit mass violence (pp. 241–242). The authors say that "very few of these boys seem to meet the physical and social ideals of masculinity—tall, handsome, muscular, athletic, and confident" and that "in three out of five cases, the shooters had suffered an attack on their masculinity . . . by being physically bullied, mercilessly teased or humiliated, sexually or physically abused, or having recently been rejected by a girl." Unable to protect themselves from attacks on their manliness, they found a bloody way to "set the record straight" (p. 242). (p. 12)

After showing that the incidences of violence in schools not only are cumulative between victim and offender but also operate across multiple levels of society, Henry identifies a common ground proximate cause: There is some threshold level of aggravation that triggers school shootings.

STEP 9: The construction of a comprehensive multilevel causal explanation of school violence must reflect the complex and systemic nature of the problem. This explanation, says Henry, will enable us to better see the interconnected processes that produce school violence. Given the complexity of the problem involving multilevel causal factors, it is to be expected that stating this more comprehensive explanation will reflect this complexity. He begins with a propositional statement that reflects the theoretical explanations of the lower levels of causation examined in STEP 8: "Lower-level and more diffuse harm production can produce victims who, over time, can come to resent their victimization and react violently against it" (p. 17). Henry then explains how this process works, noting in particular that "social exclusion can occur in multiple ways that are both evident and concealed" and that such exclusion "can be the product of social hierarchies in the social networks of peers, bolstered by societal-cultural discourses of masculinity and violence and supported by school systems through their own hierarchies of power" (p. 17).

If the focus of Henry's analysis were limited to Levels 1 and 2, this causal explanation would be sufficient. As it is, however, there are three additional levels of causation (not described earlier) that he must integrate into his multilevel causal explanation if it is to be fully interdisciplinary. This means that he must take into account "the wider framing discourses" that operate in and across all five levels and explain how these function as a system that is conducive to producing violence. Henry continues his comprehensive causal explanation:

> Although we can examine the psychological processes and situational explanations for why students acted violently, we need to step outside of the microcontexts to explore the wider framing discourses of gender and power, masculinity and violence, and social class and race that produce social exclusion, victimization, anger, and rage. We need to see how these discourses shape the school curriculum, teaching practices, the institution of education, the meaning of "school," and its associated educational policy. How do parents, both in their absence and in their presence, harm the lives of students? We need to proactively engage in the deconstruction of hierarchies of power that exclude, and in the process create, a wasted class of teenagers who feel hopeless, whose escape from hopelessness is blocked, and whose only way out are violent symbolic acts of self-destruction and other-destruction. We also need to challenge the ways in which the economic and political structure of American society reproduces and tolerates hierarchies of exclusion and structural violence. This needs to go beyond cultural

causes of school violence to see how these cultural forms are integrated with structural inequalities. Any adequate analysis of school violence, therefore, has to locate the microinteractive, institutional practices and sociocultural productions in the wider political economy of the society in which these occur. Ignoring the structural inequalities of power in the wider system reduces the cause to local and situational inequalities of power, suggesting that policies can be addressed to intervene locally, such as at the level of peer subculture or school organization. Although these levels of intervention are important, they alone are insufficient. (pp. 16–17)

Lessons for Interdisciplinary Practice. The cross-level and multilevel causal strategy that Henry uses may involve conflicting theories at one or more levels of causation. If so, common ground will have to be created at each level where these theories conflict before a comprehensive theory involving all levels can be constructed. A further possible complication is how many variables there are among the levels and how much they interact. It seems that what Henry is saying here with his talk of thresholds is that school shootings are an "emergent property" of a system of interactions: It is hard to trace them to one particular cause; rather they must be traced to the reinforcing effects of multiple causes. Once identified, this emergent property will form the basis for the more comprehensive theory.

From the Natural Sciences. van der Lecq (2012), "Why We Talk: An Interdisciplinary Approach to the Evolutionary Origin of Language." In this example of cross-level or multilevel causal integration, van der Lecq's aim is to formulate a coherent and testable theory that answers the question, "What was the primary function for which language emerged?" Her strategy is to construct an interdisciplinary model that integrates theories (identified in Chapter 12). The model is cross-level because it integrates theories from one level with theories from other levels (introduced in Chapter 11). Since the model seeks understanding that is as comprehensive as possible, all theoretical levels need to be addressed simultaneously and show parallel interactive processes.

STEP 8 Summarized: Van der Lecq creates common ground—actually three layers of common ground—among relevant theory-based explanations of why humans talk. The result of STEP 8, she explains, is that we have created two common ground concepts and one common ground assumption.

The common ground concept of *evolution* is a complex concept, because it encompasses "Darwinian" as well as "natural" evolution. The new comprehensive theory will have to distinguish between the two aspects when necessary. The common ground concept of *communication* refers to the cooperative sharing of information by producing knowledge in the minds of hearers. The common ground assumption is that man is a social and political animal. (p. 218)

Van der Lecq notes that a further common ground concept (evolution) pulls these causes together. The result of STEP 8 is a common ground proximate cause: We need to talk because we need to communicate. Then she describes about how various phenomena interact to increase our need to communicate.

STEP 9: Rather than creating a new theory from pieces of existing theories, van der Lecq begins constructing her model from a "base theory" that already incorporates both/and thinking in an interdisciplinary sense, such as niche construction theory. She explains her selection of niche construction theory as the base theory as follows:

> Niche construction theory enables us to see linguistic communication as
> one of the activities that may modify more conventional sources of natural
> selection. Group living, the theory says, depends on social and communicative
> niche construction. A *social* niche is a subset of natural selection pressures in
> an evolutionary niche that are induced by interactions with other organisms
> in the group (e.g., grooming). The construction of a *communicative* niche
> depends on the ability of living organisms to exchange meaningful information.
> Communication with evolutionary consequences typically involves learning and
> cognition. The fact that only humans use speech to communicate is explained
> by assuming that human language is a component of human culture. Cultural
> practices change rapidly and with the generation of more cultural variants better
> ways of communicating were selected for. (pp. 218–219)

Van der Lecq extends the explanatory range of the base theory by connecting a second theory, grooming and gossip theory, to it. Adding this theory helps to explain one of the base theory's key elements (niche construction).

> Using Dunbar's grooming and gossip theory, we may add that better ways of
> communicating were also necessary for humans to communicate in larger groups
> with complex networks. This theory says that language evolved to replace social
> grooming when group sizes increased. Accepting the idea that grooming contributes
> to the construction of a social niche makes it possible to include grooming and
> gossip theory in niche construction theory. From grooming and gossip theory
> we also take the social brain hypothesis and the continuity hypothesis. Together
> these hypotheses propose that language (both learning and performance) evolved
> gradually out of primate social intelligence. Research in neurology corroborates this
> idea by showing that there is an overlap in the brain between the language centers
> and the location of social intelligence (Worden, 1998). (p. 219)

She then connects a third theory, relevance-for-status, to the base theory to help it explain how another of its key elements (communication) operates.

Accepting the common ground assumption that man is a social *and* political animal and using the common ground concept of communication as a social *and* cognitive activity, it is possible to partially integrate Dessalles' relevance-for-status theory. This theory sees language (performance) as a way of "showing off," its primary function being communication of salient features. Integration of this theory is possible if we accept the idea that the (primary) function of language may have been (and still is) different for different individuals. (p. 219)

Finally, van der Lecq connects a fourth theory, complexity theory, to explain the base theory's prediction that language becomes more complex over time:

Niche construction theory suggests that growth in cultural complexity made better ways of communicating necessary. In addition complexity theory explains *how* languages evolve in complexity over time, often becoming more complex, sometimes being simplified. We must consider the possibility that in the beginning there was no selection pressure for complexity, but that once language existed it quite "naturally" evolved in complexity like other cultural institutions and then more complex language abilities were selected for. This idea is compatible with niche construction theory, when we assume that complex languages make communication about complex environmental problems and the solution of these problems possible. Language can induce changes in the environment to which organisms will adapt. (p. 219)

Thus, based on the common ground concepts of evolution and communication and the common ground assumption that humans are social and political, it is possible to extend niche construction theory such that it includes Dunbar's (1996) grooming and gossip theory and the political hypothesis exemplified by Dessalles's (2007) relevance-for-status theory. "The new theory," says van der Lecq, "is the result of niche construction theory, which in turn is an extension of conventional evolutionary theory" (p. 219).

What remains is for van der Lecq to state the more comprehensive theoretical explanation in a way that answers the question that motivated the study: What was the primary function for which language emerged? The understanding, she says, must answer three subquestions: (1) Why did language emerge? (2) Why only with humans? and (3) Why (and how) did language evolve into such a complex system? The complexity of these questions is reflected in the complexity (and length) of her integrative model and resultant understanding:

The emergence of our natural predisposition to speech can be seen as the result of gene-culture interaction. Humans belong to the family of primates. For primates, living in social groups is advantageous because it increases the fitness of individual members of the group. In other words, groups construct a necessary social context (social niche) for existence. The formation and maintenance of a social group depends on communication. Our ancestors were able to handle social problems by grooming, but when group sizes increased and social relations became more complex, grooming became time consuming. Moreover, it was advantageous to be able to share "displaced" information. This explains the emergence of language, because with language more individuals can be reached with less effort. Moreover, human group sizes and the complexity of human networks require more than grooming for social bonding. Social bonding is the most likely *original* function of language.

Once the possibility of linguistic communication had emerged, it may have served multiple purposes for different individuals, including mating, childcare, tool making, and hunting. For language to have an evolutionary impact in the Darwinian sense, a certain amount of learning and cognition must have been involved. So, most probably, it evolved as a means of advancing the social and cognitive transmission of life-skills to young hominids. But for some (male?) individuals, it may also have played a role in the "coalition game." As we know from election campaigns, language can be used as a way of showing our talent for leadership and strong leaders benefit all the members of the group. Thus, the primary function of language may differ for individuals in different situations and is not necessarily the same as the original function for which language emerged, social bonding.

Like other cultural constructs and knowledge, languages have a "natural" tendency to evolve in complexity over time. Language complexity does not seem to be advantageous in itself, but when environmental problems become more complex, we need complex linguistic skills (scientific reasoning) to solve them. But when we use language for "social grooming," one or two simple words may be enough. (p. 220)

Lessons for Interdisciplinary Practice. Van der Lecq and Repko used a similar process to identify their "base theories": Both niche construction theory and identity theory are already interdisciplinary (though narrowly so) and incorporate thinking. Therefore, they were prime candidates for extension. In making their selections, van der Lecq and Repko adhered to the principle of least action, making sure that the required changes would be the smallest possible.

Spatial Integration

Spatial integration is a strategy for researching problems of cities and suburbs studied by the fields of urban history, the new suburban history, public affairs, and public policy. Spatial integration addresses problems related to where people live and how their geographical location (in city or suburb) affects opportunity, shapes perceptions, influences political views, and influences values. Problems involving political, economic, social, and ecological conflicts because of where people live require mapping how causal processes operate across space. They also require understanding that disciplines may differ in their perspective on the spaces they typically study (Connor, 2012, p. 53). (*Note:* Spatial integration is really a type of horizontal integration—perhaps sometimes multicausal integration—but merits special treatment due to the spatial element.)

From the Humanities and the Social Sciences. Connor (2012), "The Metropolitan Problem in Interdisciplinary Perspective." Michan Connor combines spatial integration with the integrative strategy of theory extension to explain the process of metropolitan formation in the United States. His objective is to construct a comprehensive theory that fully integrates "explanatory constructs" (i.e., theories) of history, metropolitics, public choice theory, and critical legal studies concerning the complex problem of metropolitan formation (p. 71). Connor recognizes that cities and suburbs are often studied separately but can only be fully understood in relation to each other.

STEP 8 Summarized: Connor's task in STEP 8 is to leverage the theoretical insights of history, public choice theory, metropolitics, and critical legal studies by identifying a theory, already interdisciplinary but narrowly so, that could be extended "across disciplinary domains to address the spatial assumptions of each perspective," which he summarizes here:

> Although particular academic disciplines may use the concepts of spatiality and place overtly, and others implicitly, traditional disciplinary approaches fail to integrate the multiple, overlapping, and occasionally conflicting elements of the essential relationship between places and social life. In the U.S. context, suburbanization and home ownership have connected individuals and families to historical changes in social class, racial differentiation, and political interest. However, historians have typically been limited by a spatial framing of social processes that seldom transcends the presumed division between urban and suburban places. Metropolitics and public choice theory approaches in the social sciences supply frameworks for assessing social processes that cross municipal boundaries but provide starkly different assessments of which social mechanisms—politics or the market—best account for differentiation between metropolitan places. The field of critical legal studies provides a systematic critique of municipal boundaries as political, symbolic, and economic dividers between places and people. Because research in this field focuses very diligently on the processes by which legal principles of local control have evolved in legal doctrine, critical legal studies points to the need to consider the way that policy or legal doctrines have emerged in particular historical contexts and influenced the course of events in those times. (p. 84)

In spite of what he sees as its shortcomings, Connor selects Lefebvre's (1991) theory of the social production of space for modification because of its "mutually reinforcing patterns of ideas (ideologies, values, ideals), political power (over institutions and over the validation of particular ideas), and social practices" (p. 74). One of the tests he uses for gauging the appropriateness of a theory is that it must recognize (or be compatible with) "historical contingency": the way things happen to play out, and how this constrains and shapes what is possible or at least likely thereafter.

STEP 9: Theory extension for Connor involves "extending the theoretical notions of the spatial and temporal nature of social life into the domain of multiple disciplines where it is differentially acknowledged" (p. 72). Here, he explains how the extended theory increases its explanatory range or "pays attention" to additional factors:

> Integrating the theoretical insights of these fields also requires expanding the characteristic methods of local history. A more spatially and analytically open case study methodology employed here addresses the qualities of individual places, but it also pays attention to the nonlocal factors, like ideology, law, and public policy, affecting metropolitan places. In addition, it pays attention to the ways in which local places are not separate but affect each other. (p. 84)

The basis of Connor's study is his own primary research on how the theory of public choice informed the Lakewood Plan, which shaped the formation of modern-day metropolitan Los Angeles County. Connor's more comprehensive theoretical explanation of why Los Angeles County has developed in the way it has is in two parts. The first part explains the effect of the Lakewood Plan:

> The Lakewood Plan . . . shaped the metropolitan area, but that influence depended on the very particular qualities of Lakewood as a symbol of the positive values of local autonomy. Public choice theory ideas, because they have been influential in governing as well as in scholarship, merit special scrutiny in this framework. The history of metropolitan formation in Los Angeles demonstrates the partial validity of these ideas. Under the conditions of the rule of homes, suburban residents in the Lakewood Plan cities acted in ways that were consistent with rational self-interest: advocating for low taxes, demanding better services, and seeking to exclude neighbors or land uses that threatened the value of property. Under the political and cultural conditions that have prevailed in the United States and in Los Angeles in particular, it has also been rational for relatively affluent suburban areas to reject common obligations with other metropolitan places. However, this rationality is neither an inherent part of human nature nor a complete explanation for the political and social patterns of metropolitan formation. Rather, it was formed in the context of historically

produced legal, social, and political frameworks that made certain behaviors, orientations, and outlooks productive. . . . Public choice theory ideas have, in fact, been more important as shapers of public life than descriptors of it. (p. 84)

In the second part of his comprehensive explanation, Connor particularizes the "behaviors, orientations, and outlooks" resulting from these frameworks of economic incentives and political justifications:

These frameworks of incentives and justifications are not inevitable, but they are powerful, and they have been constructed and reinforced over time through the political rationality they support. In Los Angeles County, homeowners, local and state officials, judges, realtors, bankers, and builders produced new places in the county, attached certain social privileges to favored places, and protected those privileges against claims that they were unfair or unjust. These agents worked through institutional and cultural channels at local, metropolitan, and state scales to create what was visible in hindsight: a white middle class, multiple municipal jurisdictions that could insulate portions of that class from taxation or service obligations, and perhaps most important, the sense that this division was a legitimate component of politics that reflected not exclusion or segregation but "choice." (pp. 84–85)

Connor concedes that because Los Angeles County has its own particular history, making generalizations about other metropolitan areas is problematic. Even so, he contends that it should be possible to use a case study method involving Los Angeles to propose a methodology for evaluating the particular factors influencing metropolitan development elsewhere.

Metropolitan areas nationwide exhibit comparable political division, social inequality, and privilege for affluent homeowners, regardless of the particular path of development, and researchers should devote attention to the ways that institutions, ideas, and experiences (in their particular local incarnations) have interacted to shape development. It is beyond the scope of this essay to recommend policy changes that can promote greater metropolitan equity, though metropolitics scholars offer many suggestions. Rather, by demonstrating that inequality is part of a long process of metropolitan formation with roots in many dimensions of social life, this essay suggests the limits of narrow policy solutions, which are likely to provoke intense opposition and unlikely to secure political legitimacy in the current metropolitan climate of the United States. As Gerald Frug (1999) observes, a fragmented metropolitan area is "perpetuated by the kind of person this fragmentation has nurtured" (p. 80). Reform proposals for regional equity are unlikely to succeed without substantial efforts to unravel dominant values and ideals about place and community in metropolitan America, ideals which have coalesced around the priority given to local interests and the rule of homes. (p. 85)

Connor's integration and more comprehensive theory produces an analysis that is spatially integrated at the metropolitan (or regional) scale through its attention to the legal, political, and cultural dimensions of relationships among metropolitan places.

Lessons for Interdisciplinary Practice. Connor's study demonstrates that using a combination of integrative strategies may be necessary when working with highly complex processes such as metropolitan formation. Critical to such integrative efforts is first recognizing, as Connor does, the inability of disciplinary theories and their corresponding methodologies to explain the system in its complexity. This involves developing a base theory that can integrate these theories of spatial framings of metropolitan issues, and integrating these with their legal, political, and cultural dimensions.

Chapter Summary

This chapter explains STEP 9, constructing a more comprehensive understanding or theory. It defines both terms, unpacks their meaning, and explains how this STEP relates to previous STEPS in the integrative process. Since STEPS 8 and 9 both require creativity, the chapter began by discussing the nature of creativity, asserting that this capacity can be learned and practiced, and outlining some important strategies for enhancing creativity.

STEP 9 reflects the two intellectual tendencies within the academy: those disciplines that work primarily with concepts, and/or assumptions, and those disciplines that rely on the development of theories to explain the phenomena they study. Consequently, STEP 9 reflects this bifurcated approach to knowledge formation by laying out two pathways by which the understanding may be achieved. The first pathway is applicable to the humanities, the fine and performing arts, and some applied fields where the focus of integration is directly on concepts and indirectly on their underlying assumptions. In these contexts, achieving full interdisciplinarity involves consciously choosing to construct an understanding that is comprehensive and nuanced.

The second pathway is applicable to the natural and social sciences, sometimes to the humanities, and to some applied and multidisciplinary fields where the focus of knowledge formation is on the development of theories to explain the phenomena of interest. Six strategies demonstrated to achieve integration and construction of a more comprehensive theory are identified and illustrated by student and professional work. The examples by Repko (2012), Henry (2009), van der Lecq (2012), and Connor (2012) illustrate how authors creatively apply these strategies to achieve integration. Once the more comprehensive understanding or theory is constructed, the final task is to reflect on it, test it, and communicate it. This is the subject of Chapter 13.

Notes

1. The more comprehensive understanding is similar to Boix Mansilla's (2006) "complex explanation" where "aspects of the phenomenon, typically studied by different disciplines, are considered in dynamic complementary interaction" (p. 14). For a detailed discussion of the term *understanding* from a social science perspective, and an explanation of the difference between the *verstehen* (and the more recent interpretive) approach and the predictive approach, see Frankfort-Nachmias and Nachmias (2008, pp. 10–11).

2. For a detailed discussion of emergent properties and their relation to a system of causal links as a whole, see Szostak (2009, pp. 43–45).

Exercises

Defining Terms

12.1 How is the definition of *more comprehensive understanding* consistent with and an extension of the definition of *interdisciplinary studies* as a whole (see Chapter 1)?

Constructing Understandings

12.2 If you are working with concepts, explain what the understandings by Silver (2005), Bal (1999), and Delph (2005) have in common.

12.3 If you are working with theories, decide which strategy for achieving causal or propositional integration is most appropriate to the problem. Create a figure depicting the various causes, including feedback loops where these may exist.

12.4 How many levels of causality can you identify concerning the problem of the rising cost of university education, and how should they be labeled? Is there an "emergent property" involved?

12.5 What must Henry (2009) do to make his understanding of school violence fully interdisciplinary?

12.6 How is van der Lecq's (2012) "base theory" similar to Monroe and Kreidie's (1997) identity theory in Repko's (2012) example?

12.7 How would you apply spatial explanation to the complex problem of constructing a high-speed rail system somewhere in the countryside near or where you live?

Creativity

12.8 How will you seek to develop your creative capabilities?

Image by Gerd Altmann from Pixabay.

REFLECTING ON, TESTING, AND COMMUNICATING THE UNDERSTANDING OR THEORY

LEARNING OUTCOMES

By the end of this chapter, you will be able to

- Reflect on the more comprehensive understanding or theory
- Test the quality of interdisciplinary work
- Identify four ways to test the more comprehensive understanding or theory
- Identify ways to communicate the results of integration or theory

GUIDING QUESTIONS

Why and how should you reflect upon your more comprehensive understanding or theory?

Why and how should you test your more comprehensive understanding or theory?

How could you communicate your more comprehensive understanding or theory to diverse audiences?

CHAPTER OBJECTIVES

STEP 10 of the interdisciplinary research process (IRP) reflects on, tests, and communicates the more comprehensive understanding or theory constructed in STEP 9. This chapter discusses the importance of reflecting on both the new understanding itself and the utility of the IRP in its development. Since interdisciplinarians recognize that no method is perfect, this chapter advocates using four complementary ways of testing the understanding. It also explains how to test the quality of interdisciplinary work itself in a way that takes into account the literature on cognition and instruction. Finally, this chapter identifies ways to communicate the results of integration. It challenges readers at all academic levels to communicate the understanding by feeding back integrative knowledge into the scientific community.

↻ **B. INTEGRATING DISCIPLINARY INSIGHTS**

7. Identify conflicts between insights and their sources.

8. Create common ground between insights.

9. Construct a more comprehensive understanding or theory.

10. **Reflect on, test, and communicate the understanding or theory.**

   ○ **Reflect on the more comprehensive understanding or theory.**

   ○ **Test the quality of interdisciplinary work.**

   ○ **Test the more comprehensive understanding or theory.**

   ○ **Communicate the results of integration.**

REFLECT ON THE MORE COMPREHENSIVE UNDERSTANDING OR THEORY

Research of any kind calls for reflection. **Reflection in an interdisciplinary sense** is a self-conscious activity that involves thinking about why certain choices were made at various points in the research process and how these choices have affected the development of the work. Interdisciplinarians, Szostak (2009) suggests, may well have to be more reflective than disciplinary specialists for this reason: Interdisciplinarians must be able to demonstrate the ability to enter the specialized scholarly discourse (i.e., disciplinary literatures) on the problem in an interesting and persuasive manner, and explain those elements of their own research that may cause concern among the intended audience (p. 331). Brassler and Block (2017) and von Wehrden et al. (2019) are among many authors who recognize the critical importance of reflection in interdisciplinary research and education.

While reflection should occur throughout the research process, it is necessary at the end of the research process. Szostak (2009) identifies four sorts of reflection that are called for in interdisciplinary work: (1) what has actually been learned from the project, (2) STEPS omitted or compressed, (3) one's own biases, and (4) one's limited understanding of the relevant disciplines, theories, and methods (pp. 331–334).

We spoke in Chapter 12 about how to encourage creativity. Recall that creative ideas are both novel and useful. Many novel ideas may emerge from the subconscious mind into the conscious mind, but not all of these ideas will prove to be useful. The interdisciplinary researcher needs always to reflect on these ideas and not allow pride in the novelty of an idea to blind them to its shortcomings. Nor, though, should the researcher act in haste.

As we shall see, no novel idea is likely to be perfect, and thus we should be careful not to reject an idea at the first indication that there is a problem with it.

Reflect on What Has Actually Been Learned From the Project

Students at all levels need to reflect on what they have learned from the project, including its probable benefits for their careers and/or for society (see Box 13.1). The focus should be on both the *contents* of the project (i.e., what was learned about the subject of the project) and the *process* used. By "process" is meant how the project was organized, how it proceeded, positive and negative experiences, and the utility of the research process described in this book. For example, if the research project was a group project, participants can reflect on their goal to integrate different knowledge cultures and decide the ways and the extent to which this goal was achieved, as well as the ways that integration was incomplete or its results were less than fully satisfied. Participants can also reflect on when and under which circumstances common learning took place and what can be done to stimulate such moments. Another point for students to reflect on is how disciplinary and interdisciplinary knowledge were shared, and how that sharing influenced the development of the resultant understanding. Boix Mansilla, Duraisingh, Wolfe, and Haynes (2009) suggest that students ask, "Do the conclusions drawn indicate that understanding has been advanced by the integration of disciplinary views?" This question essentially asks, "Was it worth the effort? Did it yield a new, richer, deeper, broader, or more nuanced understanding of the problem?" (p. 345). With these questions, we also interrogate the content of our understanding.

BOX 13.1

Preparing for a Job Interview

Many employers are far more familiar with the traditional disciplines than with interdisciplinary programs. Yet an interdisciplinary education encourages many skills (analytical, synthetic, creative, teamwork, and more) that employers, when surveyed, say that they are looking for (Repko, Szostak, & Buchberger, 2020). So, the student can turn their interdisciplinary education to their advantage if they reflect on what they have learned and how that may be of use to a prospective employer. It is thus a useful exercise at this point to reflect on how you might describe (in just a few minutes) your interdisciplinary education in a job interview (or in a cover letter for a job application). This might be a very useful class discussion or a conversation among friends. You might speak to

(Continued)

(Continued)

- Why you chose an interdisciplinary education
- The importance of interdisciplinarity in the contemporary world
- The skills and attitudes associated with interdisciplinary studies (see our discussion of learning outcomes associated with interdisciplinarity below)
- The critical analytical and creative skills you have honed in particular
- The steps and actions you have learned to perform
- Examples of interdisciplinary analyses that you have performed

If you know what kind of job you intend to seek, you can tailor your "pitch" to the particular employer: What sort of challenges do they face and what kind of skills do they most need? If, as is generally the case, teamwork and leadership are called for, you might especially speak to what you have learned from any group projects you have undertaken.

Reflect on STEPS Omitted or Compressed

Some students may have omitted one or more STEPS for various reasons, or may not have been able to perform each STEP in the way described in this book. Or, they may have compressed some STEPS. For example, some programs and courses may view STEP 2, justifying using an interdisciplinary approach, as something that can be folded into STEP 1 or eliminated entirely. Others may compress STEP 3, identifying relevant disciplinary perspectives and their insights and theories, with STEP 4, conducting the literature search. Still others may compress STEP 5, developing adequacy in each relevant discipline, with STEP 6, analyzing the problem and evaluating each insight into it. And still others may view STEPS 8 and 9 as two parts of a single STEP.

NOTE TO READERS

For most complex problems, even graduate students and professional research teams do not always completely perform each STEP. For example, there is always more literature to read. Thus even advanced researchers should reflect on what they did not do. If the researcher decides to omit or change the order of certain STEPS, then these omissions or alterations to the research strategy should be explained and their possible cumulative effect on the final product described. Typically, these effects surface in glaring ways near the end of the research process. For example, failure to develop adequacy in each relevant discipline may result in certain theories and methods being overlooked or misunderstood.

If the more comprehensive understanding includes (as is often the case) some suggestion of a change in public policy, the researcher should reflect on the possibility of side effects. Sadly, many policies designed to further one policy goal turn out to have negative effects on other policy goals (e.g., conflicts between environmental and job-creation goals, though these can be exaggerated). In terms of interdisciplinary research, such side effects involve causal relationships that were outside the scope of the original research. It is impossible for any piece of research to fully encompass all possible side effects, but it is usually quite feasible to at least reflect on what these might be.

Reflect on One's Own Biases

Researchers at all levels should reflect on their own biases that they consciously or unconsciously bring to the problem. Szostak (2009) makes a compelling case for why reflecting on one's own biases is necessary and should even be mandatory in interdisciplinary work:

> A key guiding principle of interdisciplinary analysis is that no piece of scholarly research is perfect. If we accept that no scholarly method can guide a researcher flawlessly towards insight, then it follows that scholarly results may reflect researcher biases. This does not mean that results only reflect such biases, as some in the field of science studies have claimed. But it does mean that one way of evaluating the insights generated by research is to interrogate researcher bias. (p. 331)

Szostak makes these two points: (1) Researchers should reflect on their own biases concerning the problem, and (2) questioning researcher bias is one way of evaluating the researcher's insights.

Interrogate One's Own Biases

Practically, students should interrogate their own biases. One way is to ask these questions:

- "Why was I drawn to the problem in the first place?" Typically, students—as well as practitioners—select a topic that they have strong views on or personal experience with. For example, one student in a senior capstone course wished to examine the causes of alienation that teens often experience. She admitted that what drew her to the topic was her own experience with alienation and now, as a mother of a teen, watching her daughter experience the same struggle. Selecting a research topic on this basis need not skew the research process or its result, but might lead one to seek certain answers.

- "Did I select expert views based on whether or not I agreed with them?" Selecting insights on this basis is a common trap that many students fall into. They mistakenly believe that interdisciplinary research is similar to research papers that they have written for other courses in which they are urged to argue a particular viewpoint and support it with expert testimony. This mode of "research," of course, is antithetical to interdisciplinary work, which involves taking into account expert insights regardless of whether they run counter to the student's point of view. As noted elsewhere in this book, the role of the interdisciplinary researcher is similar to that of a marriage counselor who is attempting to bridge differences, not ignore them.

- "Did I select insights based solely or primarily on my familiarity with particular disciplines and their literatures?" The answer to this question is usually yes for undergraduates because they are generally limited to using two or three disciplines to make the research project manageable. The answer may be yes or no for graduate students, depending on course requirements and the scope of the paper or thesis. Regardless of the level, students should be able to justify the disciplines and their insights chosen and articulate in STEP 10 how any omission or limitation has affected the final understanding. For members of research teams and solo practitioners, insights should be chosen without regard for familiarity with any particular discipline and its literature.

If you find upon reflection that you were biased, you should revisit your analysis and assess whether your conclusions (more comprehensive understanding) should be adjusted.

Biases can of course affect not only a particular research project, but your entire approach to life and all of the complex challenges you will face in it. Keestra (2017) urges us all toward *metacognition*. We all necessarily develop mental representations in life that guide our understandings of phenomena, causal relationships, and people. These mental representations allow us to act in new situations without having to reflect deeply on what we are doing. Yet of course there is then a danger that our mental representations are biased, and we ignore new information that might guide us toward better mental representations. We need to balance our need to function in the world with openness to new information. Disciplinary experts are particularly likely to ignore insights from other disciplines that might change their mental representations. The antidote to bias is to consciously reflect on our mental representations and whether these reflect all relevant information. Keestra suggests asking why we do what we do, whether our existing mental representations are achieving our goals, and where we might look for information that might lead us to alter our mental representations.

Check One's Work for Biases

Students should check their work for bias. In the following composite example, an undergraduate student reveals personal bias concerning the qualities of an ideal manager in this opening paragraph to the paper's introduction.

The Ideal Manager: An Interdisciplinary Model
Good managers must have at least two essential qualities. They must be fully knowledgeable of the products that their unit is responsible for, and they must understand that a single error can produce unforeseen consequences resulting in a very dissatisfied customer. Constantly, one sees reports of higher management going terribly bad. Whether it's a case of lying, cheating, or stealing, the number of unfaithful or unreliable managers keeps increasing.

The problem with this introduction is that the student has already decided (based on her experience working under several poor managers) that an ideal manager should have these particular qualities *even before* conducting the in-depth literature search. The student's bias was also revealed (not surprisingly) in the literature search itself by preferring insights that reflected her preconceived notions of an ideal manager.

Reflect on One's Adherence to a Theoretical Approach

Reflecting on one's own bias should include reflecting on one's own epistemological attitudes (as discussed in Chapter 2). For example, it may be that one prizes personal freedom over structure and therefore objects to the very notion of following an interdisciplinary research process of any kind. In this case, one can reflect on these questions:

- How has my experience using the process changed or tempered my bias?

- Have I become more self-consciously interdisciplinary, and if so, how has this happened?

- Have I concluded that interdisciplinary studies needs some structure in order to dissuade scholars from using superficial forms of interdisciplinary analysis?

- Am I more aware that scholarly results reflect some combination of the influence of personal biases and of external reality?

Reflect on One's Limited Understanding of the Relevant Disciplines, Theories, and Methods

Students should reflect on the simple fact that all humans have limited perceptual and cognitive capabilities. Even if one has achieved adequacy in understanding the relevant insights, theories, and methods pertaining to the problem, one can hardly claim to have practitioner-level understanding of every insight produced and every theory and method used by the authors who have written about the problem. It is appropriate, therefore, to do what practitioners often do in their concluding remarks: State the ways in which you may have overestimated or underestimated the importance of some insights, theories, and methods and therefore the ways your results may reflect this possibility. Academic writing, say Boix Mansilla et al. (2009), is strengthened when authors are aware of the limitations of their work (p. 346).

Reflecting on one's limited understanding extends to one's awareness of the limitations and benefits of the contributing disciplines and how the disciplines intertwine. Interdisciplinary work, say Boix Mansilla et al. (2009), requires "a deliberate intertwining of disciplinary perspectives" and careful "evaluation of disciplinary insights for their potential contributions and limitations" (p. 345). Instead of dividing students into undergraduates and graduates, Boix Mansilla et al. (2009) categorize students according to their level of reflection. Novice-level students, they say, may limit their reflection "to a pro forma critique such as 'More research is needed on this topic.'" "Apprentice-level students" they say, "may weigh the merits and limitations of the selected disciplines in turn [i.e., serially] against alternative selections available" (pp. 345–346). Novice-level students are satisfied to name the disciplinary perspectives used and may only fleetingly refer to how each could potentially limit or advance the argument. But to move from a novice to an apprentice level, students should explicitly consider the limitations of the relevant disciplinary understandings and explain how using an interdisciplinary approach has enlarged and strengthened understanding (pp. 345–346).

TEST THE QUALITY OF INTERDISCIPLINARY WORK

Before we outline strategies for testing, it is useful to recall the purposes of interdisciplinary research and education. A pundit has compared testing the quality of interdisciplinary work to educated people being forcibly marched to the Chinese countryside during Mao's Great Cultural Revolution to dig onions, and then spending the rest of their time confessing how much they have learned by digging onions. "The biggest challenge to interdisciplinarity, particularly at the undergraduate level, is the lack of generalizable methods for judging interdisciplinary education and its direct impacts on student learning" (Rhoten, Boix

Mansilla, Chun, & Klein, 2006, Executive Summary). Veronica Boix Mansilla (2005), principal investigator of the Interdisciplinary Studies Project at Project Zero, Harvard Graduate School of Education, points to a related problem: the "lack of clarity" about interdisciplinary learning outcomes and "indicators of quality" (p. 16).

Learning Outcomes Claimed for Interdisciplinarity

The learning outcomes typically claimed for interdisciplinarity include tolerance of ambiguity or paradox, critical thinking, a balance between subjective and objective thinking, an ability to demythologize experts, increased empowerment to see new and different questions and issues, and the ability to draw on multiple methods and the knowledge to address them (Cornwell & Stoddard, 2001, p. 162; Field, Lee, & Field, 1994, p. 70). Notably, interdisciplinarians substantially agree that integration is "fundamental to any 'successful' interdisciplinary program" and consider "the ability to synthesize or integrate" as the "hallmark of interdisciplinarity" (Rhoten et al., 2006, pp. 3–4). We have stressed that creativity is essential to integration and is enhanced by interdisciplinary practices, such as perspective taking.

Some of these outcomes are also claimed for disciplinary and multidisciplinary learning. In particular, disciplines in the liberal arts and the humanities typically claim "critical thinking" as an important outcome, as do the natural sciences and the applied fields. This collective claim to critical thinking raises a question: How do interdisciplinary approaches contribute to the development of this key cognitive skill in ways that are different from or superior to single-subject approaches? "For [an interdisciplinary] learner to be truly empowered through critical thinking," says Toynton (2005), "more than one context or one discipline needs to be encountered" (p. 110). If interdisciplinarians insist on including "critical thinking" in the learning outcomes at the program level, he asserts, they should make clear that the development of this skill requires viewing "the approaches, products, and processes" of relevant disciplines "from a detached and comparative viewpoint" (p. 110).

While *critical thinking* is a broad term, few would deny that key elements include distinguishing assumptions from argument and from evidence, evaluating the quality of argument and evidence, comparing conflicting arguments and evidence, and reaching a reasoned conclusion regarding which argument or combination of arguments is best. These elements may (or may not) be addressed in a disciplinary education, but each lies at the heart of the interdisciplinary research process. Most if not all of the topics addressed in the next section may also be thought of as particular critical thinking skills.

Cognitive Abilities Attributable to Interdisciplinary Learning

The literature on cognition and instruction identifies five cognitive abilities that interdisciplinary learning fosters. These include the ability to

- Develop and apply perspective-taking techniques

- Develop structural knowledge of problems that are appropriate to interdisciplinary inquiry

- Create common ground

- Integrate conflicting insights (i.e., expert views) from two or more disciplines

- Produce a cognitive advancement or interdisciplinary understanding of the problem

We discuss the latter of these abilities here.

Produce a Cognitive Advancement or Interdisciplinary Understanding of a Problem

A **cognitive advancement** is what this book calls a more comprehensive understanding. According to Boix Mansilla (2005), four core premises underlie the concept of cognitive advancement:

- "It builds on a *performance* view of interdisciplinary understanding—one that privileges the capacity to use knowledge over that of simply having or accumulating it [italics added]. . . . From this perspective, individuals understand a concept when they are able to apply it—or think with it— accurately and flexibly in novel situations" (pp. 16–17). Using knowledge also involves effectively communicating it to others.

- "The understanding is highly *disciplined,* meaning that it is *deeply informed by disciplinary expertise* [italics added]" (p. 17). This refers to the distinction between genuine disciplinary insights and common sense. "Interdisciplinary understanding differs from naïve common sense precisely in its ability to draw on disciplinary insights" (p. 17).

- "The understanding is achieved through the *integration* of disciplinary views [italics added]. . . . In interdisciplinary work," says Boix Mansilla, "these perspectives are not merely juxtaposed," but "actively inform one another, thereby leveraging understanding" (p. 17).

- The interdisciplinary understanding "is *purposeful*" leading to a "*cognitive advancement*—e.g., a new insight, a solution, an account, an explanation [emphasis added]" (p. 17). Examples of cognitive advancements include explaining a phenomenon, creating a product, raising a new question, generating a new insight, proposing a solution, providing an account, or offering an explanation (pp. 16–17). Like all research results, the cognitive advancement must be communicated for it to be useful.

The interdisciplinary understanding, then, is new knowledge that is useful, disciplined, integrative, and purposeful. These core premises are the primary indicators of quality interdisciplinary work and form the basis for testing or assessing the quality of the understanding produced. Students can reasonably ask themselves if they aimed for useful understanding, pursued disciplinary insights or theories diligently, integrated insights or theories, and communicated their understanding in a purposeful manner.

TEST THE MORE COMPREHENSIVE UNDERSTANDING OR THEORY

Testing the more comprehensive understanding or theory is quite different from reflecting on it, though arguably testing may involve reflection. Testing is a more formal process.

Test has different meanings in different disciplinary contexts. Testing in a discipline usually means applying the method(s) favored by the discipline to the data favored by the discipline. These tests will be flawed to the extent that the methods and data used are imperfect or incomplete. Interdisciplinarians face a challenge but also an opportunity: They cannot rely on just one method but can by triangulating among different methods aspire to a less biased test. The undergraduate is unlikely to do any formal testing but can reflect on what different methods might produce (and might refer back to the earlier discussion in Chapter 6 of strengths and limitations of methods in doing so).

There are four ways to test an interdisciplinary understanding. Three of these look at the results and ask in turn whether there seem to be useful policy implications (Newell, 2007a), whether others find the results interesting (Tress, Tress, & Fry, 2006), and whether the interdisciplinary understanding is indeed better in some identifiable way (Szostak, 2009). The fourth test looks at the process used to produce the result and asks whether this was appropriate (Boix Mansilla et al., 2009). This fourth test bears some similarity to the practice within disciplines of judging whether disciplinary methods were applied properly in generating a certain result.

The Newell (2007a) Test

Newell's (2007a) concern with the *utility* or practical application of the understanding is balanced by his concern with the *process* used to produce it. He identifies three of the most common sources of "failure" or "inadequacy" in producing the understanding. The first is failure to perform each step adequately. The second is doing insufficient work in identifying connections between the aspects of the problem studied by different disciplines, and thus the connections between the variables or concepts used by those disciplines. Relatively little is known about connections, he says, because most scholars are disciplinarians and are not interested in other aspects of the problem or how other disciplines explain them. The third common source of failure or inadequacy is overlooking one or more perspectives (Newell, 2007a, p. 262). Students may test their understanding by responding to one or more of Newell's questions:

- Does it allow for more effective action?

- Does it help solve the problem, resolve the issue, or answer the research question?

- Is it useful to practitioners, the public, or policy makers who are concerned with that particular complex problem, issue, or question? (p. 262)

The value of these questions, he says, is that they serve as a bridge between the ivory tower of the academy and the real world. If the pragmatic judgments of practitioners are that the more comprehensive understanding lacks utility, or if it has limited value because of a serious weakness, then the interdisciplinarian must correct the weakness by revisiting the earlier steps in the interdisciplinary process (Newell, 2007a, p. 262).

The Tress et al. (2006) Test

Another test of an interdisciplinary understanding's utility is whether others are interested in it. Tress et al. (2006) emphasize the importance of communicating one's research to the scientific community (we will provide advice on how to do so below). By "research," they mean "original investigation undertaken specifically to gain knowledge and understanding" (p. 21). Gathering data, recording observations, collecting experiences, developing plans, discussing with stakeholders, and even solving real-world problems, though valuable, are not necessarily research. Rather, integrative "knowledge creation becomes research when all data and information we have gathered are systematized, analyzed and *fed back* into academic communities [italics added]" (p. 21). Theories of organization learning stress the importance of researchers developing formal and informal mechanisms to engage in information exchanges within their institutions. "If knowledge remains implicit and is not integrated and shared at the institutional level, it can easily get lost" (p. 23).

Integrative research, they say, requires creating new knowledge, which must be transformed from "tacit knowledge" (not directly accessible to others) into "explicit knowledge" (accessible to others). Explicit knowledge "is fixed on some kind of medium such as a book, scientific journal, CD, video or a web site that moves it into the wider context of the public domain" (Tress et al., 2006, p. 22). Subjecting the new knowledge to peer review and publishing it is one of the main pillars of scientific progress (p. 22). Only when this happens, they say, can we "speak of research activity, since it is through this . . . feedback that generic knowledge is created" (p. 23).

Though Tress et al. (2006) have in mind graduate students, professionals, and interdisciplinary teams, undergraduates can also communicate their research in these ways:

- *As feedback into the academy:* Students can share the results of their research inside the academy. Commonly used communication strategies include public poster sessions, student panels, symposia, student journals, and student portfolios. Some graduate programs host an annual interdisciplinary research conference that involves outside speakers, poster sessions, and other activities to highlight interdisciplinarity and the quality of their work.

- *As feedback into the community:* Students can share their research with appropriate public or private entities and solicit their feedback. Students at one undergraduate program have been successfully leveraging their senior research projects to secure professional positions and gain entry into graduate programs.

We stress here that scholarship is a conversation. The books and papers we read did not just happen but were contributions to an ongoing conversation: The authors evaluated and built upon preceding arguments either within a discipline or across disciplines. Over time, some insights are judged to be particularly valuable (often these are combinations of preceding insights). Any researcher should be open to the comments they receive, and ready to revise their ideas in response to criticism. All scholars need to learn that criticism is valuable: It not only helps to clarify thinking, but also serves as a signal that their peers found their ideas worthy of critique. For interdisciplinarians, responding positively to criticism provides an opportunity to gain further attention for your now-improved understandings.

The Szostak (2009) Test

Szostak (2009) offers a two-part approach to testing the understanding. The first part asks two questions of the understanding:

1. Does it give us better insight into the problem than if there was no new understanding (i.e., is something better than nothing)? If so, how is it better? (*Note:* This is a low-level question that asks the student, "Is the new understanding better than no new understanding?")

2. Does it explain some aspect of a causal relationship (or set of causal relationships) better than any alternative explanation? If so, how (p. 335)? (*Note:* This is a higher-level question that asks the student to compare the understanding [or some part of it] to the various alternative [disciplinary] understandings and explain how the student's understanding is better than each of these.)

If the new understanding seems to provide some sort of insight, but it turns out that some other understanding explains everything captured by the new understanding *and more* (and/or seems to provide a more plausible explanation), then, says Szostak (2009), we should be very wary of attributing any importance to the new understanding. But a more likely result is that the new understanding will explain the problem (or at least some aspects of it) better than other (i.e., disciplinary) understandings (Szostak, 2009, p. 335). If so, the student should make this contribution clear.

The test of the understanding generally should be to see *not* whether particular disciplinary insights were "right," but whether the student has accorded them too little or too much emphasis in the new understanding. Szostak (2009) admits that such a test "may seem both more difficult and more arbitrary," especially for undergraduates. Still, "no method or piece of scholarly research is perfect, and thus no test of any hypothesis is perfect. Scholarly judgment always has to be exercised in determining how important a particular argument is" (p. 336).

The second part of Szostak's approach to testing the new understanding is to subject it to what he calls "the holistic test." This test is appropriate in situations where new understanding is aimed at shaping public policy. The understanding passes the "holistic test" if it proves useful to policy makers. In this respect, it is similar to Newell's (2007a) utility test that relies on the pragmatic judgment of practitioners. Szostak (2009) cautions against neglecting the possible negative side effects of any policy-oriented understanding. "Interdisciplinary analysis," he says, "naturally tends to highlight the potential weakness of each theory or disciplinary perspective. Attention to the full range of causal relationships should further reduce the probability of this type of error" (p. 337).

More advanced students, as well as interdisciplinary teams and solo practitioners, may wish to subject their understanding (if it is policy oriented) to Szostak's (2009) more demanding "holistic" test. This involves submitting the understanding to a second set of questions that are broader and more probing:

1. Does the interdisciplinary analysis neglect possible negative side effects of the understanding if it is implemented?

2. Does the interdisciplinary analysis pay attention to the full range of causal relationships involved?

3. Is the understanding and the problem it addresses adequately contextualized?

4. Do policy makers find the understanding and the insights the study provides useful? (p. 337)

The Boix Mansilla et al. (2009) Test

Boix Mansilla et al. (2009) have developed a rubric to test the quality of interdisciplinary work. Their approach differs from the approaches offered by Newell (2007a) and

Szostak (2009) in two important respects: (1) Whereas Newell's and Szostak's focus is on *students testing their own work,* the focus of Boix Mansilla et al. is on *faculty evaluating student work*; (2) whereas Newell's and Szostak's focus is on *the utility of the understanding,* the focus of Boix Mansilla et al. is on *showing developmental differences* between the interdisciplinary writing produced by upper- and lower-division undergraduate students. Note that the Boix Mansilla et al. test thus evaluates both the understanding itself and the process employed in developing the understanding. However, faculty can adapt the Boix Mansilla et al. rubric so that students can judge the quality of their own work and reflect on ways in which they can develop it further (Brough & Pool, 2005; Huber & Hutchings, 2004; Walvoord & Anderson, 1998).[1]

The Boix Mansilla et al. (2009) rubric identifies four defining qualities that apply to student work at any level and that students can use to assess their own work. These qualities (which are identical to the core premises noted earlier) include the work's *purposefulness,* its *disciplinary grounding,* its *integration,* and its *critical awareness.* Each quality is connected to one or more of the STEPS in the IRP. Producing a quality interdisciplinary understanding requires that students demonstrate the following:

- *Purposefulness:* Have clarity about the purpose of their inquiry and the intended audience (STEP 1) and an explicit rationale for taking an interdisciplinary approach (STEP 2). Two questions relating to purposefulness are these:

  (1) "Does the student's framing of the problem invite an interdisciplinary approach?

  (2) Does the student use the writing genre effectively to communicate with his or her intended audience?" (Boix Mansilla et al., 2009, p. 342)

- *Disciplinary grounding:* Show understanding of the chosen disciplinary insights, modes of thinking, or disciplinary perspective (i.e., the discipline's preferred concepts, units of analysis, methods, and forms of communication in a discipline). This is similar to STEPS 3, 4, and 5. Two questions relating to disciplinary grounding are these:

  (1) "Does the student use disciplinary *knowledge* accurately and effectively (e.g., concepts, theories, perspectives, findings, examples)?

  (2) Does the student use disciplinary *methods* accurately and effectively (e.g., experimental design, philosophical argumentation, textual analysis)?" (Boix Mansilla et al., 2009, p. 343)

- *Integration:* Identify "points where insights from different disciplines are brought together and articulate the cognitive advantage enabled by the combination of these insights" (Boix Mansilla et al., 2009, p. 338). This involves STEPS 8, 9, and 10. Four questions to be asked that relate to integration are these:

(1) "Does the student include selected disciplinary perspectives or insights from two or more disciplinary traditions that are relevant to the purpose of the paper?

(2) Is there an integrative device or strategy (e.g., a model, metaphor, analogy)?

(3) Is there a sense of balance in the overall composition of the piece with regard to how the student brings the disciplinary perspectives or insights together to advance the purpose of the piece?

(4) Do the conclusions drawn by the student indicate that understanding has been advanced by the integration of disciplinary views?" (Boix Mansilla et al., 2009, pp. 344–345)

- *Critical awareness:* "Engage in a process of considered judgment and critique: weighing disciplinary options, making informed adjustments to achieve their proposed aims, recognizing the limitations of the work produced" (Boix Mansilla et al., 2009, pp. 338–339). A question relating to critical awareness is this: "Does the student show awareness of the limitations and benefits of the contributing disciplines and how the disciplines intertwine?" (Boix Mansilla et al., 2009, p. 346)

For each criterion, Boix Mansilla et al. (2009) describe four qualitatively distinct levels of student achievement: *naïve, novice, apprentice,* and *master.* These levels are described here:

- *Naïve:* "A project can be characterized as naïve if it lacks clarity of purpose and audience, is built primarily on common sense or folk beliefs about the topic at hand, fails to draw on disciplinary insights, and makes no effort to integrate them because disciplinary perspectives themselves are not considered as such. These students may benefit from discussions about why the topic matters, what would be gained by understanding it in depth, how the topic connects with and expands personal experience, or how their intuitions or beliefs about the topic may be challenged" (Boix Mansilla et al., 2009, pp. 339–340).

- *Novice:* "A project exemplifies a novice understanding when it exhibits a superficial understanding of the nature of interdisciplinary academic work. The purpose of the project may be too broad or unviable, disciplinary concepts and theories are uncritically presented as matters of fact, and integrative language may be mechanistic and pro forma. These students grasp the nature of and differences between disciplinary work and the distinctive process of interdisciplinary inquiry. Students at this level may benefit from analyzing examples of work by experts in which the process of knowledge construction in and across disciplines is apparent as well as to reflect critically on the strengths and limitations of the different disciplinary insights" (Boix Mansilla et al., 2009,

p. 340). (*Note:* There is no mention of integration itself [i.e., the quality of the more comprehensive understanding], though Boix Mansilla et al. do mention the language used in the integration.)

- *Apprentice:* Projects at this level approach professional interdisciplinary work. It "exhibits a clear and viable purpose and a sense of the multiple audiences for the work." These students adequately use disciplinary elements (e.g., concepts, theories, assumptions) and modes of thinking, and support key claims with examples and sources. "Integration is reached through a metaphor, conceptual framework, causal explanation, or other device that contributes to a deepening understanding of the topic" (Boix Mansilla et al., 2009, p. 340). Student projects that fall within this category may still fail to strengthen an argument or critically probe an insight's benefit or shortcoming. Nevertheless, they attain "a robust understanding of disciplinary foundations" and understand "how and why integration can deepen understanding of the topic at hand" (p. 340).

- *Master:* Projects at the master level are characterized by their creativity, analytical sophistication, and self-reflection. Students "demonstrate comfortable understanding of disciplinary foundations and interdisciplinary integration" (Boix Mansilla et al., 2009, p. 341). Their work exhibits a clear sense of purpose and the need for an interdisciplinary approach. Students at this level "have mastered multiple expressive genres," introduce "new insightful examples to support disciplinary claims," and integrate insights "elegantly and coherently" while not overlooking opportunities to advance the argument. Undergraduate students performing at this level are ready to move to a new topic. Graduate students are ready to consider new criteria "such as originality, potential impact of the work, and whether scholarly precedent and contributions have been accounted for" (p. 341).

Integrating These Tests

From the varying approaches of Newell (2007a), Tress et al. (2006), Szostak (2009), and Boix Mansilla et al. (2009) to testing an interdisciplinary understanding, it is possible to identify seven key indicators of quality. Each of the following indicators, with the exception of "communication," which only Tress et al. advocate, is supported by at least two authors. Certainly, scholars all advocate communicating the results of their research to appropriate audiences; they just don't think of it as a test (but should). The decision as to which indicators are most appropriate for a particular project must be made locally on the basis of a course's academic level, purpose, and content.

- *Usefulness:* Applicability to practitioners who are concerned with the problem (Boix Mansilla et al., Newell, Szostak).

- *Disciplinary grounding:* Adequacy is achieved in relevant disciplines; linkages between disciplines are identified; careful evaluation is made of disciplinary insights for their potential contributions and limitations; all relevant perspectives, insights, and theories are addressed; balance is achieved in using disciplinary insights; strengths and limitations of author's insight(s) and understanding are addressed; personal bias is recognized and suspended so as not to skew analysis (Boix Mansilla et al., Newell, Szostak, Tress et al.).

- *Integration:* Synthesis of varying or conflicting viewpoints is explicit (Boix Mansilla et al., Newell, Szostak, Tress et al.).

- *Process:* Which steps proved advantageous or problematic and why, which steps were modified or not followed and why, and which steps proved most necessary to producing a better understanding (Newell, Szostak, Tress et al.).

- *Comparison:* Insight(s) into the problem are better than any alternative explanations (Boix Mansilla et al., Newell, Szostak).

- *Self-reflection:* Author is aware of the limitations of his or her work (Boix Mansilla et al., Szostak, Tress et al.).

- *Communication:* Subject to peer review feedback (Tress et al.).

BOX 13.2

Counteracting a Bias Against Novelty

This book has noted that there is sometimes a tradeoff between creativity and feasibility. That tradeoff arises again when we test a more comprehensive understanding. It is harder to judge a more novel understanding, precisely because it is so different. It is much easier to evaluate an understanding that deviates only a little from previous insights. Some of the above tests, such as asking practitioners and disciplinary scholars for advice, may be particularly biased against novelty for they will compare the novel idea to existing ideas that have already been subjected to critical revision. Having appreciated that scholarship is a conversation, we can imagine that a very novel idea will, if not ignored completely from the outset by other scholars or practitioners, be subjected to a variety of criticisms from scholars or practitioners that will likely result in important modifications. In evaluating a novel understanding, some allowance needs to be made for the probability that it will be improved through time. We should be particularly careful not to discard a novel understanding just because of a criticism that might possibly be addressed.

COMMUNICATE THE RESULTS OF INTEGRATION

Advanced undergraduate or graduate courses typically encourage or even require students to use metaphors, models, or narratives to capture creatively the new understanding in all of its richness. However one chooses to communicate the new understanding, it should be inclusive of each discipline's insights but beholden to none of them. That is, each relevant insight, theory, or concept should contribute to the understanding but not dominate it. The objective of this part of STEP 10 is to achieve unity, coherence, and balance among the disciplinary influences that have contributed to the understanding (Newell, 2007a, p. 261). In effect, the use of metaphors, models, and narratives to communicate the understanding constitutes a test of whether it is coherent, unified, and balanced and, thus, truly interdisciplinary. Other ways to communicate results of integration include a new process, a new product, a critique of existing policy and/or a proposed new policy, and a new question or avenue of scientific inquiry. The results of integration may take several forms, or some combination of these, as discussed below.

The literature on creativity stresses the importance of *persuasion*. Some authors go so far as to suggest that the key characteristic of a creative person—in both art and science—is their ability to persuade others of the value of their creations, rather than their ability to create in the first place. These authors note that the histories of both art and science are riddled with examples of discoveries that were only appreciated decades or centuries later. We cannot depend on our creative ideas to sell themselves, but must be prepared to actively "sell" our ideas to others. Many scholars achieve limited attention because of failures in persuasion, which often reflect limited self-confidence. The researcher should see their more comprehensive understanding as the starting point from which to launch an act(s) of persuasion. While the arguments and evidence marshaled in developing the more comprehensive understanding will be important, a telling anecdote or metaphor or picture that illustrates the more comprehensive understanding may prove far more important in persuading others (Szostak 2017a).[2]

Recall that scholarship is a conversation. If you wish to persuade others of the value of your ideas, you need first to reflect on what their interests are. If you can say, "You care about X. I have an idea that will help deal with X," you will catch their attention. (Or perhaps, "Your argument suffered from weakness Y. My idea can fix Y.") You need then to connect your idea directly to their concern. A story or metaphor or diagram that makes this connection will be powerful. Since different disciplines have different concerns, you may need to construct different stories or metaphors or diagrams for different audiences. You want to convince different disciplines that you understand their internal debates and can add something useful to these.

A Metaphor

A metaphor "brings out the defining characteristics of the understanding without denying the remaining conflict that underlies it" (Newell, 2007a, p. 261). Metaphors allow readers to connect new information to what they already know. They are particularly useful in the humanities, where meaning cannot be adequately expressed using quantitative and empirical approaches. The social sciences and even the natural sciences also make use of metaphors. They help us to understand one thing in terms of another. An interdisciplinary understanding has been reached when the metaphor is consistent with (1) the contributing disciplinary insights as modified to create common ground, (2) the interdisciplinary linkages found, and (3) the patterns observable in the overall behavior of the complex system (Newell, 2007a, p. 261).

Visual or physical metaphors can be effective and powerful integrative devices. They "frame reality in terms of similarities between constructs pertaining to different realms" (Boix Mansilla, 2009, p. 10). A visual metaphor combines a *"vehicle concept"* with the *topic*. An example of a visual metaphor is Maya Lin's Vietnam Veterans Memorial in Washington, DC. She uses the vehicle concept of a scar—a cut in the earth to be healed by time—to highlight certain features of the topic—the devastating consequences of the Vietnam War on American society. Boix Mansilla (2009) explains Lin's approach:

> Framing the Vietnam War as a scar sheds light on the personal emotional experience of war and its long-lasting impact. It does not illuminate, for instance, the political and military connundra that the war presented to American administrations at different points in time. To the extent that the mind can explicate the tacit analogy presented by a [visual] metaphor, the metaphor offers parsimony and impact in our representation of reality. (p. 10)

Visual metaphors "create a holistic synthesis and operate in a physical medium—in this case, the landscape, the stone, the engravings [of names]" (Boix Mansilla, 2009, p. 10). As an integrative device, the visual metaphor itself constitutes the more comprehensive understanding of the topic or event. The metaphor brings out the defining characteristics of the understanding without denying the remaining ambiguity or conflict that underlies it (Newell, 2007a, p. 261).

A Model

A model may be a physical object, a performance, or a written work set before others for guidance or imitation. Serving as a visual aid to comprehension, a model may capture the unity, coherence, and balance contained in the interdisciplinary understanding.

Examples of Models

In the first example, Nagy, an advanced undergraduate, writes about sustainable development in Costa Rica that integrates economic development, environmental concerns, and indigenous social and cultural values.

From the Natural Sciences. Nagy* (2005), *Anthropogenic Forces Degrading Tropical Ecosystems in Latin America: A Costa Rican Case Study.* Nagy produces a model of "sustainable development" for the Costa Rican coastal region that is designed to meet the immediate needs of the people, but in a way that will not mortgage the ability of future generations to secure their basic needs. Sustainable development, she says, considers not just the environment and economic development, but social and cultural aspects as well. The components of her model, though she does not map relationships among them, include "resource conservation, ecosystem protection, economic motivation, cultural celebration and protection, and social considerations" (p. 107). Nagy believes that her integrative model makes it easier to incorporate cultural diversity and protection of indigenous peoples for the simple reason that the practices of these groups, often linked myopically to endangered ecosystems, are also in need of economic development to ensure their survival. Within her continuum, indigenous culture (and thus the needs of indigenous peoples) would fall between economic development and environmental protection. Perhaps, she says, by focusing on conservation rather than preservation, sustainability could be achieved for these people as well. This insight is drawn from recognizing that it is unlikely countries like Costa Rica, which need more economic development to meet the needs of their growing populations, will continue to set aside large tracts of land for these small groups as they have been doing. "Rather than emphasizing the preservation of land in a pristine and untouched condition, it is more realistic that activities allowing for multi-purpose land use (for ecosystem and cultural protection and economic development) will progress towards sustainability" (p. 108). (*Note:* A diagram outlining the type of causal arguments made in this paragraph can be very persuasive by showing disciplinary scholars that the phenomena they study are an important part of the understanding.)

In the second example, Foisy, an advanced undergraduate, integrates various theories on how to create meaning in everyday life.

From the Social Sciences. Foisy* (2010), *Creating Meaning in Everyday Life: An Interdisciplinary Understanding.* In this threaded example, Foisy refers to the Meaning Construction Model (introduced in Chapter 12) that she developed as the basis for integrating the various theories on how to create meaning in everyday life. In the conclusion to her paper, reproduced here, she applies the Meaning Construction Model to the case of Andrew the Volunteer.

CONCLUSION

A Return to the Case Study of Andrew the Volunteer

I will conclude by illustrating the usefulness of the Meaning Construction Model. Andrew's first step (according to the Meaning Construction Model) was making the conscious decision to volunteer in Africa (goal-setting stage). He made his goal specific ("I want to volunteer in Moshi, Tanzania") and realistic ("I will spend one year planning this trip and will double-check that I am able to get the time off work"). Andrew's goal was challenging, adjustable, and personally important (for example, he originally wanted to volunteer for six months, but could not get that much time off work, so he adjusted the length of his volunteer position to three months). His volunteer trip fell into several different "goal categories": traveling, giving, loving, and working.

Once Andrew set his goal, he began working towards it by booking his plane ticket, fund-raising, getting vaccinated, and speaking with his boss to get time off work. One year later, Andrew's plane landed in Africa: He had reached his goal.

Upon arrival, Andrew began working towards several *new* goals, one of which was to "ensure that every child I treat makes a full recovery." Andrew eventually found himself in many situations where the child did not get better, and he had to acknowledge that his goal was not realistic. Andrew therefore modified his goal to the following: "treat every child with compassion and respect while giving them high-quality medical care."

Now, because Andrew's goal is realistic, he is able to fully immerse himself in the moment and dedicate his full attention to each child. Andrew spends much of each day in a state of flow, where each moment is meaningful in and of itself. Because he is actively engaged in work that he finds meaningful, he is also able to provide better medical care.

When Andrew returns home at the end of his volunteer period, he will likely feel a sense of accomplishment and purpose. The sense of meaning that Andrew felt in Africa will likely transfer to his own medical practice, and he might find himself experiencing a state of flow while treating patients in his home country. Upon his return, Andrew might decide to set entirely new goals ("teach my daughter to ride her bike this summer"), or he might set a goal that *builds* from his volunteer experience ("send monthly care packages to the clinic where I volunteered"). It is likely that he will make both kinds of goals and pursue them concurrently.

Because of its interdisciplinary nature, the Meaning Construction Model can account for a variety of real-life goals that Andrew could set for himself. While this model is by no means flawless, it takes important concepts from four separate theories and integrates them into a practical model that can be applied to everyday situations.

The Meaning Construction Model is by no means the one correct answer to the question, "How do people create meaning in their lives?" If three other students along with myself had researched this same topic, it is likely that the four of us would have reached very different conclusions. However, I believe that there would be true value in this type of "side-by-side" interdisciplinary work: If each of us had examined four different theories on meaning construction and reached four different conclusions, another interdisciplinarian could then take our four interdisciplinary papers and apply Repko's 10-STEP research process to our *interdisciplinary* papers. The interdisciplinarian would have then integrated 16 different theories on meaning construction. In this way, I believe that it is possible for interdisciplinary research to "build upon" itself, and this process might allow us to approach answers that we currently cannot even conceive.

A Narrative

A narrative is a written or spoken account or story. Because humans naturally think in terms of stories, narratives are essential components to integrative learning and research, and are the most common way in which students and practitioners communicate the understanding. Narratives may be short or extended, depending on the complexity of the subject. The purpose of a narrative is not just to explain the more comprehensive understanding, but to convince the reader that this understanding might be useful. *Stories can be a particularly powerful form of narrative.* The author can use a story to show how their more comprehensive understanding works in real life. If the reader can identify with the story, they are persuaded of the value of the more comprehensive understanding.

Examples of Narratives

In the first example, J. Lewis (2009), an undergraduate student, has written a paper on why the U.S. public education system is failing to educate an increasingly diverse student population.

From the Social Sciences. J. Lewis* (2009), *The Failure of Public Education to Educate a Diverse Population.* Lewis's narrative is as follows:

The interdisciplinary understanding that resulted from this integration is that it's going to take more than just those involved in education to begin the process of revitalizing public education. The massive veil that is held in place over the public education system must be removed to reveal how subtractive schooling and the increasing attrition rates are negatively impacting [the United States]. Sociological facets of education must be examined and given more attention to stop the blatant curtailing of innovative and adaptive teaching methods needed to satisfy the diverse needs of our children. Working together, change can be implemented to avoid the bleak outlook that economists are predicting for the future. Their input should be valued as much as everyone else. The hegemony of the current public education system must be challenged by more than one discipline to create a more relevant and adaptive curriculum. New and integrative legislation is needed to ensure the problems created by current mandates are not allowed to continue to push our education system and its benefactors towards peril. (p. 1)

In the second example, Gary Blesser and Linda-Ruth Salter (2007), pioneers of the interdisciplinary field of aural architecture, provide a book-length study of this rapidly evolving field that is fully interdisciplinary.

From the Interdisciplinary Field of Aural Architecture. Blesser and Salter (2007), *Spaces Speak, Are You Listening?* Aural architecture refers to using digital simulations to build public spaces such as an auditorium or amphitheater that are sonically complex. The concept of aural architecture, say Blesser and Salter, "is an intellectual edifice built from bricks of knowledge, borrowed from dozens of disciplinary subcultures and thousands of scholars and researchers." When integrated into a single concept, however, "the marriage of aural architecture and auditory spatial awareness provides a way to explore our aural connection to the spaces built by humans and those provided us by nature" (p. 8). Their comprehensive understanding of how aural architecture has and continues to influence social cohesion in different historical periods and cultures consists of three points:

- "Over the centuries, aural spaces have been created or selected to provide environments for a variety of groups and individuals. And the aural qualities of these spaces can either impede or support social cohesion over social distances that range from intimate to public" (p. 363).

- "Because the nature of social cohesion shifts as culture evolves, so, too, does aural architecture. Historically, small towns in warm climates actively encouraged cohesion by embracing aural connections through open windows, public commons, large churches, and outdoor living. Currently, modern advanced cultures embrace independence and privacy, supporting cohesion by means of the telephone and the Internet, the electronic fusion of remote acoustic arenas. The current generation frequently experience aural architecture with its visual spaces of manufactured music. The difference between then and now is nothing more than an evolution of cultural values" (p. 363).

- "Unlike other art forms, we cannot escape the influence of aural architecture because we live inside it. Whether intentionally designed or accidentally selected, our aural spaces influence our moods and behavior. Learning to appreciate aural architecture by closely attending to auditory spatial awareness is one way we can control, and thus improve, our personal environment" (p. 364).

Lessons for Readers

J. Lewis's (2009) undergraduate paper and Blesser and Salter's (2007) book conform, to varying degrees, to Boix Mansilla et al.'s (2009) criteria for quality disciplinary work:

1. Both build on a performance or practical view of understanding. In the case of Lewis, the understanding is a narrative describing a framework for a more comprehensive approach to curriculum design to solve the dropout problem. In the case of Blesser and Salter, the understanding is grounded in theory (which they helped to develop) and field experience concerning how public spaces (both man-made and natural) should be developed. Lewis's paper was written as a class project with the goal of using it to support his candidacy for a teaching position in a particular public school with a primarily minority student body. Blesser and Salter have multiple specialized audiences in view: academics, technicians, and public officials who are involved in the design of public spaces.

2. Both are informed by disciplinary expertise and draw on the insights of relevant disciplines to create something new and more comprehensive. However, the disciplinary depth and breadth of the two projects varies considerably. J. Lewis, an undergraduate, was restricted to using a few disciplines and a handful of insights from each of them. By contrast, Blesser and Salter's research is exhaustive and draws on *all* relevant disciplines and utilizes *all* important concepts and theories, including the authors' own fieldwork.

3. Both integrate disciplinary views. J. Lewis follows the IRP described in this book and achieves partial integration, whereas Blesser and Salter compress many of the steps of the IRP (unconsciously so) and produce a fully integrated result.

4. Both projects lead to cognitive advancements, although to varying degrees. J. Lewis's understanding attempts to integrate important theories from sociology and education that, if applied, would add increased creativity and flexibility to the learning process. Blesser and Salter's understanding is a cognitive advancement in that it is the first attempt to produce a fully interdisciplinary approach to integrating the aural into the design of public spaces.

5. Both narratives speak to the importance of their topics and thus their understandings. The reader should be left with no doubt as to why they should care.

A New Process to Achieve New Outcomes

A process is a way of making the understanding immediately useful (and comprehensible) to practitioners. Processes include new approaches to solving particular sorts of problems or new modes of intellectual analysis that yield new insights. Here the author is trying to convince the reader that a particular process will be useful in clearly identified situations. A key audience is practitioners that face such situations.

Examples of New Processes

In the first example, Delph (2005), an undergraduate student, presents a new integrative process that she believes will increase the success rate of homicide investigations and reduce the number of unsolved murders.

From the Social Sciences. Delph* (2005), *An Integrative Approach to the Elimination of the "Perfect Crime."* Criminal profiling, says Delph, could achieve its greatest potential if profilers from forensic psychology and forensic science would integrate their analytical techniques and share them with local criminal investigators. If adopted, she believes her approach would produce four likely outcomes: (1) quickly reduce the list of possible suspects, (2) predict the prime suspect's future behavior, (3) offer investigative avenues that have been overlooked by police, and (4) empower local law enforcement agencies to use these integrated profiling techniques themselves (p. 32).

In the second example, Bal (1999), a professional, shows how using cultural analysis, an interdisciplinary mode of analysis that she pioneered, can be applied to an enigmatic note or letter painted in yellow letters on a red brick wall in post–World War II Amsterdam.

From the Humanities. Bal (1999), *"Introduction," The Practice of Cultural Analysis: Exposing Interdisciplinary Interpretation.* Cultural analysis, as noted earlier, stands for an interdisciplinarity that is primarily analytical and seeks to discover new meaning in objects and texts that traditional approaches are unable to achieve. The note or short letter (i.e., "graffito") written on a wall, Bal says, is "a good case for the kind of objects at which cultural analysis looks, and—more importantly—*how* it goes about doing so [emphasis added]" (p. 2). The graffito, she says, is a letter both visually and linguistically. Though the literal translation of the opening from the Dutch is "Note," the more usual address that comes to Bal's mind is "Dearest" or "Sweety" (p. 3). "This implied other word fits in with the beginning of the rest of the text that says something like 'I love you'" (p. 3).

With the "discourse of the love letter" firmly in place, the graffito shifts to epistemic philosophy by continuing with "I did not invent you" or "I did not make you up" or "I have not thought you up." "The past tense, the action negated, the first-person speaking," observes Bal, "all indicate the discourse of narrative only to make a point about what's real and what is not" (p. 3). What is striking is that the address changes a real person, the anonymous writer's beloved, into a self-referential description of the note: A referential "Dearest" becomes a self-referential "Note" or short letter. "This turns the note into fiction," says Bal, and the addressee into a made-up "you," after all.

Yet, by the same token, this inscription of literariness recasts the set of characters, for the identity of the "you" has by now come loose from the implied term of endearment that personalized him or her. So, the passerby looks again, tripping over this word that says, "YOU! . . . Addressed as beloved and not as a guilty citizen, the city dweller gets a chance to reshape her or his identity, gleaming in the light of this anonymous affection. But is it real?" (p. 13).

One possible meaning of the note or letter, and one Bal considers plausible, is that the addressee is real even if the beloved cannot be found; "she or he is irretrievably lost, and the graffito mourns that absence" (p. 4). For Bal, the letter is an autographic note or letter. "Moreover, it is publicly accessible, semantically dense, pragmatically intriguing, visually appealing and insistent, and philosophically profound. Just like poetry" (p. 4).

Bal's understanding extends to linking the graffito to the interdiscipline of cultural analysis itself, thus emphasizing why people in that field should care about her analysis. It is the interest in more than the public self-exposure of the subject (author) and object (the lost beloved) of the note or letter that makes the exposition an exposure of the self. Such exposure, says Bal, is an act of producing meaning (1999, p. 2). Cultural analysis is also interested in moving from a literal meaning of the note or letter to its broader and metaphorical meaning. Exposing meaning, as Bal does, creates a subject/object dichotomy. The dichotomy enables the subject (in this case, the author of the note or letter) to make a statement about the object (in this case, the lost beloved). The object is there, explains Bal (1999), to substantiate the statement and enable the statement to come across (p. 3). There is an addressee for the statement: the reader. In expositions like the note or letter, a "first person," the exposer, tells a "second person," the reader, about a third person, "the object" or lost beloved, who does not participate in the conversation. But, Bal says, "unlike many other constative speech acts, the object, although mute, is present" (1999, p. 4). In this sense, the note or letter is a *sign*. A sign stands for a thing or an idea in some capacity, or for someone. In this instance, the note or letter is a sign for the writer's beloved, who is now lost. It is also a sign for the culture that produced it, the reader that reads it, and the field that examines it.

A New Product

Products are both immediately useful and comprehensible. They may be of two kinds: (1) technological innovations or manufactured products and (2) works of art such as plays, poems, sculptures, paintings, or media productions. Technological innovations or manufactured products are the result of extracting relevant information from disciplines and applied fields and integrating their contributions (Blesser & Salter, 2007, p. x). Works of art operate differently, but are useful to communicate insights that are hard to express in words (insights that have an emotional or intuitive content).

The late Steve Jobs of Apple Corporation was widely celebrated for introducing new products such as the computer mouse and the smartphone. These products were always integrative, combining not only multiple technical ideas, but synthesizing these with ideas from the field of industrial design. (He was not the first to use a mouse or think of smartphones, but he also developed both into mass market devices.) Importantly, the devices always reflected something that a large number of consumers wanted to do, even if this consumer desire was not widely understood beforehand. The products persuaded consumers that they wanted these products, and thus valued the more comprehensive understanding on which they were based.

A Critique of an Existing Policy and/or a Proposed New Policy

An interdisciplinary understanding can also take the form of a critique of an existing policy to show how it is failing to meet a societal need because of its disciplinary or conceptual narrowness. The critique may be followed by a proposed new policy, plan, program, or schema that, because of its inclusiveness, is more likely to solve the problem.

Examples of Critiques

In the first example, Szostak (2009) uses his integrative understanding of the causes of economic growth as a basis for critiquing government preference for a "one-size-fits-all" approach to facilitate economic growth in underdeveloped countries.

From the Social Sciences. Szostak (2009), *The Causes of Economic Growth: Interdisciplinary Perspectives.* In his statement of understanding (his "Concluding Remarks"), Szostak's main insight is that we should not expect our understanding of economic growth to yield "one true model" or "one secret formula for achieving growth everywhere, anytime" (p. 341). The reason, he explains, is because economic growth "is generated by a wide array of causes involving a host of economic and non-economic factors" (p. 341). He finds that insights "from diverse theories, methods, and disciplines into what causes growth can be integrated in order to produce more accurate and nuanced insights than could be generated by any one discipline" (p. 341). He identifies two policy implications of his understanding. The first is that scholars should jettison the "misguided presumption" that "the goal of scholarly analysis of complex questions is some simple grand theory" (p. 341). The second is that

> we should not give advice to any country without looking at its unique situation. What institutions has it inherited from the past? What is the fiscal capacity of its government? What is its infrastructure and education and health system like? Does the government have legitimacy? Does the culture favor hard work and thrift? . . . [T]he set of best policies depends very much on their answers. (p. 343)

Szostak could have been even more persuasive in 2009. In 2012, he developed a complex diagram of the various causal relationships he had identified in 2009. This diagram communicates the simple fact that there are many interdependent influences on economic growth much more powerfully than a narrative alone could do.

In the second example, Henry (2009) critiques narrow disciplinary policies that deal with the causes of school violence.

From the Social Sciences. Henry (2009), *School Violence Beyond Columbine: A Complex Problem in Need of an Interdisciplinary Analysis.* In this threaded example, Henry presents his more comprehensive understanding of the causes of school violence.

> I have argued here that to understand the genesis of school violence, we need to adopt an interdisciplinary, multilevel analytical approach. In this way, we are able to better see the interconnected processes that produce school violence. Such an approach sensitizes us to the ways lower-level and more diffuse harm production can produce victims who, over time, can come to resent their victimization and react violently against it. In particular, social exclusion can occur in multiple ways that are both evident and concealed. In particular, they can be the product of social hierarchies in the social networks of peers, bolstered by societal-cultural discourses of masculinity and violence and supported by school systems through their own hierarchies of power. Although we can examine the psychological processes and situational explanations for why students acted violently, we need to step outside of the microcontexts to explore the wider framing discourses of gender and power, masculinity and violence, and social class and race that produce social exclusion, victimization, anger, and rage. We need to see how these discourses shape the school curriculum, teaching practices, the institution of education, the meaning of "school," and its associated educational policy. How do parents, both in their absence and in their presence, harm the lives of students? We need to proactively engage in the deconstruction of hierarchies of power that exclude, and in the process create, a wasted class of teenagers who feel hopeless, whose escape from hopelessness is blocked, and whose only way out are violent symbolic acts of self-destruction and other destruction. We also need to challenge the ways in which the economic and political structure of American society reproduces and tolerates hierarchies of exclusion and structural violence. This needs to go beyond cultural causes of school violence to see how these cultural forms are integrated with structural inequalities. Any adequate analysis of school violence, therefore, has to locate the microinteractive, institutional practices and sociocultural productions in the wider political economy of the society in which these occur. Ignoring the structural inequalities of power in

the wider system reduces the cause to local and situational inequalities of power, suggesting that policies can be addressed to intervene locally, such as at the level of peer subculture or school organization. Although these levels of intervention are important, they alone are insufficient. (pp. 16–17)

BOX 13.3

Columbia River Salmon

We have referred multiple times in this book to Dietrich's (1995) study of the Columbia River. In 2018, an agreement was reached between state and local authorities and tribal agencies to enhance the ability of salmon to travel upstream past the many dams that have been built on the river. The authority that runs the dams will allow increased spill during the spring migration season at times of day when power demand is low. It is hoped that this will have a limited negative impact on electricity rates. State governments will adjust water quality regulations to allow the increased spill. Though some remain opposed to the agreement, and there is still some possibility of continued litigation, an official with the Bonneville Power Administration spoke of how "people 'who historically have been on opposite sides of the table' found common ground" (Mapes & Berton, 2018).

A New Question or Avenue of Scientific Inquiry

Because so much interdisciplinary work is conducted at the boundaries and intersections of disciplines, researchers are well positioned to detect gaps in existing knowledge. Detecting new lines of inquiry and pursuing these is part of normal research activity. The new understanding may include a new question or a new avenue of research that builds upon and extends the work just completed.

Recall yet again that scholarship is a conversation. You want to show others how you have built constructively upon their work and answered questions raised by their research. You may be even more persuasive if you can show them how they can in turn build upon your work. Hypothetically, a more comprehensive understanding that answered all questions in a field might gain little attention: Researchers might then just move on to other research questions. The best outcome may be an understanding that clearly advances the field but leaves many opportunities for further research. This is not uncommon: The big breakthroughs in scholarly understanding are hailed not just for the answers they gave to previous questions but for the research avenues they opened up. For example, Einstein's theory of relativity explained much *and* provided a basis for continued research in nuclear physics over the last century. (Recall from above that novel understandings invite clarification.) Most scholarly articles contain a section in which the scholar suggests avenues for further research: This may be a critical component of the act of persuasion. Students, and especially graduate students, should also devote much attention to how their understanding might inspire research by others.

The Value of Communicating Back to Disciplines

There is value to communicating the product of interdisciplinary work back to the disciplines. Szostak (2009) says that this communication challenges the disciplines to do the following:

- Appreciate the value of interdisciplinarity and of integrating disciplinary insights in order to gain a new insight into the problem, intellectual question, or object.

- Appreciate the value of theoretical and methodological flexibility.

- Appreciate that each theory has a particular range of applicability.

- Appreciate the value of integrating the widest array of empirical analysis within an organizing structure of causal links and emergent properties. (p. 344)

- In addition to these general influences, communicating a particular interdisciplinary understanding about a particular issue will hopefully encourage the discipline to be more flexible in its approach to that issue.

There are also challenges involved in communicating back to disciplines. Most obviously, one cannot communicate to a discipline unless one uses its terminology and publishes in its journals. Less obviously, one's communication will be ignored unless it is grounded in the discipline's literature, showing how the interdisciplinary insight is useful for the sort of question(s) the discipline investigates.

Chapter Summary

STEP 10 of the IRP includes reflecting on the understanding produced in STEP 9, testing it, and then communicating it. This chapter argues that interdisciplinarians should be more reflective than disciplinary specialists, must demonstrate the ability to enter the specialized scholarly discourse (i.e., disciplinary literatures) on the problem in an interesting and persuasive manner, and must be willing to explain those elements of their own research that may cause concern among the intended audience. The four sorts of reflection that are called for in interdisciplinary work include what has actually been learned from the project, which STEPS of the IRP were omitted, what are one's own biases, and what are the strengths and limitations of the insights, theories, and methods used, including the IRP itself.

(Continued)

(Continued)

In addition to detailing the cognitive abilities distinctive to interdisciplinarity, the chapter explains how these abilities can (and should) form the basis for testing the quality of the interdisciplinary understanding, identifies four approaches to testing the quality of interdisciplinary work, and integrates these.

Finally, the chapter stresses the importance of communicating the understanding in multiple ways to multiple audiences regardless of academic level. The activity of communicating the results of integrative work is, in fact, another way of testing its coherence, unity, and balance, and thus whether it constitutes partial or full interdisciplinarity. Persuasion is an important component of scholarly activity and of creative activity more generally.

Notes

1. For a basic introduction to rubrics and their utility for students, see Popham (1997).

2. Werder et al. (2018) argue that the emerging field of "strategic communication," which explores how actors of all types can communicate effectively in pursuit of their goals, should pursue an interdisciplinary worldview and the sort of research process recommended in this book.

Exercises

What Has Been Learned

13.1 Reflect on the project and ask, "What have I learned about (1) the content of the topic, (2) the process involved, and (3) whether or not the conclusions drawn indicate that understanding has been advanced by the integration of disciplinary views?"

STEPS Omitted or Compressed

13.2 If you decided to omit or compress some of the STEPS of the IRP, what was your reason for doing so? What was the cumulative effect of this/these decision(s) on the final product?

Self-Interrogation

13.3 Reflect on your own bias at the outset of the project and ask how your view of the topic has changed as a result of examining views that conflict with your own.

13.4 Referencing the introduction to the paper, "The Ideal Manager: An Interdisciplinary Model," how might the paragraph be revised to eliminate personal bias while retaining the goal of the project: developing a model of the ideal manager?

Reflecting on One's Limited Understanding of the Relevant Disciplines, Theories, and Methods

13.5 Assuming you are at the apprentice level, consider the limitations of the relevant disciplinary understandings you used in your project and explain how using an interdisciplinary approach has enlarged and strengthened your understanding.

Cognitive Abilities

13.6 How is the ability to engage in critical thinking developed differently in interdisciplinary contexts than in disciplinary contexts?

Testing for Quality

13.7 Of the five cognitive abilities that interdisciplinary learning fosters, which has been the most difficult for you to develop and why?

Holistic Approaches to Testing the Understanding

13.8 If the result of your research is policy oriented, test it by applying either the Newell (2007a) or the Szostak (2009) test.

13.9 If the result of your research is not policy oriented, test it by applying the Boix Mansilla et al. (2009) rubric, which features four defining traits of quality interdisciplinary work.

13.10 Referencing the four qualitatively distinct levels of student achievement (i.e., *naïve, novice, apprentice,* and *master*), decide which level best describes your project. Why?

Metaphor, Model, Process, Product, or Avenue of Scientific Inquiry

13.11 Can you think of a metaphor or model that could depict your understanding visually?

13.12 If your understanding does not lend itself to either a metaphor or a model, write a narrative describing the new or improved process or product that you have created.

13.13 If your understanding is policy oriented, use it as the basis for critiquing an existing policy.

13.14 If your work revealed a gap in knowledge, identify it and describe what steps need to be taken to address it.

Creativity

13.15 What were the key challenges to creativity identified in this chapter, and how can these be addressed?

CONCLUSION

Interdisciplinarity for the New Century

This book invites researchers at all levels to practice interdisciplinary research in a more explicit, self-conscious, knowledgeable, rigorous, and nuanced way. It also asks them to think about eight interrelated issues: (1) the integrated definition of interdisciplinary studies introduced in Chapter 1 and its implications for education and research; (2) the ramifications of interdisciplinary studies having achieved the status of a maturing academic field; (3) the latitude and utility of the research model, which enable it to address complex problems that cut across multiple knowledge domains; (4) the body of theory that undergirds interdisciplinarity; (5) the refinement of our understanding of integration as a hallmark of interdisciplinarity; (6) the purpose and product of the interdisciplinary research process (IRP); (7) the cognitive outcomes fostered by interdisciplinary education and research; and (8) the variations in how the IRP plays out when disciplinary insights are drawn from the natural sciences, the social sciences, or the humanities.

THE DEFINITION OF INTERDISCIPLINARY STUDIES

The book dwells heavily on defining interdisciplinary studies and interdisciplinarity for two practical reasons. If interdisciplinary programs are fuzzy in their conception of what interdisciplinarity is, then they are unlikely to provide the distinctive educational outcomes that interdisciplinary education potentially offers. Developing a sustainable, rigorous, and coherent interdisciplinary studies program begins with having a clear notion of what interdisciplinarity is. Therefore, interdisciplinary programs should develop a local conception of the unique characteristics of their program (which might, for example, focus on complex problems of local interest). This conception should be informed by the integrated definition of interdisciplinarity presented in Chapter 1.

The second reason for dwelling at length on definition is that interdisciplinarity is still a widely misunderstood approach to education and research. Indeed, it is commonplace to hear well-meaning but uninformed academics claim, "It's in the air we breathe," and

"We are already doing it." Perhaps this is true, but according to what standard and informed by what body of theory? Accompanying this claim is the occasional call to fold the local interdisciplinary program into disciplinary units now that interdisciplinarity is supposedly ubiquitous. These academics, it seems fair to say, should be obliged to show how their discipline-based conception of interdisciplinarity is informed by their familiarity with the field's literature and theory. Without such grounding, their claim of "doing interdisciplinarity" is suspect.

One of the most promising developments in the academy at both the undergraduate and graduate levels is the increase in border-crossing activity. This makes interdisciplinary programs more necessary than ever because they can provide the intellectual center of gravity for interdisciplinary education and research on campus. The danger is that programs will be introduced that falsely claim to be interdisciplinary and thus fail to provide the understandings that guide interdisciplinary research. Even a cursory reading of the field's extensive literature shows that interdisciplinarity in all its breadth and complexity cannot be ignored or folded into narrow disciplinary structures (see Szostak, 2019).

INTERDISCIPLINARY STUDIES AS A MATURING ACADEMIC FIELD

This book shows that interdisciplinary studies can rightfully stake its claim as a maturing academic field that deserves its place in the academy alongside the disciplines. The criteria necessary to substantiate this claim are already in place and identified in Chapter 1: There is a consensus understanding of what interdisciplinarity is; there is an integrated model of the interdisciplinary research process; there is a body of theory underlying the field's approach to education and research; there is a growing community of interdisciplinary experts; and there is an extensive and growing body of literature on best practices. This literature addresses administration, assessment, curriculum design, pedagogy, research process, theory, student learning, faculty development, and research on specific problems. Those claiming the mantle of interdisciplinarity should familiarize themselves, at least cursorily, with this literature.

The emergence of interdisciplinary studies as an academic field has far-reaching implications that can only be summarized here: (1) There now exists a paradigm of knowledge production of increasing sophistication that is compatible with, but essentially different from, that of the disciplines; (2) this paradigm is essentially incongruent with the discipline-based administrative structure of the academy; and (3) this paradigm's flexibility provides new and far-reaching opportunities to savvy administrators, innovative scholars, and enterprising students.

THE RESEARCH MODEL

The book presents an integrated model of the interdisciplinary research process that rests on several assumptions: (1) Interdisciplinarians should identify such a process, (2) the disciplines are foundational to the interdisciplinary enterprise, (3) the accumulation of knowledge cannot and should not be limited to existing paradigms, and (4) integration is a cognitive process that is achievable.

By presenting the research model, the book addresses the sometimes contentious issue of what role the disciplines should play in interdisciplinary work. What we have in fact shown is that interdisciplinarians draw upon specialized research. The book does not establish that the existing disciplines are the only or the best way of organizing specialized research. The disciplines provide the depth while interdisciplinarity provides the breadth and the integration. Interdisciplinarity stands as a counterweight to the narrow focus of the disciplines. This book is one way that interdisciplinarians can communicate to their disciplinary colleagues that interdisciplinarity is not about competing with the disciplines, or about replacing them, but rather about working with them to transcend their limitations while building on their strengths. A proper understanding of interdisciplinarity might lead to more outward-looking disciplines. On this point, the literature is clear: Interdisciplinarity needs the disciplines, and the disciplines for relevance to complex issues need interdisciplinarity. This mutual dependence (symbiosis) warns against an interdisciplinarity that is overenthusiastic, overconfident, and overwhelming.

THEORY

The book presents the body of theory that undergirds the field and the research model. This includes theories on complexity, perspective taking, common ground, cognitive interdisciplinarity, and creativity. The field also draws on leader-member exchange theory and critical theory (not presented) as well as theories associated with learning, linguistics, and criminal justice. In addition, groundbreaking work has been done on the role of intuition and its connection to creativity and insight in understanding complexity, performing integration, and applying the interdisciplinary research process. Of particular importance is the development (in progress) of a philosophically grounded theory of interdisciplinarity that fully situates interdisciplinarity in the history of ideas and establishes an epistemology and ontology of complexity.

UNDERSTANDING INTEGRATION

In the controversy over integration, the book sides with "integrationists" who assert that integration is the key distinguishing characteristic of interdisciplinarity and that the goal of interdisciplinary work should be full integration whenever possible. Minimizing, obscuring, or eliminating integration from a conception of interdisciplinarity hollows out the concept of interdisciplinarity and makes it far easier for critics to argue that interdisciplinarity "is whatever we say it is." It also makes problematic the production and assessment of quality interdisciplinary work.

The book advances our understanding of integration in at least two ways. First, it pries open a nebulous concept that has long resisted definition and shows it to be at root a cognitive process that is a native tendency of the human brain, which is designed to process information integratively, allowing us to adapt to the complexities of reality. We have shown that it involves both conscious and subconscious mental processes. More particularly, the book introduces the concept and theory of common ground as a prerequisite for integration. We can create common ground between conflicting disciplinary concepts, assumptions, or theories using identifiable techniques. Common ground enables us to integrate insights or theories concerning a problem and construct a more comprehensive understanding or theoretical explanation of it. Second, since the process of integration is now transparent, it invites reflection and testing.

THE PURPOSE AND PRODUCT OF THE RESEARCH PROCESS

The book explains the purpose and product of the interdisciplinary research process, which is to construct a more comprehensive understanding of the problem. This STEP of the research process and its underlying theory is of major importance to interdisciplinary research for two reasons. First, it makes possible a more rigorous and more granular evaluation of interdisciplinary work through the use of rubrics based upon course and program learning outcomes. Second, the enlarged conception of interdisciplinary understanding effectively connects interdisciplinary work to the real world in new and creative ways because the products of the research effort include work that is practical, purposeful, and performance oriented.

THE COGNITIVE OUTCOMES OF INTERDISCIPLINARITY

This book briefly discusses the cognitive abilities, outcomes, and underlying theory associated with interdisciplinary education and research. Students, academics, and administrators should be aware of how the interdisciplinary approach to problem solving and decision making involved in the research process differs from the learning that occurs in many traditional disciplinary contexts. A major recent finding of learning theory, for example, shows that with repeated exposure to interdisciplinary thought, students develop more mature epistemological beliefs, enhanced critical thinking ability, metacognitive skills, and an understanding of the relations among perspectives derived from different disciplines. These are critical cognitive abilities that students need when entering the professions or pursuing graduate study. The skills they have learned are useful in almost any job and in life more generally, enabling them to consciously reflect on how to grapple with any complex problem that they may face (see Repko, Szostak, & Buchberger, 2020).

THE VARIATIONS IN HOW THE INTERDISCIPLINARY RESEARCH PROCESS PLAYS OUT

Lastly, the book explains the variations in how the interdisciplinary research process plays out when disciplinary insights are drawn from the natural sciences, the social sciences, or the humanities. Insights produced especially by the social and physical sciences generally rely on theories to explain the relations among the phenomena they study. Integration in these contexts involves presenting one integrative theory that combines elements of several competing theories to achieve the goal of full integration. Rather than providing one integration, scholars in the fine and performing arts, and often the humanities that study them, prefer setting up a range of alternative integrations for readers or viewers to consider.

LOOKING TOWARD THE FUTURE

For interdisciplinary studies to fulfill its potential to advance knowledge in the new century, several developments need to occur, each of which this book encourages. One is the need to develop a course on how to conduct interdisciplinary research. This course should be included in each program's required core to be taken immediately *after* completing an introduction to the field of interdisciplinary studies course and *before* taking advanced courses requiring substantive research. Such a course on the interdisciplinary research process (and its underlying theory) would show students how their interdisciplinary work differs from disciplinary work and prepare them for doing research and writing in advanced theme-based or problem-based courses. It would allow later courses to go into more depth about the nature of interdisciplinarity in various contexts and address conceptual, theoretical, and methodological issues with greater sophistication. Adding such a course would enhance the program's academic standing among disciplinarians who highly value research methods courses. At a minimum, focusing more intently on how to do interdisciplinary research would mute disciplinary criticism of interdisciplinary work for its lack of rigor and achieve balance between disciplinary depth and interdisciplinary breadth. (This book is used widely in graduate education, often coupled with works on multimethod research. It can usefully guide graduate students on how to perform interdisciplinary research.)

A second need is to inform faculty who are new to the field or who have been doing interdisciplinarity on their own but have not had time to immerse themselves in the literature or to keep up with that literature about the interdisciplinary research process. With a few exceptions (e.g., American studies), most faculty teaching in interdisciplinary programs were themselves trained in a discipline and picked up an interdisciplinary approach later in their careers. Some may have developed their own idiosyncratic style of interdisciplinary research because the professional literature, until recently, had little to offer them. Some may not have reexamined the professional literature in recent years and may have missed its dramatic increase in sophistication, depth of analysis, and utility. Their graduate students, however, are likely to seek out the professional literature in conducting interdisciplinary research. (As noted above, this book is often used in graduate programs.) Increasingly, some faculty may be directing the research of graduate students who are more familiar with the professional literature on interdisciplinarity than they are. This gap needs to close.

Finally, there is need for more research that is explicitly interdisciplinary and that creatively applies the research model (or some version of it) presented in this book. The publication of journal articles and books, most notably *Case Studies in Interdisciplinary Research* (Repko, Newell, & Szostak, 2012), is encouraging because they demonstrate the utility of using an interdisciplinary approach to address a wide range of complex problems. Indeed, breakthroughs of lasting importance are increasingly the product of cross-fertilization between different knowledge formations and research cultures. What the interdisciplinarian brings to a complex problem is a toolkit of cognitive abilities and skills, a process to achieve integration, and techniques to construct a more comprehensive solution. It is hoped that the research model presented in this book will inform and challenge a new generation of students and scholars to engage in this much-needed work.

APPENDIX

Interdisciplinary Resources

Authors' Note: Some of the information in this Appendix is adapted from Klein, J. T. (2003). Thinking about interdisciplinarity: A primer for practice. *Colorado School of Mines Quarterly, 103*(1), 101–114; Klein, J. T., & Newell, W. H. (2002). Strategies for using interdisciplinary resources across K–16. *Issues in Integrative Studies, 20,* 139–160; and Klein, J. T. (2010). Resources. In *Creating interdisciplinary campus cultures: A model for strength and sustainability* (pp. 161–180). We have also drawn on Fiscella, J. B., & Kimmel, S. E. (Eds.). (1999). *Interdisciplinary education: A guide to resources.* New York, NY: The College Board. We thank Julie Thompson Klein for making available her "Interdisciplinary Searching Module," and C. Diane Shepelwich, former interdisciplinary librarian at the University of Texas at Arlington, for providing the information on databases and online resources.

ASSOCIATIONS

American Studies Association (ASA), www.theasa.net

For a current summary of the state of American studies as an interdisciplinary academic field, please refer to Bronner, S. (2008). The ASA Survey of Departments and Programs, 2007: Findings and projections. *ASA Newsletter,* March. Retrieved July 14, 2011, from http://www.theasa.net/images/uploads/Final_Copy_Simon_Bronner_Article_PDF.pdf.

Association of American Colleges and Universities (AAC&U), www.aacu.org/

The AAC&U is the leading national association concerned with the quality, vitality, and public standing of undergraduate liberal education. Integrative learning is a recurring theme in many of its conferences and publications.

Association for General and Liberal Studies (AGLS), http://web.oxford.emory.edu/

AGLS is a community of learners—faculty, students, administrators, and alumni—intent upon improving general and liberal education at 2-year and 4-year institutions.

AGLS identifies and supports the benefits of students' liberal education attained through general education programs. As an advocate, AGLS tracks changes in general education and liberal studies and sponsors professional activities that promote successful teaching, curricular innovation, and effective learning.

Association for Interdisciplinary Studies (AIS), www.oakland.edu/ais/

AIS, formerly the Association for Integrative Studies, is an interdisciplinary professional organization founded in 1979 to promote the interchange of ideas among scholars and administrators in all of the arts and sciences on intellectual and organizational issues related to furthering interdisciplinary studies. The website has materials that may be downloaded at no charge, including its journal, newsletter, peer-reviewed course syllabi, directories of master's and doctoral programs, advice on best practices in interdisciplinary teaching research and administration, interdisciplinary assessment tools, and a list of core publications with tables of contents. The AIS journal *Issues in Interdisciplinary Studies,* formerly *Issues in Integrative Studies,* publishes articles on a wide range of interdisciplinary topics, including assessment, pedagogy, program development, research process, theory, special reports on the status of and challenges to interdisciplinarity, and topics of current interest.

Integration and Implementation Sciences (I2S), http://i2s.anu.edu.au/

I2S emphasizes the role of integration in interdisciplinary research. It also stresses the translation of interdisciplinary research into public policy. Its website, based in Australia, hosts a wide variety of resources, including tools for performing interdisciplinary research and links to a variety of other websites.

Network for Transdisciplinary Research (td-net), www.transdisciplinarity.ch/index/Aktuell/News.html

Funded by the Swiss Academy of Arts and Sciences, td-net is an organization involving scholars across Europe and beyond that supports the study of transdisciplinarity in both research and teaching. It defines transdisciplinarity in a way that is very similar to the definition of interdisciplinarity utilized in this book, but with an emphasis on involving people outside the academy in collaborative research. The td-net website has many resources, including a useful bibliography.

Science of Team Science, www.scienceofteamscience.org/

As the title suggests, this group focuses on the challenges of team research: both the cognitive challenges of integrating insights and communicating across diverse perspectives, but also the personal challenges of collaboration and leadership. It hosts an annual conference at different locations in the United States. Its website provides many resources, including a toolkit of strategies for successful team research.

DATABASES

Academic Search Complete

The world's largest scholarly, multidiscipline, full-text database, Academic Search Complete offers critical information from many sources found in no other database, including peer-reviewed full-text articles for almost 4,600 periodical titles in more than 100 scholarly journals dating back to 1965, or the first issue published (whichever is more recent). Areas of study include social sciences, humanities, education, computer sciences, engineering, language and linguistics, arts and literature, medical sciences, and ethnic studies.

ERIC

The ERIC (Education Resources Information Center) database is sponsored by the U.S. Department of Education to provide extensive access to education-related literature. The database corresponds to two printed journals: *Resources in Education* (RIE) and *Current Index to Journals in Education* (CIJE). Both journals provide access to some 14,000 documents and over 20,000 journal articles per year. In addition, ERIC provides coverage of conferences, meetings, government documents, theses, dissertations, reports, audiovisual media, bibliographies, directories, books, and monographs.

H-Net, www.h-net.org

H-Net is a self-described "international interdisciplinary organization" that provides teachers and scholars with forums for the exchange of ideas and resources in the arts, humanities, and social sciences. The database includes over 100 free, edited listservs and websites that coordinate communication in a wide variety of disciplinary and interdisciplinary fields as well as subjects and topics.

JSTOR

JSTOR (Journal Storage) is an archive collection of over 620 full-text scholarly journals primarily from university presses and professional society publishers. Subject areas include African American studies, anthropology, Asian studies, botany, ecology, economics, education, finance, folklore, history, history of science technology, language literature, mathematics, philosophy, political science, population studies, public policy administration, science, Slavic studies, sociology, and statistics.

ProQuest Dissertation Abstracts International

The ProQuest Dissertation Abstracts International database contains citations for dissertations and theses from institutions in North America and Europe. Citations for dissertations published from 1980 forward also include abstracts. Citations for master's theses from 1988 forward include abstracts. Titles published from 1997 forward have 24-page previews and are available as full-text PDF documents.

Web of Knowledge

Web of Knowledge contains abstracts and citations from Thompson Reuters's Science Citation Index Expanded, Social Sciences Citation Index, and Arts & Humanities Citation Index. The database allows for searches by subject term, author name, journal title, or author affiliation, as well as for articles that cite an author or a work. Along with an article's abstract, its cited references (bibliography) are listed for further searching.

WorldCat

WorldCat contains more than 32 million records describing books and other materials owned by libraries around the world.

ONLINE RESOURCES

Carleton Interdisciplinary Science and Math Initiative (CISMI), http://serc.carleton.edu/cismi/index.html

This site has compiled literature and resources to support interdisciplinary and integrative teaching activities, with an emphasis on science and math. Areas include research on expert interdisciplinary thinking and practice, assessing interdisciplinary work in college, strategies for interdisciplinary teaching, integrative learning, national reports, and books.

Interdisciplinary Studies Project, www.interdisciplinarystudiespz.org

The project examines the challenges and opportunities of interdisciplinary work carried out by experts, faculty, and students in well-recognized research and education contexts. Building on an empirical understanding of cognitive and social dimensions of interdisciplinary work, the project develops practical tools to guide quality interdisciplinary education.

JOURNALS

There are many interdisciplinary journals, a few of the most useful of which appear here. Articles on interdisciplinary topics such as interdisciplinary resources and interdisciplinary curriculum design are scattered across professional journals, some of which periodically devote special issues to interdisciplinarity. Locating these requires searching databases using keyword searching and Boolean search strategy.

Academic Exchange Quarterly, www.rapidintellect.com/AEQweb/

History of Intellectual Culture, www.ucalgary.ca/hic/homepage

The Integral Review, http://integral-review.org/

Issues in Interdisciplinary Studies, wwwp.oakland.edu/ais/publications/

Journal of General Education, http://muse.jhu.edu/journals/journal_of_general_education/

Journal of Interdisciplinary History, http://muse.jhu.edu/journals/jih/

Journal of Research Practice, http://jrp.icaap.org/index.php/jrp

SEARCH STRATEGIES

Locating resources for interdisciplinary purposes is typically not a straightforward process for interdisciplinary students. A metaphor descriptive of the challenge of identifying and locating relevant resources is suggested in the title of a special issue of *Library Trends,* 45(2), "Navigating Among the Disciplines: The Library and Interdisciplinary Inquiry." Students must "navigate" across multiple knowledge forums to locate relevant information. Before using the following approaches, students must state the problem or question they are investigating as clearly and as concisely as possible.

There are four approaches or strategies to navigation. The first is to use the traditional method of keyword searching. The typical search box options are author, title, keyword, and subject. This approach works well when the author and title of the article are already known. If this information is not known, then keyword and subject searching should be used. For example, if the problem under investigation is "The Causes of Childhood Obesity: An Interdisciplinary Analysis," the primary search term is *obesity*. The search will identify numerous articles written by experts from several disciplines. The student must identify the disciplines these writers represent because interdisciplinary research projects typically involve analyzing a problem from three or more disciplinary perspectives.

The second approach is the Boolean search strategy. This strategy is useful when it is necessary to narrow the number of references to those that are most relevant to the problem. For example, a keyword search of the problem "The Causes of Suicide Terrorism: An Interdisciplinary Analysis" would focus on *terrorism* and *suicide* because these terms are at the heart of the problem. However, navigating databases using just one, or even both, of them will yield an overwhelming number of references. The value of using the Boolean search strategy when faced with an abundance of resources is this: It refines the search by creating a "string" of terms that frame the search more precisely. The more one refines the search, the better the results.

The Boolean search strategy is based on the words AND, OR, and NOT. The basic formula for combining keyword and Boolean search strategy is as follows:

_____ AND _____ AND [OR] _____ *[Fill in your search terms.]*

Inserting the terms *suicide* and *terrorism* connects the two terms and narrows the number of relevant resources. A more refined search can be achieved by adding another term, say, *Islamic*, to the string.

Another way to narrow the number of resources is to use the NOT feature of the Boolean search strategy. For example, if the problem is youth gun violence, inserting the key terms *youth, gun,* and *violence* into the string will produce a substantial number of references. Adding a restrictive term, such as the name of a city, or a type of violence, such as homicide, into the string after NOT will produce a more precise result.

_____ AND _____ AND [OR] _____ NOT _____
[Fill in your search terms.]

Klein (2003) notes that different databases respond in different ways, so students should be prepared to "play" with the terms in the search string. If a particular term is not yielding good results, use a synonym, consult the thesaurus of a particular database, or check the list of common terms in the Library of Congress Classification (LCC) system. If the student encounters further problems, a librarian should be consulted.

The third approach is federated searching. This tool is a boon for interdisciplinarians because it enables access to multiple databases at a single keystroke. For example, the database ABI/INFORM allows for federated searching.

Regardless of the strategy used, achieving the most relevant results requires that students use precise keywords and Boolean logic, and this requires a clear statement of the problem or question.

Last but not least, it is possible to search using the subject headings assigned to books by library classification systems. The advantage here is that classifiers carefully attach one or more subjects to a particular book: This subject may be poorly captured by the book's title (which is what a keyword search in a library catalog will focus on). These controlled subject headings reduce some of the terminological ambiguity that hampers searching: Books using different words to talk about the same thing should still receive the same subject heading (though subject headings may still reflect disciplinary jargon, since all major library classifications are organized around disciplines). The disadvantage of subject searching—beyond the important fact that it captures only books and not journal articles—is that one first must learn what subject headings are employed. Library catalogs will often guide the searcher here. Another strategy is to locate one or two useful resources using a keyword search and then use the subject headings assigned to those works for future searching. Unfortunately, many library catalogs focus on keyword searching and make subject searching difficult or impossible (See Martin 2017; Szostak, Gnoli, & Lopez-Huertas 2016; and Mack & Gibson, 2012, for more information on how library classifications are unnecessarily challenging for interdisciplinary researchers.)

CORE RESOURCES ON INTERDISCIPLINARY STUDIES

This section identifies core resources under subject headings for ease of access. In a few cases, a publication may appear under more than one heading.

Assessment and Evaluation

Boix Mansilla, V. (2005). Assessing student work at disciplinary crossroads. *Change, 37*, 14–21.

Boix Mansilla, V. (2010). Learning to synthesize: The development of interdisciplinary understanding. In R. Frodeman, J. T. Klein, & C. Mitcham (Eds.), *The Oxford handbook of interdisciplinarity* (pp. 288–291). New York, NY: Oxford University Press.

Boix Mansilla, V., Duraising, E. D., Wolfe, C. R., & Haynes, C. (2009). Targeted assessment rubric: An empirically grounded rubric for interdisciplinary writing. *The Journal of Higher Education, 80*(3), 334–353.

Brooks, B., & Widders, E. (2012) interdisciplinary studies and the real world: A practical rationale for and guide to postgraduation evaluation and assessment. *Issues in Integrative Studies, 30*, 75–98.

Field, M., Lee, R., & Field, M. E., Assessing Interdisciplinary Learning. In J. T. Klein & W. Doty (Eds.), *Interdisciplinary studies today* (pp. 69–84). San Francisco, CA: Jossey-Bass.

Field, M. & Stowe, D. (2002), Transforming Interdisciplinary Teaching and Learning through Assessment. In C. Haynes (ed.), *Innovations in interdisciplinary teaching* (pp. 256–274). Westport, CT: Oryx Press.

Huutoniemi, K. (2010). Evaluating interdisciplinary research. In R. Frodeman, J. T. Klein, & C. Mitcham (Eds.), *The Oxford handbook of interdisciplinarity* (pp. 309–320). New York, NY: Oxford University Press.

Ivanitskaya, L., Clark, D., Montgomery, G., & Primeau, R. (2002). Interdisciplinary learning: Process and outcomes. *Innovative Higher Education, 27*(2), 95–111.

Klein, J. T. (2002). Assessing interdisciplinary learning K–16. In J. T. Klein (Ed.), *Interdisciplinary education in K–16 and college: A foundation for K–16 dialogue* (pp. 179–196). New York, NY: The College Board.

Klein, J. T. (2006). Afterword: The emergent literature on interdisciplinary and transdisciplinary research evaluation. *Research Evaluation, 15*(1), 75–80.

Klein, J. T. (2008). Evaluation of interdisciplinary and transdisciplinary research: A literature review. *American Journal of Preventative Medicine, 35*(2S), S116–S123.

Lyall, C., Bruce, A. Tait, J., & Meagher, L. (2011). *Interdisciplinary research journeys.* London, UK: Bloomsbury Academic.

Popham, W. J. (1997, October). What's wrong—and what's right—with rubrics. *Educational Leadership,* 72–75.

Repko, A. F. (2008). Assessing interdisciplinary learning outcomes. *Academic Exchange Quarterly, 12*(3), 171–178.

Research Evaluation. (2006). [Special issue devoted to evaluating interdisciplinary research.] *15*(1), 1–80. Retrieved July 14, 2011, from http://www.ingentaconnect.com/content/beech/rev/2006/000 00015/00000001;jsessionid=72tg4b5q5k61e.alice.

Seabury, M. B. (2004). Scholarship about interdisciplinarity: Some possibilities and guidelines. *Issues in Integrative Studies, 22,* 52–84.

Wolfe, C., & Haynes, C. (2003). Interdisciplinary assessment profiles. *Issues in Integrative Studies, 21,* 126–169.

Bibliographies and Literature Review

Chettiparamb, A. (2007). *Interdisciplinarity: A literature review.* Southampton, UK: University of Southampton. Retrieved July 14, 2011, from http://www.heacademy.ac.uk/assets/York/-documents/ourwork/sustainability/interdisciplinarity_literature_review.pdf.

Dubrow, G. L. (2007). *Interdisciplinary approaches to teaching, research, and knowledge: A bibliography.* Retrieved July 14, 2011, from http://www.grad.umn.edu/oii/Leadership/inter disciplinary_bibliogra phy.pdf.

Holly, K. A. (2009). Understanding interdisciplinary challenges and opportunities in higher education. *ASHE Higher Education Report, 35*(2). San Francisco, CA: Jossey-Bass.

Klein, J. T. (2006, April). Resources for interdisciplinary studies. *Change,* 52–56, 58.

Collaboration

Amey, M. J., & Brown, D. F. (2004). *Breaking out of the box: Interdisciplinary collaboration and faculty work.* Greenwich, CT: Information Age.

Derry, S. J., Schunn, C. D., & Gernsbacher, M. A. (Eds.). (2005). *Interdisciplinary collaboration: An emerging cognitive science.* Mahwah, NJ: Lawrence Erlbaum Associates.

Hall, K. L., Vogel, A. L., Huang, G. C., Serrano, K. J., Rice, E. L., Tsakraklides, S. P., & Fiore, S. M. (2018). The science of team science: A review of the empirical evidence and research gaps on collaboration in science. *American Psychologist, 73*(4), 532–548.

O'Rourke, M., Crowley, S., Eigenbrode, S. D., & Wulfhorst, J. D. (Eds.). (2013). *Enhancing communication and collaboration in interdisciplinary research.* Thousand Oaks, CA: Sage.

Comparative National Perspectives

Lenoir, Y., & Klein, J. T. (2010). Interdisciplinarity in schools: A comparative view of national perspectives. *Issues in Integrative Studies, 28.*

Creativity

Darbellay, F. Moody, Z., & Lubart, T. (Eds.). (2017). *Creative design thinking from an interdisciplinary perspective.* Berlin, Germany: Springer.

Curriculum Design

Association for Interdisciplinary Studies. (2017). *Interdisciplinary general education.* Retrieved from https://oakland.edu/ais/publications/. [Click on "Resources"]

Augsburg, T. (2003). Becoming interdisciplinary: The student portfolio in the Bachelor of Interdisciplinary Studies Program at Arizona State University. *Issues in Integrative Studies, 21,* 98–125.

Borrego, M., & Newswander, L. K. (2010). Definitions of interdisciplinary research: Toward graduate-level interdisciplinary learning outcomes. *The Review of Higher Education, 34,* 1.

Linkon, S. (2004). *Understanding interdisciplinarity: A course portfolio.* Retrieved July 14, 2011, from http://www.educ.msu.edu/cst/events/2004/linkon.htm.

Newell, W. H. (1994). Designing interdisciplinary courses. In J. T. Klein & W. G. Doty (Eds.), *New directions for teaching and learning: Vol. 58. Interdisciplinary studies today* (pp. 35–51). San Francisco, CA: Jossey-Bass.

Repko, A. F. (2006). Disciplining interdisciplinarity: The case for textbooks. *Issues in Integrative Studies, 24,* 112–142.

Repko, A. F. (2007). Interdisciplinary curriculum design. *Academic Exchange Quarterly, 11*(1), 130–137.

Spelt, E.J.H., Biemans, J.A.H., Pieternel, H. T., Luning, A., & Mulder, M. (2009). Teaching and learning in interdisciplinary higher education: A systematic review, *Educational Psychology Review, 21,* 365–378.

Szostak, R. (2003). "Comprehensive" curricular reform: Providing students with an overview of the scholarly enterprise. *Journal of General Education 52*(1), 27–49.

Vess, D. (2000–2001). *Interdisciplinary learning, teaching, and research.* Retrieved July 14, 2011, from http://www.faculty.de.gcsu.edu/~dvess/ids/courseportfolios/front.htm.

Definitions of Interdisciplinarity

About Interdisciplinarity (n.d.). *Retrieved from* http://wwwp.oakland.edu/ais/ [Click on "Resources"]

Committee on Facilitating Interdisciplinary Research. (2004). *Facilitating interdisciplinary research.* Washington, DC: National Academies Press.

Klein, J. T. (1996). *Crossing boundaries: Knowledge, disciplinarities, and interdisciplinarities.* Charlottesville: University Press of Virginia.

Klein, J. T., & Newell, W. H. (1997). Advancing interdisciplinary studies. In J. G. Gaff, J. L. Ratcliff, & Associates (Eds.), *Handbook of the undergraduate curriculum: A comprehensive guide to purposes, structures, practices, and change* (pp. 393–415). San Francisco, CA: Jossey-Bass.

Szostak, R. (2015). Extensional definition of interdisciplinarity. *Issues in Interdisciplinary Studies, 33,* 94–117.

DOMAINS OF PRACTICE

Science and Technology

Committee on Facilitating Interdisciplinary Research. (2004). *Facilitating interdisciplinary research*. Washington, DC: National Academies Press.

Culligan, P. J., & Pena-Mora, F. (2010). Engineering. In R. Frodeman, J. T. Klein, & C. Mitcham (Eds.), *The Oxford handbook of interdisciplinarity* (pp. 147–160). New York, NY: Oxford University Press.

Oberg, G. (2011). Interdisciplinary environmental studies: A primer. Oxford, UK: Wiley-Blackwell.

Weingart, P., & Stehr, N. (2000). *Practising interdisciplinarity*. Toronto, Canada: University of Toronto Press.

Social Sciences

Calhoun, C., & Rhoten, D. (2010). Integrating the social sciences: Theoretical knowledge, methodological tools, and practical applications. In R. Frodeman, J. T. Klein, & C. Mitcham (Eds.), *The Oxford handbook of interdisciplinarity* (pp. 103–118). New York, NY: Oxford University Press.

Fuchsman, K. (2012). Interdisciplines and interdisciplinarity: Political psychology and psychohistory compared. *Issues in Integrative Studies, 30,* 128–154.

Kessel, F., Rosenfield, P. L., & Anderson, N. B. (Eds.). (2003). *Expanding the boundaries of health and social sciences: Case studies in interdisciplinary innovation*. New York, NY: Oxford University Press.

Klein, J. T. (2007). Interdisciplinary approach. In S. Turner & W. Outhwaite (Eds.), *Handbook of social science methodology* (pp. 32–50). Thousand Oaks, CA: Sage.

Miller, R. C. (2018). *International political economy: Contrasting world views* (2nd ed.). New York, NY: Routledge.

Smelser, N. J. (2004). Interdisciplinarity in theory and practice. In C. Camic & H. Joas (Eds.), *The dialogic turn: New roles for sociology in the postdisciplinary age* (pp. 34–46). Lanham, MD: Rowman & Littlefield.

Humanities

Bal, M. (2002). *Traveling concepts in the humanities*. Toronto, Canada: University of Toronto Press.

Fredericks, S. E. (2010). Religious studies. In R. Frodeman, J. T. Klein, & C. Mitcham (Eds.), *The Oxford handbook of interdisciplinarity* (pp. 161–173). New York, NY: Oxford University Press.

Klein, J. T. (2005). *Humanities, culture, and interdisciplinarity: The changing American academy*. Albany: State University of New York Press.

Klein, J. T., & Parncutt, R. (2010). Art & music research. In R. Froedeman, J. T. Klein, & C. Mitcham (Eds.), *The Oxford handbook of interdisciplinarity* (pp. 133–146). New York, NY: Oxford University Press.

History of Interdisciplinarity

Klein, J. T. (1990). *Interdisciplinarity: History, theory and practice*. Detroit, MI: Wayne State University Press.

Klein, J. T. (1999). *Mapping interdisciplinary studies: The academy in transition series* (Vol. 2). Washington, DC: Association of American Colleges and Universities.

Moran, J. (2002). *Interdisciplinarity*. London, UK: Routledge.

Newell, W. H. (2008). The intertwined history of interdisciplinary undergraduate education and the Association for Integrative Studies: An insider's view. *Issues in Integrative Studies, 26,* 1–59.

Wernli, D., & Darbellay, F. (2017) *Interdisciplinarity and the 21st century research—intensive university. Report for the League of European Research Universities.* Retrieved from https://www.leru.org/publications/interdisciplinarity-and-the-21st-century-research-intensive-university.

Information Research

Mack, D. C., & Gibson, C. (Eds.). (2012). *Interdisciplinarity and academic libraries.* Chicago: Association of Research and College Libraries.

Martin, V. (2017). *Transdisciplinarity revealed: What librarians need to know.* Santa Barbara: Libraries Unlimited.

Palmer, C. L. (1999). Structures and strategies of interdisciplinary science. *Journal of the American Society for Information Science, 50*(3), 242–253.

Palmer, C. L. (2001). Work at the boundaries of science: Information and the interdisciplinary research process. Boston, MA: Kluwer Academic.

Palmer, C. L. (2010). Information research on interdisciplinarity. In R. Frodeman, J. T. Klein, C. Mitcham, & J. B. Holbrook (Eds.), *The Oxford handbook of interdisciplinarity* (pp. 174–188). New York, NY: Oxford University Press.

Palmer, C. L., & Neuman, L. J. (2002). The information work of interdisciplinary humanities scholars: Exploration and translation. *The Library Quarterly, 72*(1), 85–117.

Palmer, C. L., Teffeau, L. C., & Pirmann, C. M. (2009). *Scholarly information practices in an online environment: Themes from the literature and implications for library service development.* Dublin, OH: Online Computer Library Center.

Szostak, R. (2008). Classification, interdisciplinarity, and the study of science. *Journal of Documentation, 64*(3), 319–332.

Szostak, R., Gnoli, C., & Lopez-Huertas, M. (2016). *Interdisciplinary knowledge organization.* Berlin, Germany: Springer.

Integration

McDonald, D., Bammer, G., & Deane, P. (2009). *Research integration using dialogue methods.* Canberra, Australia: ANU Press.

Newell, W. H. (2006). Interdisciplinary integration by undergraduates. *Issues in Integrative Studies, 24,* 89–111.

Newell, W. H. (2007). Decision making in interdisciplinary studies. In G. Morçöl (Ed.), *Handbook of decision making* (pp. 245–264). New York, NY: Marcel-Dekker.

O'Rourke, M., Crowley, S., & Gonnerman, C. (2016). On the nature of cross-disciplinary integration: A philosophical framework. *Studies in History and Philosophy of Biological and Biomedical Sciences, 56,* 62–70.

Piso, Z., O'Rourke, M., & Weathers, K. (2016) Out of the fog: Catalyzing integrative capacity in interdisciplinary research. *Studies in History and Philosophy of Science, 56,* 84–94.

Pohl, C., van Kirkhoff, L., Hadorn, G. H., & Bammer, G. (2008). Integration. In G. H. Hadorn, H. Hoffman-Riem, S. Biber-Klemm, W. Grossbacher-Mansuy, D. Joye, C. Pohl, U. Wiesmann, & E. Zemp (Eds.), *Handbook of transdisciplinary research* (pp. 411–426). Berlin, Germany: Springer.

Repko, A. F. (2007). Integrating interdisciplinarity: How the theories of common ground and cognitive interdisciplinarity are informing the debate on interdisciplinary integration. *Issues in Integrative Studies, 27,* 1–31.

Sill, D. (1996). Integrative thinking, synthesis, and creativity in interdisciplinary studies. *Journal of General Education, 45*(2), 129–151.

Spooner, M. (2004). Generating integration and complex understanding: Exploring the use of creative thinking tools within interdisciplinary studies. *Issues in Integrative Studies, 22*, 85–111.

LITERATURE AND RESOURCE GUIDES

Ackerson, L. G. (Ed.). (2007). *Literature search strategies for interdisciplinary research: A sourcebook for scientists and engineers.* Lanham, MD: Scarecrow Press.

Fiscella, J. (1996). Bibliography as an interdisciplinary service. *Library Trends, 45*(2), 280–295.

Fiscella, J. B., & Kimmel, S. E. (Eds.). (1999). *Interdisciplinary education: A guide to resources.* New York, NY: College Entrance Examination Board.

Klein, J. T. (2006). Resources for interdisciplinary studies. *Change,* April, 52–56, 58.

Klein, J. T., & Newell, W. H. (2002). Strategies for using interdisciplinary resources across K–16. *Issues in Integrative Studies, 20*, 139–160.

Newell, W. H. (2007). Distinctive challenges of library-based research and writing: A guide. *Issues in Integrative Studies, 25*, 84–110.

Palmer, C. L. (2001). *Work at the boundaries of science: Information and the interdisciplinary research process.* Boston, MA: Kluwer Academic.

Palmer, C. L., & Neumann, L. J. (2002). The information work of interdisciplinary humanities scholars: Exploration and translation. *Library Quarterly, 72*, 85–117.

Pedagogy

Arvidson, P. S. (2015). Cultivating integrity: Balancing autonomy and discipline in integrative programs. In Hughes, P., Muñoz, J. & Tanner, M. N. (Eds.), *Perspectives in interdisciplinary and integrative studies* (pp. 95–116). Lubbock: Texas Tech University Press.

Davis, J. (1995). *Interdisciplinary courses and team teaching: New arrangements for learning.* Phoenix, AZ: Oryx Press.

Foshay, R. (Ed.). (2011). *Valences of interdisciplinarity: Theory, practice, pedagogy.* Edmonton, AB, Canada: Athabasca University Press.

Haynes, C. (Ed.). (2002). *Innovations in interdisciplinary teaching.* American Council on Education. Series on Higher Education. Westport, CT: Oryx Press/Greenhaven Press.

Kain, D. L. (2005). Integrative learning and interdisciplinary studies. *Peer Review, 7*(4), 8–10.

Klein, J. T. (Ed.). (2002). *Interdisciplinary education in K–12 and college: A foundation for K–16 dialogue.* New York, NY: The College Board.

Newell, W. H. (2001). Powerful pedagogies. In B. L. Smith & J. McCann (Eds.), *Reinventing ourselves: Interdisciplinary education, collaborative learning and experimentation in higher education* (pp. 196–211). Bolton, MA: Anker Press.

Newell, W. H. (2006). Interdisciplinary integration by undergraduates. *Issues in Integrative Studies, 24*, 89–111.

Repko, A. F. (2006). Disciplining interdisciplinary studies: The case for textbooks. *Issues in Integrative Studies, 24*, 112–142.

Seabury, M. B. (Ed.). (1999). *Interdisciplinary general education: Questioning outside the lines.* New York, NY: The College Board.

Szostak, R. (2007). How and why to teach interdisciplinary research practice. *Journal of Research Practice, 3*(2), Article M17.

van der Lecq, R. (2016). Self-authorship characteristics of learners in the context of an interdisciplinary curriculum. Evidence from reflections. *Issues in Integrative Studies* 34, 79–108.

Program Development and Sustainability

Augsburg, T., & Henry, S. (Eds.). (2009). *The politics of interdisciplinary studies: Interdisciplinary transformation in undergraduate American higher education.* Jefferson, NC: McFarland.

Carmichael, T. S. (2004). *Integrated studies: Reinventing undergraduate education.* Stillwater, OK: New Forum Press.

Chandramohan, B., & Fallows, S. (Eds.). (2009). *Interdisciplinary learning and teaching in higher education: Theory and practice.* London, UK: Routledge.

Holley, K. A. (2009). Interdisciplinary strategies as transformative change in higher education. *Innovations in Higher Education, 34,* 331–344

Klein, J. T. (2010). *Creating interdisciplinary campus cultures: A model for sustainability and growth.* San Francisco, CA: Jossey-Bass.

Klein, J. T. (2013). The state of the field: Institutionalization of interdisciplinarity. *Issues in Interdisciplinary Studies,* 66–74.

Lyall, C., Bruce, A., Tait, J., & Meagher, L. (2011). *Interdisciplinary research journeys.* London, UK: Bloomsbury Academic.

Seabury, M. B. (Ed.). (1999). *Interdisciplinary general education: Questioning outside the lines.* New York, NY: The College Board.

Smith, B. L., & McCann, J. (Eds.). (2001). *Reinventing ourselves: Interdisciplinary education, collaboration, learning, and experimentation in higher education.* San Francisco, CA: Anker/Jossey-Bass.

Thew, N. (2007). *The impact of the internal economy of higher education institutions on interdisciplinary teaching and learning.* England: University of Southampton.

Research Practice

Atkinson, J., & Crowe, M. (Eds.). (2006). *Interdisciplinary research: Diverse approaches in science, technology, health and society.* West Sussex, England: Wiley.

Bergmann, M., Jahn, T., Knobloch, T. Krohn, W., Pohl, C., & Schramm, E. (2012). *Methods for transdisciplinary research: A primer for practice.* Berlin, Germany: Campus.

Boix Mansilla, V. (2006). Interdisciplinary work at the frontier: An empirical examination of expert interdisciplinary epistemologies. *Issues in Integrative Studies, 24,* 1–31.

Committee on Facilitating Interdisciplinary Research. (2004). *Facilitating interdisciplinary research.* Washington, DC: National Academies Press.

Hadorn, G. H., Hoffmann-Riem, H., Biber-Klemm, S., Grossenbacher-Mansuy, W., Joye, D., Pohl, C., . . . & Zemp, E. (Eds.). (2008). *Handbook of transdisciplinary research.* New York, NY: Springer.

Hesse-Bulber, S., & Johnson, R. B. (Eds.). (2015). *Oxford Handbook of multimethod and mixed method research.* Oxford, UK: Oxford University Press.

Hughes, P.C., Muñoz, J. S., & Tanner, M. N. (Eds.). (2015). *Perspectives in interdisciplinary and integrative studies*. Lubbock: Texas Tech University Press.

Newell, W. H. (2007). Decision making in interdisciplinary studies. In G. Morçöl (Ed.), *Handbook of decision making* (pp. 245–264). New York, NY: Marcel-Dekker.

Palmer, C. L. (2010). Information research on interdisciplinarity. In R. Froedeman, J. T. Klein, & C. Mitcham (Eds.), *The Oxford handbook of interdisciplinarity* (pp. 174–188). New York, NY: Oxford University Press.

Palmer, C. L., & Neumann, L. J. (2002). The information work of interdisciplinary humanities scholars: Exploration and translation. *Library Quarterly, 72,* 85–117.

Repko, A. F., Newell, W. H., & Szostak, R. (2012). *Case studies in interdisciplinary research*. Thousand Oaks, CA: Sage. (*Note:* This book provides several applications of the interdisciplinary research process described in Repko, A. F. (2008). *Interdisciplinary research: Process and theory*. Thousand Oaks, CA: Sage.)

Rowe, J. W. (2003). Approaching interdisciplinary research. In F. Kessel, P. L. Rosenfield, & N. B. Anderson (Eds.), *Expanding the boundaries of health and social science* (pp. 3–12). New York, NY: Oxford University Press.

Szostak, R. (2009). *The causes of economic growth: Interdisciplinary perspectives.* Berlin, Germany: Springer. (*Note:* This is the only book-length application of the research process described in this book.)

Szostak, R. (2012). The interdisciplinary research process. In A. F. Repko, W. H. Newell, & R. Szostak (Eds.), *Case studies in interdisciplinary research* (pp. 3–20). Thousand Oaks, CA: Sage.

Szostak, R. (2013). The state of the field: Interdisciplinary research. *Issues in Interdisciplinary Studies, 31,* 44–65.

Wallis, S. E. (2014). Existing and emerging methods for integrating theories within and between disciplines. *Journal of Organisational Transformation & Social Change, 11,* 3–24.

Weingart, P., & Stehr, N. (2000). *Practising interdisciplinarity.* Toronto, Canada: University of Toronto Press.

Theory

Bhaskar, R., Danermark, B., & Price, L. (2016). *Interdisciplinarity and wellbeing: A critical realist general theory of interdisciplinarity.* London: Routledge.

Boix Mansilla, V. (2006). Interdisciplinary work at the frontier: An empirical examination of expert epistemologies. *Issues in Integrative Studies, 24,* 1–31.

Henry, S. (2018). Beyond interdisciplinary Theory: Revisiting William H. Newell's Integrative theory from a Critical Realist Perspective. *Issues in Interdisciplinary Studies 36:2,* 68–107.

Newell, W. H. (2001a). A theory of interdisciplinary studies. *Issues in Integrative Studies, 19,* 1–25.

Newell, W. H. (2001b). Reply to respondents to "A theory of interdisciplinary studies." *Issues in Integrative Studies, 19,* 135–146.

Newell, W. H. (2010). Educating for a complex world. *Liberal Education, 96*(4), 6–11.

Newell, W. H. (2013). State of the field: Interdisciplinary theory. *Issues in Interdisciplinary Studies, 31,* 22–43.

Repko, A. F. (2007). Integrating interdisciplinarity: How the theories of common ground and cognitive interdisciplinarity are informing the debate on interdisciplinary integration. *Issues in Integrative Studies, 27,* 1–31.

Spooner, M. (2004). Generating integration and complex understanding: Exploring the use of creative thinking tools within interdisciplinary studies. *Issues in Integrative Studies, 22*, 85–111.

Szostak, R. (2004). *Classifying science: Phenomena, data, theory, method, practice.* Dordrecht, The Netherlands: Springer.

Szostak, R. (2007). Modernism, postmodernism, and interdisciplinarity. *Issues in Integrative Studies, 25*, 32–83.

Weingart, P., & Stehr, N. (2000). *Practising interdisciplinarity.* Toronto, Canada: University of Toronto Press.

Welch, J., IV. (2007). The role of intuition in interdisciplinary insight. *Issues in Integrative Studies, 25*, 131–155.

Welch, J., IV. (2009). Interdisciplinarity and the history of western epistemology. *Issues in Integrative Studies, 27*, 35–69.

Welch, J., IV. (2011). The emergence of interdisciplinarity from epistemological thought. *Issues in Integrative Studies, 29*, 1–39.

Welch, J., IV. (2018). The impact of Newell's "A Theory of Interdisciplinary Studies": Reflection and analysis. *Issues in Interdisciplinary Studies, 36*(2), 193–211.

Other Works

Augsburg, T. (2006). *Becoming interdisciplinary: An introduction to interdisciplinary studies* (2nd ed.). Dubuque, IA: Kendall/Hunt.

Czechowski, J. (2003). An integrated approach to liberal learning. *Peer Review, 5*(4), 4–7.

Graff, G. (1991, February 13). Colleges are depriving students of a connected view of scholarship. *The Chronicle of Higher Education*, p. 48.

OVERVIEWS OF THE FIELD

About Interdisciplinarity. Retrieved from http://wwwp.oakland.edu/ais/. [Click on "Resources"]

Bammer, G. (2013). *Disciplining interdisciplinarity.* Canberra, Australia: ANU E-Press.

Bergmann, M., Jahn, T., Knobloch, T. Krohn, W. Pohl, C., & Schramm, E. (2012). *Methods for transdisciplinary research: A primer for practice.* Berlin, Germany: Campus.

Committee on Facilitating Interdisciplinary Research. (2004). *Facilitating interdisciplinary research.* Washington, DC: National Academies Press.

Foshay, R. (Ed.). (2011). *Valences of interdisciplinarity: Theory, practice, pedagogy.* Edmonton, AB: Athabasca University Press.

Frodeman, R., Klein, J. T., & Mitcham, C. (Eds.). (2010). *The Oxford handbook of interdisciplinarity.* New York, NY: Oxford University Press.

Graff, G. (1991). Colleges are depriving students of a connected view of scholarship. *The Chronicle of Higher Education*, February 13, p. 48.

Huber, M. T., Hutchings, P., & Gale, R. (2005). Integrative learning for liberal education. *Peer Review, 7*(4), 4–7.

Kain, D. L. (1993). Cabbages—and kings: Research directions in integrated/interdisciplinary curriculum. *Journal of Educational Thought/Revue de la Pensee Educative, 27*(3), 312–331.

Klein, J. T. (1999). *Mapping interdisciplinary studies.* Washington, DC: Association of American Colleges and Universities.

Klein, J. T., & Newell, W. H. (1997). Advancing interdisciplinary studies. In J. Gaff & J. Ratcliff (Eds.), *Handbook of the undergraduate curriculum: A comprehensive guide to purposes, structures, practices, and change* (pp. 393–415). San Francisco, CA: Jossey-Bass.

Lattuca, L. R. (2001). *Creating interdisciplinarity: Interdisciplinary research and teaching among college and university faculty.* Nashville, TN: Vanderbilt University Press.

Roberts, J. A. (2004). *Riding the momentum: Interdisciplinary research centers to interdisciplinary graduate programs.* Paper presented at the July 2004 Merrill conference, University of Kansas.

Szostak, R. (2019). *Manifesto of interdisciplinarity.* Retrieved from https://sites.google.com/a/ualberta.ca/manifesto-of-interdisciplinarity/manifesto-of-interdisciplinarity.

Walker, D. (1996). *Integrative education.* Eugene, OR: ERIC Clearinghouse on Educational Management.

TEXTBOOKS FOR STUDENTS

Augsburg, T. (2016). *Becoming interdisciplinary: An introduction to interdisciplinary studies* (3rd ed.). Dubuque, IA: Kendall/Hunt.

Menken, S., & Keestra, M. (Eds.). (2016). *An introduction to interdisciplinary research.* Amsterdam, the Netherlands: Amsterdam University Press.

Repko, A. F. (2012). *Interdisciplinary research: Process and theory* (2nd ed.). Thousand Oaks, CA: Sage.

Repko, A. F., Szostak, R., & Buchberger, M. (2020). *Introduction to interdisciplinary studies* (3rd ed.). Thousand Oaks, CA: Sage.

Repko, A. F., Szostak, R., & Newell, W. (Eds.). (2012). *Case studies in interdisciplinary research.* Thousand Oaks, CA: Sage.

GLOSSARY OF KEY TERMS

Adequacy (in interdisciplinary sense): An understanding of each discipline's cognitive map sufficient to identify its perspective on the problem, epistemology, assumptions, concepts, theories, and methods in order to understand its insights into the problem. [6:150]

Antidisciplinary: Preferring a more "open" understanding of "knowledge" and "evidence" that would include "lived experience," testimonials, oral traditions, and interpretation of those traditions by elders (Vickers, 1998, pp. 23–26). [1:12]

Applied fields: These include business (and its many subfields such as finance, marketing, management), communications (and its various subfields including advertising, speech, and journalism), criminal justice and criminology, education, engineering, law, medicine, nursing, social work. [1:5]

Assumptions: The principles that underlie the discipline as a whole and its overall perspective on reality. As the term implies, these principles are accepted as the truths upon which the discipline's theories, concepts, methods, and curriculum are based. [2:53]

Causal arguments: Examine the underlying cause for any particular situation or argument, and analyze what causes a trend, an event, or a certain outcome. [12:334]

Causal or propositional integration: Refers to combining truth claims from disciplinary theoretical explanations to form an integrated theory—that is, a new proposition that is interdisciplinary and more comprehensive. [12:334]

Causal relationship or link: Refers to how a change in one variable produces or leads to change in another variable. Theories explain causal relationships. [11:305]

Close reading: A method of modern criticism that calls for careful analysis of a text and close attention to individual words, syntax, and the order in which sentences and ideas unfold. [1:5]

Cognitive discord: Disagreement among a discipline's practitioners over the defining elements of the discipline. [2:36]

Cognitive advancement: A more comprehensive understanding. [13:366]

Common ground: The shared basis that exists between conflicting disciplinary insights or theories and makes integration possible. [1:20]

Common ground integrator: The *concept, assumption,* or *theory*—by which these conflicting insights (whether disciplinary or stakeholder) can be integrated. [8:228]

Complexity: Refers to the parts of a phenomenon or problem that interact in surprising/unexpected ways. [1:15]

Complexity (operational definition): The problem has multiple parts studied by different disciplines. [3:94]

Concept: A symbol expressed in language that represents a phenomenon or an abstract idea generalized from particular instances. [2:58]

Concept or principle map: Organizes information about the problem showing meaningful relationships between the parts of the problem that requires thinking through all the parts of the problem and anticipating how these behave or function. [4:114]

Conceptual integration (theory of): Explains the innate human ability to create new meaning by blending concepts and create new ones. [1:22]

Contextualization: The practice of placing "a text, or author, or work of art into context, to understand it in part through an examination of its historical, geographical, intellectual, or artistic location" (Newell, 2001, p. 4). [1:16]

Creating: Involves putting elements together—integrating them—to produce something that is new and useful. [1:21]

Critical interdisciplinarity: A society-driven approach that "interrogates the dominant structure of knowledge and education with the aim of transforming them, while raising epistemological and political questions of value and purpose" (Klein, 2010, p. 30). [1:10]

Defining elements of a discipline's perspective: The phenomena it studies, its epistemology or rules about what constitutes evidence, the assumptions it makes about the natural and human world, its basic concepts or vocabulary, its theories about the causes and behaviors of certain phenomena, and its methods (the way it gathers, applies, and produces new knowledge). [2:34]

Disciplinarity: The system of knowledge specialties called disciplines. [2:36]

Disciplinary: Relating "to a particular field of study" or specialization. [1:4]

Disciplinary bias: To state the problem using words and phrases that connect it to a particular discipline. [3:87]

Disciplinary inadequacy: The view that the disciplines by themselves are inadequate to address complex problems. [1:12]

Disciplinary jargon: Using technical terms and concepts that are not generally understood outside a particular discipline. [3:87]

Disciplinary method: The particular procedure or process or technique used by a discipline's practitioners to conduct, organize, and present research. [6:162]

Disciplinary perspective: A discipline's view of reality in a general sense, which embraces and in turn reflects the ensemble of its defining elements that include phenomena, epistemology, assumptions, concepts, theories, and methods. [2:35]

Disciplinary theory: Explains a behavior or phenomenon that falls within a discipline's traditional research domain and *may* have a specified range of applicability. [11:302]

Discipline: A particular branch of learning or body of knowledge such as physics, psychology, or history. [1:4]

Disciplines: Scholarly communities that specify which phenomena to study, advance certain central concepts and organizing theories, embrace certain methods of investigation, provide forums for sharing research and insights, and offer career paths for scholars. [2:36]

Epistemic norms of a discipline: Agreements about how researchers should select their evidence or data, evaluate their experiments, and judge their theories. [2:47]

Epistemological pluralism: Refers to the diverse attitudes that disciplines have about how to know and describe reality. [1:11]

Epistemological self-reflexivity: Awareness of how epistemological choices tend to influence one's selection of research methods that, in turn, influence research outcomes (Bell, 1998, p. 101). [2:53]

Epistemology: The branch of philosophy that studies how one knows what is true and how one validates truth. [2:47]

Extension: A technique used to increase the scope of the "something" that we are talking about. [10:283]

Feedback: Corrective information about a decision, an operation, an event, or a problem that compels the researcher to revisit an earlier STEP. [3:80]

Feedback loops: Central elements of systems thinking that describe a reciprocal relationship among variables. [3:80]

Full integration: Where all relevant disciplinary insights have been integrated into a new, single, coherent, and comprehensive understanding or theory that is consistent with the available empirical evidence. [8:226]

Generalist interdisciplinarians: Understand interdisciplinarity loosely to mean "any form of dialog or integration between two or more disciplines" while minimizing, obscuring, or rejecting altogether the role of integration (Moran, 2010, p. 14). [1:20]

Heuristic: An aid to understanding or discovery or learning. [3:79]

Holistic thinking: The ability to understand how ideas and information from relevant disciplines relate to each other and to the problem (Bailis, 2002, pp. 4–5). [1:18]

Ideographic: Referring to a theory that is applicable to only a narrow range of phenomena and under a constrained set of circumstances. [6:162]

Ideographic theory: Posits a relationship only under specified conditions. [6:162]

Insight: A scholarly contribution to the understanding of a problem based on research. Insights into a problem can be produced either by disciplinary experts or by interdisciplinarians. [1:4]

Instrumental interdisciplinarity: A pragmatic problem-driven approach that focuses on research, borrowing (supplemented by integration), and practical problem solving in response to the external demands of society. [1:10]

Integration: The cognitive process of critically evaluating disciplinary insights and creating common ground among them to construct a more comprehensive understanding of the problem. [1:23]

Integrationist interdisciplinarians: Believe that integration should be the *goal* of interdisciplinary work because integration addresses the challenge of complexity. [1:20]

Integrationist position: That integration is achievable and that researchers should achieve the greatest degree of integration possible given the problem under study and the disciplinary insights at their disposal. [1:20]

Integrative studies: Seeks to integrate various elements of student experience such as coursework and residential life. [1:27]

Integrative wisdom: The synthetic interaction between inspiration, intellect, and intuition. [10:275]

Intellectual center of gravity: That which enables each discipline to maintain its identity and have a distinctive overall perspective. [2:37]

Interdisciplinarity: The intellectual essence of the field of interdisciplinary studies which integrates insights from multiple disciplines after evaluating these in the context of disciplinary perspective. [1:10]

Interdisciplinary common ground: Involves modifying one or more concepts or theories and their underlying assumptions. [10:270]

Interdisciplinary complexity theory: States that interdisciplinary study is necessitated when the problem or question is multifaceted and functions as a "system." [1:15]

Interdisciplinary integration: The cognitive process of critically evaluating disciplinary insights and creating common ground among them to construct a more comprehensive understanding. [8:223]

Interdisciplinary research: A decision-making process that is heuristic, iterative, and reflexive. [3:78]

Interdisciplinary research process (IRP): A practical and demonstrated way to make decisions about how to approach problems, decide which ones are appropriate for interdisciplinary inquiry, and construct comprehensive understandings of them. [3:78]

Interdisciplinary studies: A process of answering a question, solving a problem, or addressing a topic that is too broad or complex to be dealt with adequately by a single discipline, and draws on the disciplines with the goal of integrating their insights to construct a more comprehensive understanding. [1:9]

Interdisciplines: Fields of study that cross traditional disciplinary boundaries and whose subject matter is taught by informal groups or scholars or by well-established research and teaching faculties. [1:6]

In-text evidence of disciplinary adequacy: Statements about the disciplinary elements that pertain to the problem, the disciplinary affiliation of leading theorists, and the disciplinary methods used. [6:176]

Intuition: The natural ability to understand or perceive something immediately without consciously using reason, analysis, or inference. [10:274]

Iterative: Procedurally repetitive. [3:80]

Literature search: The process of gathering scholarly information on a given topic. [5:128]

Meaning: Important concept in the humanities where it is often equated with the intent of the author or artist, or the effect on the audience. [1:9]

Method: Concerns how one conducts research, analyzes data or evidence, tests theories, and creates new knowledge. [2:60]

Model: A representation that serves to visualize and communicate a theory. [11:303]

More comprehensive understanding: The result of the integration of *insights* to produce a new and more complete and perhaps nuanced understanding. [12:328]

Most relevant disciplines: Those disciplines, often three or four, which are most directly connected to the problem, have generated the most important research on it, and have advanced the most compelling theories to explain it. These disciplines, or parts of them, provide information about the problem that is essential to developing a comprehensive understanding of it. [4:118]

Multidisciplinarity: The placing side by side of insights from two or more disciplines. [1:24]

Multidisciplinary research: "Involves more than a single discipline in which each discipline makes a *separate* contribution (National Academies, 2005, p. 27). [1:25]

Multidisciplinary studies: Merely bringing insights from different disciplines together in some way but failing to engage in the additional work of integration. [1:25]

Narrow interdisciplinarity: It draws on disciplines that are epistemologically close (e.g., physics and chemistry). [10:273]

New humanities: "Interrogates the dominant structure of knowledge and education with the aim of transforming them" with the "explicit intent of deconstructing disciplinary knowledge and boundaries" (Klein, 2010, p. 30). [2:56]

Nomothetic: Referring to a theory that is applicable to a broad range of phenomena. [6:162]

Nomothetic theory: Posits a general relationship among two or more phenomena. [6:162]

Organization: A technique that creates common ground by clarifying how certain phenomena interact and mapping their causal relationships. [10:288]

Partial integration: When only some insights have been integrated and it applies to only some part(s) of the problem. [8:226]

Peer review: Subjecting an author's scholarly paper or book manuscript to the scrutiny of experts in the field who evaluate it according to certain academic standards that are viewed as fair and rigorous by the discipline's members. [5:129]

Personal bias: To inject one's own point of view when introducing the problem. [3:88]

Perspectival approach: The relying on each discipline's unique perspective on reality as presented in Table 2.2. [2:43]

Perspective taking: Involves analyzing the problem from the standpoint or perspective of each interested discipline and identifying their commonalities and differences. [1:16]

Phenomena: Enduring aspects of human existence that are of interest to scholars and are susceptible to scholarly description and explanation. [2:41]

Potentially relevant discipline: One whose research domain includes at least one *phenomenon* involved in the problem or research question, whether or not its community of scholars has recognized the problem and published its research. [4:104]

Problem-based research: Focuses on unresolved societal needs, practical problem solving, and intellectual problems that are the focus of the humanities, such as the meaning of some artifact. [3:95]

Process: Following a procedure or strategy. [3:78]

Qualitative approach: Method focusing on evidence that cannot easily be quantified, such as cultural mannerisms and personal impressions of a musical composition. [2:61]

Qualitative research strategies: Focus on the what, how, when, and where of a thing—its essence and its ambiance. Qualitative research, then, refers to meanings, concepts, definitions, characteristics, metaphors, symbols, and descriptions of things or people that are not measured and expressed numerically. [6:165]

Quantitative approach: Method emphasizing that evidence can be expressed numerically over a specified time frame. [2:61]

Quantitative research strategies: Emphasize evidence that can be quantified, such as the number of atoms in a molecule, the flow rate of water in a river, or the amount of energy derived from a windmill. [6:165]

Redefinition: A technique used to modify or redefine *concepts* in different texts and contexts to bring out a common meaning. [10:279]

Reflection in an interdisciplinary sense: A self-conscious activity that involves thinking about why certain choices were made at various points in the research process and how these choices have affected the development of the work. [13:358]

Reflexive: To become self-conscious or self-aware of our disciplinary or personal biases that may influence our inquiry and possibly skew our analysis of insights and thus bias the end product. [3:81]

Research map: Helps those new interdisciplinary research visualize the research process from beginning to end. [4:114]

Researchable in an interdisciplinary sense: Authors from at least two disciplines have written on the topic or at least on some aspect of it. [3:84]

Scholarly knowledge: Knowledge that has been vetted by a discipline's community of scholars through its peer review process. [5:128]

Scholarship: A contribution to knowledge that is "public, susceptible to *critical review and evaluation*, and accessible for *exchange and use* by other members of one's scholarly community" (Shulman, 1998, p. 5). [1:8]

Scientific method: Method (*idealized*) of producing new knowledge, has four steps: (1) observation and description of phenomena and processes; (2) formulation of a hypothesis to explain the phenomena; (3) use of the hypothesis to predict the existence of other phenomena, or to predict quantitatively the result of new observations; (4) execution of properly performed experiments to test those hypotheses or predictions. The scientific method is based on beliefs in empiricism (whether the observation is direct or indirect), quantifiability (including precision in measurement), replicability or reproducibility, and free exchange of information (so that others can test or attempt to replicate or reproduce). [2:65]

Scope: Refers to the parameters of what you intend to include and exclude from consideration. [3:86]

Skewed understanding: The degree to which an insight reflects the biases inherent in the discipline's perspective, and thus the way the author understands the problem. [7:193]

Studies: Typically refers to cultural groups (including women, Hispanics, and African Americans) and also appears in a host of contexts in the natural sciences and social sciences. In fact, "studies" programs are proliferating in the modern academy. [1:6]

Subdiscipline: A subdivision of an existing discipline. [1:8]

System map: A visual that shows all the parts of a system (complex problem) and illustrates the causal relationships among them to help the researcher to visualize the system as a complex whole. [4:108]

Systems thinking: A method of visualizing interrelationships within a complex problem or system by (1) breaking it down into its constituent parts, (2) identifying which parts different disciplines address, (3) evaluating the relative importance of different causal linkages, and (4) recognizing that a system of linkages is much more than the sum of its parts. [3:80]

Taxonomy: A systematic and orderly classification of selected disciplines and their perspectives. [2:40]

Theory: A generalized scholarly explanation about some aspect of the natural or human world, how it works, and why specific facts are related that is supported by data and research. [1:14]

Theory extension: Involves taking well-known facts normally treated as exogenous (i.e., external) to the disciplinary theory (such as organisms altering their environment) and making them endogenous (i.e., internal) to and mutually interactive with it. [11:312]

Theory map: Describes a theory's supporting evidence and importance and compares it to other theories. [4:114]

Transdisciplinarity: Complementary to interdisciplinarity, involves the integration also of *insights generated outside the academy*, a team approach to research, the active involvement of non-academic participants in research design, and a "case study" approach. [1:27]

Transformation: A technique used to modify concepts or assumptions that are not merely different (e.g., love, fear, selfishness) but opposite (e.g., rational, irrational) into continuous variables. [10:285]

Triangulation of research methodology: Uses multiple data-gathering techniques to investigate the same problem/system/process. In this way, findings can be cross-checked, validated, and confirmed. [6:175]

Variable: "Something that can take on different values or assume different attributes— it is something that varies" (Remler & van Ryzin, 2011, p. 31). [11:304]

Wide interdisciplinarity: It draws on disciplines that are epistemologically much further apart (e.g., art history and mathematics). [10:273]

REFERENCES

Ackerson, L. G. (2007). Introduction. In L. G. Ackerson (Ed.), *Literature search strategies for interdisciplinary research: A sourcebook for scientists and engineers* (pp. vii–xvii). Lanham, MD: Scarecrow Press.

Adams, L. S. (1996). *The methodologies of art: An introduction.* Boulder, CO: Westview Press.

Agger, B. (1998). *Critical social theories: An introduction.* Boulder, CO: Westview Press.

Akers, R. (1994). *Criminological theories: Introduction, evaluation and application.* Los Angeles, CA: Roxbury.

Alford, R. R. (1998). *The craft of inquiry: Theories, methods, evidence.* New York, NY: Oxford University Press.

Alliance for Childhood. (1999). *Fool's gold: A critical look at computers in childhood.* Retrieved July 14, 2011, from http://www.allianceforchildhood.net.

Alvesson, M. (2002). *Postmodernism and social research.* Philadelphia, PA: Open University Press.

American Sociological Association. (n.d.). *Society and social life.* Retrieved July 18, 2011, from http://www.asanet.org/employment/society.cfm.

Anderson, L. W., Krathwohl, D. R., Airasian, P. W., Cruikshank, K. A., Mayer, R. E., Pintrich, P. R., . . . Wittrock, M. C. (2000). *Taxonomy for learning, teaching, and assessing: A revision of Bloom's taxonomy of educational objectives* (2nd rev. ed.). Boston, MA: Allyn & Bacon.

Arms, L. A. (2005). *Mathematics and religion: Processes of faith and reason.* Unpublished manuscript, Western College Program, Miami of Ohio University.

Arvidson, S. (2016). Interdisciplinary research and phenomenology as parallel processes of consciousness. *Issues in Interdisciplinary Studies, 34,* 30–51.

Atkinson, J., & Malcolm Crowe, M. (Eds.). (2006). *Interdisciplinary research: Diverse approaches in science, technology, health and society.* West Sussex, England: John Wiley & Sons.

Atran, S. (2003a, March 7). Genesis of suicide terrorism. *Science, 299,* 1534–1539.

Atran, S. (2003b). *Genesis and future of suicide terrorism.* Retrieved August 14, 2006, from http://interdisciplines.org/terrorism/papers/1.

Bailis, S. (2001). Contending with complexity: A response to William H. Newell's "A Theory of Interdisciplinary Studies." *Issues in Integrative Studies, 19,* 27–42.

Bailis, S. (2002). Interdisciplinary curriculum design and instructional innovation: Notes on the social science program at San Francisco State University. In C. Haynes (Ed.), *Innovations in interdisciplinary teaching* (pp. 3–15). Westport, CT: Oryx Press.

Bal, M. (1999). Introduction. In M. Bal (Ed.), *The practice of cultural analysis: Exposing interdisciplinary interpretation* (pp. 1–14). Stanford, CA: Stanford University Press.

Bal, M. (2002). *Traveling concepts in the humanities: A rough guide.* Buffalo, NY: University of Toronto Press.

Bal, M., & Bryson, N. (1991). Semiotics and art history. *The Art Bulletin, 73*(2), 174–208.

Bammer, G. (2005). Integration and integration sciences: Building a new specialization. *Ecology and Society, 10*(2), 6.

Bandura, A. (1998). Mechanism of moral disengagement. In W. Reich (Ed.), *Origins of terrorism: Psychologies, ideologies, theologies, states of mind* (pp. 161–191). Washington, DC: Woodrow Wilson Center Press.

Barnet, S. (2008). *A short guide to writing about art* (9th ed.). Upper Saddle River, NJ: Pearson Prentice Hall.

Beauchamp, T. L., & Childress, J. F. (2001). *Principles of biomedical ethics* (5th ed.). Oxford, UK: Oxford University Press.

Becher, T., & Trowler, P. R. (2001). *Academic tribes and territories* (2nd ed.). Buckingham, UK: The Society for Research into Higher Education & Open University Press.

Bechtel, W. (2000). From imagining to believing: Epistemic issues in generating biological data. In R. Creath & J. Maienschein (Eds.), *Biology and epistemology* (pp. 138–163). Cambridge, UK: Cambridge University Press.

Bell, J. A. (1998). Overcoming dogma in epistemology. *Issues in Integrative Studies, 16,* 99–119.

Berg, B. L. (2004). *Qualitative research methods for the social sciences* (5th ed.). Boston, MA: Pearson Education.

Bergmann, M., Jahn, T., Knobloch, T., Krohn, W., Pohl, C., & Schramm, E. (2012). *Methods for transdisciplinary research: A primer for practice.* Berlin, Germany: Campus.

Bernard, H. R. (2002). *Research methods in anthropology: Qualitative and quantitative methods* (3rd ed.). New York, NY: AltaMira Press.

Bhaskar, R., Danermark, B., & Price, L. (2016). *Interdisciplinarity and wellbeing: A critical realist general theory of interdisciplinarity.* London, UK: Routledge.

Blackburn, S. (1999). *Think: A compelling introduction to philosophy.* Oxford, UK: Oxford University Press.

Blesser, B., & Salter, L. R. (2007). *Spaces speak, are you listening?* Cambridge, MA: The MIT Press.

Boix Mansilla, V. (2005, January/February). Assessing student work at disciplinary crossroads. *Change, 37,* 14–21.

Boix Mansilla, V. (2006). Interdisciplinary work at the frontier: An empirical examination of expert interdisciplinary epistemologies. *Issues in Integrative Studies, 24,* 1–31.

Boix Mansilla, V. (2009). *Learning to synthesize: A cognitive-epistemological foundation for interdisciplinary learning.* Retrieved September 22, 2019, from https://www.semanticscholar.org/paper/Learning-to-synthesize-%3A-A-foundation-for-learning-Mansilla/aca3b2fde63096b305379491b6b6d0143939089d.

Boix Mansilla, V., Duraisingh, E. D., Wolfe, C., & Haynes, C. (2009, May/June). Targeted assessment rubric: An empirically grounded rubric for interdisciplinary writing. *The Journal of Higher Education, 80*(3), 334–353.

Boix Mansilla, V., Miller, W. C., & H. (2000). On disciplinary lenses and interdisciplinary work. In S. Wineburg & P. Gossman (Eds.), *Interdisciplinary curriculum: Challenges to implementation* (pp. 17–38). New York, NY: Teachers College, Columbia University.

Boon, M., & Van Baalen, S. (2019). Epistemology for interdisciplinary research—shifting philosophical paradigms of science. *European Journal of Philosophy of Science, 9*(16). Retrieved from https://doi.org/10.1007/s13194-018-0242-4.

Booth, W. C., Columb, G. G., & Williams, J. M. (2003). *The craft of research* (2nd ed.). Chicago, IL: University of Chicago Press.

Boulding, K. (1981). *A preface to grants economics: The economy of love and fear.* New York, NY: Praeger.

Bradsford, J. D., Brown, A. L., & Cocking, R. R. (Eds.). (1999). *How people learn: Brain, mind, experience, and school.* Washington, DC: National Academy Press.

Brassler M., & Block M. (2017). Interdisciplinary teamwork on sustainable development—The top ten strategies based on experience of student-initiated projects. In W. Leal Filho, U. Azeiteiro, F. Alves, & P. Molthan-Hill (Eds.), *Handbook of theory and practice of sustainable development in higher education.* Berlin, Germany: Springer,

Bressler, C. E. (2003). *Literary criticism: An introduction to theory and practice* (3rd ed). Upper Saddle River, NJ: Pearson Education.

Bromme, R. (2000). Beyond one's own perspective: The psychology of cognitive interdisciplinarity. In P. Weingart & N. Stehr (Eds.), *Practising interdisciplinarity* (pp. 115–133). Toronto, Canada: University of Toronto Press.

Brough, J. A., & Pool, J. E. (2005). Integrating learning and assessment: The development of an assessment culture. In J. Etim (Ed.), *Curriculum Integration K–12: Theory and practice* (pp. 196–204). Lanham, MD: University Press of America.

Brown, R. H. (1989). Textuality, social science, and society. *Issues in Integrative Studies, 7,* 1–19. Cited in S. Henry & N. L. Bracy (2012). Integrative theory in criminology applied to the complex social problem of school violence. In A. F. Repko, W. H. Newell, & R. Szostak (Eds.), *Case studies in interdisciplinary research* (pp. 259–282). Thousand Oaks, CA: Sage.

Bryman, A. (2004). *Social research methods* (2nd ed). New York, NY: Oxford University Press.

Buzan, T. (2010). *The mindmap book.* London: BBC Books.

Calhoun, C. (Ed.). (2002). *Dictionary of the social sciences.* Oxford, UK: Oxford University Press.

Carey, S. S. (2003). *A beginner's guide to scientific method* (2nd ed.). Belmont, CA: Wadsworth.

Carp, R. M. (2001). Integrative praxes: Learning from multiple knowledge formations. *Issues in Integrative Studies, 19,* 71–121.

Choi S., & Richards K. (2017). *Interdisciplinary Discourse: Communicating Across Disciplines.* London: Palgrave Macmillan.

Clark, H. H. (1996). *Using language.* Cambridge, MA: Cambridge University Press.

Connor, M. A. (2012). The metropolitan problem in interdisciplinary perspective. In A. F. Repko, W. H. Newell, & R. Szostak (Eds.), *Case studies in interdisciplinary research* (pp. 53–90). Thousand Oaks, CA: Sage.

Cooke, N. J., & Hilton, M. L. (Eds.). (2015). *Enhancing the effectiveness of team science.* Washington, DC: National Research Council.

Cornwell, G. H., & Stoddard, E. W. (2001). Toward an interdisciplinary epistemology: Faculty culture and institutional change. In B. L. Smith & J. McCann (Eds.), *Reinventing ourselves: Interdisciplinary education, collaborative learning, and experimentation in higher education* (pp. 160–178). Bolton, MA: Anker.

Crenshaw, M. (1998). The logic of terrorism: Terrorist behavior as a product of strategic choice. In W. Reich (Ed.), *Origins of terrorism: Psychologies, ideologies, theologies, states of mind* (pp. 7–24). Washington, DC: Woodrow Wilson Center Press.

Creswell, J. W. (1997). *Qualitative inquiry and research design: Choosing among five traditions.* Thousand Oaks, CA: Sage.

Creswell, J. W. (2002). *Research design: Qualitative, quantitative, and mixed methods approaches* (2nd ed.). Thousand Oaks, CA: Sage.

Csikszentmihalyi, M., & Sawyer, K. (1995). Creative insight: The social dimension of a solitary moment. In R. Sternberg & J. Davidson (Eds.), *The nature of insight* (pp. 329–363). Cambridge, MA: The MIT Press.

Cullenberg, S., Amariglio, J., & Ruccio, D. (2001). Introduction. In S. Cullenberg, J. Amariglio, & D. Ruccio (Eds.), *Postmodernism, economics and knowledge* (pp. 3–57). New York, NY: Routledge.

Czuchry, M., & Dansereau, D. F. (1996). Node-link mapping as an alternative to traditional writing assignments in undergraduate courses. *Teaching of Psychology, 23*, 91–96.

Darbellay, F., Moody, Z., Sedooka, A., & Steffen, G. (2014). Interdisciplinary research boosted by serendipity. *Creativity Research Journal, 26*(1), 1–10.

Davis, J. R. (1995). *Interdisciplinary courses and team teaching: New arrangements for learning.* Phoenix, AZ: American Council on Education, Oryx.

Delph, J. B. (2005). *An integrative approach to the elimination of the "perfect crime."* Unpublished manuscript, University of Texas at Arlington.

Denzin, N. K., & Lincoln, Y. S. (Eds.). (2005). *The SAGE handbook of qualitative research* (3rd ed.). Thousand Oaks, CA: Sage.

Dessalles, J.-L. (2007). Why we talk: The evolutionary origins of language. Oxford, UK: Oxford University Press. 2007.

Dietrich, W. (1995). *Northwest passage: The great Columbia River.* Seattle, WA: University of Washington Press.

Dogan, M., & Pahre, R. (1989). Fragmentation and recombination of the social sciences. *Studies in Comparative International Development, 24*, 56–73.

Dogan, M., & Pahre, R. (1990). *Creative marginality: Innovation at the intersections of the social sciences.* Boulder, CO: Westview Press.

Dominowski, R. L., & Ballob, P. (1995). Insights and problem solving. In R. Sternberg & J. Davidson (Eds.), *The nature of insight* (pp. 33–62). Cambridge, MA: The MIT Press.

Donald, J. (2002). *Learning to think: Disciplinary perspectives.* San Francisco, CA: Jossey-Bass.

Dorsten, L. E., & Hotchkiss, L. (2005). *Research methods and society: Foundations of social inquiry.* Upper Saddle River, NJ: Prentice-Hall.

Dow, S. (2001). Modernism and postmodernism: A dialectical analysis. In S. Cullenberg, J. Amariglio, & D. F. Ruccio (Eds.), *Postmodernism, economics and knowledge* (pp. 61–101). New York, NY: Routledge.

Dunbar, R. (1996). *Grooming, gossip and the evolution of language.* Cambridge, MA: Harvard University Press.

Education for Change, Ltd., SIRU at the University of Brighton, and the Research Partnership. (2002). *Researchers' use of libraries and other information sources: Current patterns and future trends.* London, UK: Higher Education Funding Council for England. Retrieved January 25, 2010, from http://www.rslg.ac.uk/research/libuse/LUrep1.pdf.

Eilenberg, S. (1999). Voice and ventriloquy in "The Rime of the Ancient Mariner." In P. H. Fry (Ed.), *Samuel Taylor Coleridge: The rime of the ancient mariner* (pp. 282–314). Boston, MA: Bedford/St. Martin's.

Elliott, D. J. (2002). Philosophical perspectives on research. In R. Colwell & C. Richardson (Eds.), *The new handbook of research on music teaching and learning* (pp. 85–102). Oxford, UK: Oxford University Press.

Englehart, L. (2005). *Organized environmentalism: Towards a shift in the political and social roles and tactics of environmental advocacy groups.* Unpublished manuscript, Miami of Ohio University.

Etzioni, A. (1988). *The moral dimension: Towards a new economics.* New York, NY: Free Press.

Fauconnier, G. (1994). *Mental spaces: Aspects of meaning construction in modern language.* Cambridge, UK: Cambridge University Press.

Ferguson, F. (1999). Coleridge and the deluded reader: "The Rime of the Ancient Mariner." In P. H. Fry (Ed.), *Samuel Taylor Coleridge: The rime of the ancient mariner* (pp. 113–130). Boston, MA: Bedford/St. Martin's.

Fernie, E. (1995). Glossary of concepts. In E. Fernie (Ed.), *Art history and its methods: A critical anthology* (pp. 323–368). London, UK: Phaidon Press.

Field, M., Lee, R., & Field, M. L. (1994). Assessing interdisciplinary learning. *New Directions in Teaching and Learning, 58*, 69–84.

Fiscella, J. B., & Kimmel, S. E. (Eds.). (1999). *Interdisciplinary education: A guide to resources.* New York, NY: The College Board.

Fischer, C. C. (1988). On the need for integrating occupational sex discrimination theory on the basis of causal variables. *Issues in Integrative Studies, 6*, 21–50.

Foisy, M. (2010). *Creating meaning in everyday life: An interdisciplinary understanding.* Unpublished manuscript, University of Alberta, Canada.

Foster, A. (2004). A nonlinear model of information-seeking behavior. *Journal of the American Society for Information Science and Technology, 55*(3), 228–237.

Foster, H. (1998). Trauma studies and the interdisciplinary: An overview. In A. Coles & A. Defert (Eds.), *The anxiety of interdisciplinarity* (pp. 157–168). London, UK: BACKless Books.

Frankfort-Nachmias, C., & Nachmias, D. (2008). *Research methods in the social sciences* (7th ed.). New York, NY: Worth.

Frodeman, R., Klein, J. T., Mitcham, C., & Holbrook, J. B. (Eds.). (2010). *The Oxford handbook of interdisciplinarity.* New York, NY: Oxford University Press.

Frug, G. E. (1999). *City making: Building communities without building walls.* Princeton, NJ: Princeton University Press.

Fry, P. H. (Ed.). (1999). *Samuel Taylor Coleridge: The rime of the ancient mariner.* Boston, MA: Bedford/St. Martin's.

Fuchsman, K. (2009). Rethinking integration in interdisciplinary studies. *Issues in Integrative Studies, 27*, 70–85.

Fuchsman, K. (2012). Interdisciplines and interdisciplinarity: Political psychology and psychohistory compared. *Issues in Integrative Studies, 30*, 128–154.

Fussell, S. G., & Kraus, R. M. (1991). Accuracy and bias in estimates of others' knowledge. *European Journal of Social Psychology, 21*, 445–454.

Fussell, S. G., & Kraus, R. M. (1992). Coordination of knowledge in communication: Effects of speakers' assumptions about what others know. *Journal of Personality and Social Psychology, 62*, 378–391.

Galinsky, A. D., & Moskowitz, G. B. (2000). Perspective-taking: Decreasing stereotype expression, stereotype accessibility, and in-group favoritism. *Journal of Personality and Social Psychology, 78*(4), 708–724.

Garber, M. (2001). *Academic instincts*. Princeton, NJ: Princeton University Press.

Gauch, H. G., Jr. (2002). *Scientific method in practice*. Cambridge, UK: Cambridge University Press.

Geertz, C. (1983). *Local knowledge: Further essays in interpretative anthropology*. New York, NY: Basic Books.

Geertz, C. (2000). The strange estrangement: Charles Taylor and the natural sciences. In C. Geertz (Ed.), *Available light: Anthropological reflections on philosophical topics* (pp. 143–159). Princeton, NJ: Princeton University Press.

Gerber, R. (2001). The concept of common sense in the workplace learning and experience. *Education + Training, 43*(2), 72–81.

Gerring, J. (2001). *Social science methodology: A critical framework*. Boston, MA: Cambridge University Press.

Giere, R. N. (1999). *Science without laws*. Chicago, IL: University of Chicago Press.

Goldenberg, S. (1992). *Thinking methodologically*. New York, NY: Harper Collins.

Goodin, R. E., & Klingerman, H. D. (Eds.). (1996). *A new handbook of political science*. New York, NY: Oxford University Press.

Griffin, G. (2005). Research methods for English studies: An introduction. In G. Griffin (Ed.), *Research methods for English studies* (pp. 1–16). Edinburgh, Scotland: Edinburgh University Press.

Hacking, I. (2004). *The complacent disciplinarian*. Retrieved July 18, 2011, from https://apps.lis.illinois.edu/wiki/download/attachments/2656520/Hacking.complacent.pdf.

Hagan, F. E. (2005). *Essentials of research methods in criminal justice and criminology*. Boston, MA: Allyn & Bacon.

Hall, D. J., & Hall, I. (1996). *Practical social research: Project work in the community*. Basingstoke, UK: Macmillan.

Halpern, D. F. (1996). *Thought and knowledge* (3rd ed.). Mahwah, NJ: Erlbaum.

Harris, J. (2001). *The new art history: A critical introduction*. New York, NY: Routledge.

Hart, C. (1998). *Doing a literature review: Releasing the social science research imagination*. Thousand Oaks, CA: Sage.

Hatfield, E., & Rapson, R. (1996). *Love and sex: Cross-cultural perspectives*. Boston, MA: Allyn & Bacon.

Hemminger, B. M., Lu, D., Vaughn, K. T. L., & Adams, S. J. (2007). Information seeking behavior of academic scientists. *Journal of the American Society for Information Science and Technology, 58*(14), 2205–2225.

Henry, S. (2009). School violence beyond Columbine: A complex problem in need of an interdisciplinary analysis. *American Behavioral Scientist, 52*(8), 1–20.

Henry, S. (2018). Beyond interdisciplinary theory: Revisiting William H. Newell's integrative theory from a Critical Realist Perspective. *Issues in Interdisciplinary Studies, 36*(2), 68–107.

Henry, S., & Bracy, N. L. (2012). Integrative theory in criminology applied to the complex social problem of school violence. In A. F. Repko, W. H. Newell, & R. Szostak (Eds.), *Case studies in interdisciplinary research* (pp. 259–282). Thousand Oaks, CA: Sage.

Hesse-Biber, S. N., & Johnson, R. B. (Eds.). *The Oxford handbook of multimethod and mixed methods research inquiry*. Oxford, UK: Oxford University Press.

Holmes, F. L. (2000). The logic of discovery in the experimental life sciences. In R. Creath & J. Maienschein (Eds.), *Biology and epistemology* (pp. 167–190). Cambridge, UK: Cambridge University Press.

Hooker, B. (2000). *Ideal code, real world: A rule-consequentialist theory of morality*. Oxford, UK: Oxford University Press.

Howell, M., & Prevenier, W. (2001). *From reliable sources: An introduction to historical methods*. Ithaca, NY: Cornell University Press.

Huber, M. T., & Hutchings, P. (2004). *Integrative learning: Mapping the terrain*. Washington, DC: The Association of American Colleges and Universities.

Huber, M. T., & Morreale, S. P. (2002). Situating the scholarship of teaching and learning: A cross-disciplinary conversation. In M. T. Huber & S. P. Morreale (Eds.), *Disciplinary styles in the scholarship of teaching and learning: Exploring common ground* (pp. 1–24). Stanford, CA: The Carnegie Foundation.

Hughes, P. C., Muñoz, J. S., & Tanner, M. N. (Eds.). (2015). *Perspectives in interdisciplinary and integrative studies*. Lubbock: Texas Tech University Press.

Hyland, K. (2004). *Disciplinary discourses: Social interactions in academic writing*. Ann Arbor: University of Michigan Press.

Hyneman, C. S. (1959). *The study of politics: The present state of American political science*. Champaign: University of Illinois Press.

Iggers, G. G. (1997). *Historiography in the twentieth century: From scientific objectivity to postmodern challenges*. Middletown, CT: Wesleyan University Press.

Karlqvist, A. (1999). Going beyond disciplines: The meanings of interdisciplinarity. *Policy Sciences, 32*, 379–383.

Keestra, M. (2012). Understanding human action: Integrating meanings, mechanisms, causes, and contexts. In A. F. Repko, W. H. Newell, & R. Szostak (Eds.), *Case studies in interdisciplinary research* (pp. 225–258). Thousand Oaks, CA: Sage.

Keestra, M. (2017). Metacognition and reflection by interdisciplinary experts: Insights from cognitive science and philosophy. *Issues in Interdisciplinary Studies, 35,* 121–169. [He summarizes his arguments in a blog post at https://i2insights .org/2019/02/05/metacognition-and-interdisciplinarity/.]

Kelly, J. S. (1996). Wide and narrow interdisciplinarity. *The Journal of Education, 45*(2), 95–113.

Klein, J. T. (1996). *Crossing boundaries: Knowledge, disciplinarities, and interdisciplinarities.* Charlottesville: University Press of Virginia.

Klein, J. T. (1999). *Mapping interdisciplinary studies.* Number 13 in the Academy in Transition series. Washington, DC: Association of American Colleges and Universities.

Klein, J. T. (2003). Unity of knowledge and transdisciplinarity: Contexts and definition, theory, and the new discourse of problem-solving. *Encyclopedia of Life Support Systems.* Retrieved July 15, 2011, from http://www.eolss.net/.

Klein, J. T. (2005a). *Humanities, culture, and interdisciplinarity: The changing American academy.* Albany: State University of New York Press.

Klein, J. T. (2005b). Interdisciplinary teamwork: The dynamics of collaboration and integration. In S. J. Derry, C. D. Schunn, & M. A. Gernsbacher (Eds.), *Interdisciplinary collaboration: An emerging cognitive science* (pp. 23–50). Mahwah, NJ: Erlbaum.

Klein, J. T. (2010). *Creating interdisciplinary campus cultures: A model for strength and sustainability.* San Francisco, CA: Jossey-Bass.

Klein, J. T. (2012). Research integration: A comparative knowledge base. In A. F. Repko, W. H. Newell, & R. Szostak (Eds.), *Case studies in interdisciplinary research* (pp. 283–298). Thousand Oaks, CA: Sage.

Klein, J. T., & Newell, W. H. (1997). Advancing interdisciplinary studies. In J. G. Gaff, J. L. Ratcliff, & Associates (Eds.), *Handbook of the undergraduate curriculum: A comprehensive guide to purposes, structures, practices, and change* (pp. 393–415). San Francisco, CA: Jossey-Bass.

Kockelmans, J. J. (1979). Why interdisciplinarity. In J. J. Kockelmans (Ed.), *Interdisciplinarity and higher education* (pp. 123–160). University Park and London: The Pennsylvania State University Press.

Kuhn, T. (1996). *The structure of scientific revolutions* (3rd ed.). Chicago, IL: University of Chicago Press.

Lakoff, G. (1987). *Women, fire, and dangerous things: What categories reveal about the mind.* Chicago, IL: University of Chicago Press.

Lanier, M. M., & Henry, S. (2004). *Essential criminology* (3rd ed.). Boulder, CO: Westview.

Lattuca, L. (2001). *Creating interdisciplinarity: Interdisciplinary research and teaching among college and university faculty.* Nashville, TN: Vanderbilt University Press.

Larner, J. (1999). *Marco Polo and the discovery of the world.* New Haven CT: Yale University Press.

Leary, M. R. (2004). *Introduction to behavioral research methods* (4th ed.). Boston, MA: Pearson Education.

Lefebvre, H. (1991). *The production of space.* Oxford and Cambridge, UK: Blackwell.

Lenoir, Y., & Klein, J.T. (Eds.) (2010). Interdisciplinarity in schools: A comparative view of national perspectives. *Issues in Integrative Studies, 28.* [Special Issue]

Lewis, B. (2002, January). What went wrong? *The Atlantic Monthly, 289,* 1. Retrieved July 25, 2002, from http://www.the atlantic.com/doc/200201/lewis.

Lewis, J. (2009). *The failure of public education to educate a diverse population.* Unpublished manuscript.

Li, T. (2000). *Social science reference sources.* Westport, CT: Greenwood Press.

Long, D. (2002). *Interdisciplinarity and the English school of international relations.* Paper presented at the International Studies Association Annual Convention, New Orleans, March 25–27.

Longo, G. (2002). The constructed objectivity of the mathematics and the cognitive subject. In M. Mugur-Schachter & A. van der Merwe (Eds.), *Quantum mechanics, mathematics, cognition and action* (pp. 433–462). Boston, MA: Kluwer Academic.

Looney, C., Donovan, S., O'Rourke, M., Crowley, S., Eigenbrode, S. D., Totschy, L., . . .& Wulfhorst, J. D. (2014). Seeing through the eyes of collaborators: Using Toolbox workshops to enhance cross-disciplinary collaboration. In Michael O'Rourke, Stephen Crowley, Sanford D. Eigenbrode, & J. D. Wulfhorst (Eds.), *Enhancing communication and collaboration in interdisciplinary research* (pp. 220–243). Thousand Oaks, CA: Sage.

Mack, D. C., & Gibson, C., eds. (2012). *Interdisciplinarity and academic libraries.* Chicago: Association of Research and College Libraries.

Magnus, D. (2000). Down the primrose path: Competing epistemologies in early twentieth-century biology. In R. Creath & J. Maienschein (Eds.), *Biology and epistemology* (pp. 91–121). Cambridge, UK: Cambridge University Press.

Maienschein, J. (2000). Competing epistemologies and developmental biology. In R. Creath & J. Maienschein (Eds.), *Biology and epistemology* (pp. 122–137). Cambridge, UK: Cambridge University Press.

Manheim, J. B., Rich, R. C., Willnat, L., & Brians, C. L. (2006). *Empirical political analysis: Research methods in political science* (6th ed.). Boston, MA: Pearson Education.

Mapes, L.V., & Bernton, H. (2018). Changes to dams on Columbia, Snake rivers to benefit salmon, hydropower and orcas. *Seattle Times,* Originally published December 18, 2018.

Retrieved from https://www.seattletimes.com/seattle-news/environment/a-new-day-for-fish-hydropower-on-the-columbia-and-snake-rivers/.

Marsh, D., & Furlong, P. (2002). A skin, not a sweater: Ontology and epistemology in political science. In D. Marsh & G. Stoker (Eds.), *Theory and methods in political science* (2nd ed., pp. 17–41). New York, NY: Palgrave Macmillan.

Marshall, C., & Rossman, G. B. (2006). *Designing qualitative research*. Thousand Oaks, CA: Sage.

Marshall, D. G. (1992). Literary interpretation. In J. Gibaldi (Ed.), *Introduction to scholarship in modern languages and literatures* (pp. 159–182). New York, NY: The Modern Language Association of America.

Martin, R., Thomas, G., Charles, K., Epitropaki, O., & McNamara, R. (2005). The role of leader-member exchanges in mediating the relationship between locus of control and work reactions. *Journal of Occupational and Organizational Psychology, 78*, 141–147.

Martin, V. (2017). *Transdisciplinarity Revealed: What Librarians need to Know*. Santa Barbara: Libraries Unlimited.

Mathews, L. G., & Jones, A. (2008). Using systems thinking to improve interdisciplinary learning outcomes. *Issues in Integrative Studies, 26*, 73–104.

Maurer, B. (2004). Models of scientific inquiry and statistical practice: Implications for the structure of scientific knowledge. In M. L. Taper & S. R. Lee (Eds.), *The nature of scientific evidence: Statistical, philosophical, and empirical considerations* (pp. 17–31). Chicago, IL: University of Chicago Press.

Mayr, E. (1997). *This is biology*. Cambridge, MA: Harvard University Press.

McDonald, D., Bammer, G., & Deane, P. (2009). Research integration using dialogue methods. ANU E-Press. Retrieved July 15, 2011, from http://press.anu.edu.au/titles/dialogue_methods_citation/.

McKim, V. R. (1997). Introduction. In V. R. McKim, S. P. Turner, & S. Turner (Eds.), *Causality in crisis? Statistical methods and the search for causal knowledge in the social sciences*. Notre Dame, IN: University of Notre Dame Press.

Mepham, B. (2000). A framework for the ethical analysis of novel foods: The ethical matrix. *Journal of Agricultural and Environmental Ethics, 12*(2), 165–176.

Merari, A. (1998). The readiness to kill and die: Suicidal terrorism in the Middle East. In W. Reich (Ed.), *Origins of terrorism: Psychologies, ideologies, theologies, states of mind* (pp. 192–210). Washington, DC: Woodrow Wilson Center Press.

Miles, M. B., & Huberman, M. (1994). *Qualitative data analysis: An expanded sourcebook* (2nd ed.). Thousand Oaks, CA: Sage.

Miller, A. I. (1996). *Insights of genius: Imagery and creativity in science and art*. Cambridge, UK: The MIT Press.

Miller, R. C. (1982). Varieties of interdisciplinary approaches in the social sciences. *Issues in Integrative Studies, 1*, 1–37.

Miller, R. C. (2018). *International political economy: Contrasting world views* (2nd ed). New York. NY: Routledge.

Misra, S., Hall, K., Feng, A., Stipelman, B., & Stokols, D. (2011). Collaborative processes in transdisciplinary work. In M. J. Kirst et al. (Eds.), *Converging disciplines: A transdisciplinary research approach to urban health problems* (pp. 97–110). New York, NY: Springer.

Modiano, R. (1999). Sameness or difference? Historicist readings of "The Rime of the Ancient Mariner." In P. H. Fry (Ed.), *Samuel Taylor Coleridge: The rime of the ancient mariner* (pp. 187–219). Boston, MA: Bedford/St. Martin's.

Mokari Yamchi, A., Alizadeh-sani, M., Khezerolou, A., Zolfaghari Firouzsalari, N., Akbari, Z., & Ehsani, A. (2018). Resolving the food security problem with an interdisciplinary approach. *Journal of Nutrition, Fasting and Health, 6*(3), 132–138.

Monroe, K. R., & Kreidie, L. H. (1997). The perspective of Islamic fundamentalists and the limits of rational choice theory. *Political Psychology, 18*(1), 19–43.

Moran, J. (2010). *Interdisciplinarity* (2nd ed.). New York, NY: Routledge.

Motes, M. A., Bahr, G. S., Atha-Weldon, C., & Dansereau, D. F. (2003). Academic guide maps for learning psychology. *Teaching of Psychology, 30*(3), 240–242.

Murfin, R. C. (1999a). Deconstruction and "The Rime of the Ancient Mariner." In P. H. Fry (Ed.), *Samuel Taylor Coleridge: The rime of the ancient mariner* (pp. 261–282). Boston, MA: Bedford/St. Martin's.

Murfin, R. C. (1999b). Marxist criticism and "The Rime of the Ancient Mariner." In P. H. Fry (Ed.), *Samuel Taylor Coleridge: The rime of the ancient mariner* (pp. 131–147). Boston, MA: Bedford/St. Martin's.

Murfin, R. C. (1999c). The new historicism and "The Rime of the Ancient Mariner." In P. H. Fry (Ed.), *Samuel Taylor Coleridge: The rime of the ancient mariner* (pp. 168–186). Boston, MA: Bedford/St. Martin's.

Murfin, R. C. (1999d). Psychoanalytic criticism and "The Rime of the Ancient Mariner." In P. H. Fry (Ed.), *Samuel Taylor Coleridge: The rime of the ancient mariner* (pp. 220–238). Boston, MA: Bedford/St. Martin's.

Murfin, R. C. (1999e). Reader-response criticism and "The Rime of the Ancient Mariner." In P. H. Fry (Ed.), *Samuel Taylor Coleridge: The rime of the ancient mariner* (pp. 97–113). Boston, MA: Bedford/St. Martin's.

Myers, D. G. (2002). *Intuition: Its powers and perils*. New Haven & London: Yale University Press.

Nagy, C. (2005). *Anthropogenic forces degrading tropical ecosystems in Latin America: A Costa Rican case study*. Unpublished manuscript, Miami of Ohio University.

National Academy of Sciences, National Academy of Engineering, & Institute of Medicine. (2005). *Facilitating interdisciplinary research*. Washington, DC: National Academies Press.

Neuman, W. L. (2006). *Social research methods: Qualitative and quantitative approaches* (6th ed.). Boston, MA: Pearson Education.

Newell, W. H. (1990). Interdisciplinary curriculum development. *Issues in Integrative Studies, 8,* 69–86.

Newell, W. H. (1992). Academic disciplines and undergraduate interdisciplinary education: Lessons from the school of interdisciplinary studies at Miami University, Ohio. *European Journal of Education, 27*(3), 211–221.

Newell, W. H. (1998). Professionalizing interdisciplinarity: Literature review and research agenda. In W. H. Newell (Ed.), *Interdisciplinarity: Essays from the literature.* New York, NY: College Entrance Examination Board.

Newell, W. H. (2001). A theory of interdisciplinary studies. *Issues in Integrative Studies, 19,* 1–25.

Newell, W. H. (2004). Complexity and interdisciplinarity. In L. Douglas Kiel (Ed.), *Encyclopedia of life support systems* (EOLSS). Oxford, UK: EOLSS. Retrieved October 20, 2006, from http://www.eolss.net.

Newell, W. H. (2007a). Decision making in interdisciplinary studies. In G. Morçöl (Ed.), *Handbook of decision making* (pp. 245–264). New York, NY: Marcel-Dekker.

Newell, W. H. (2007b). Distinctive challenges of library-based research and writing: A guide. *Issues in Integrative Studies, 25,* 84–110. Retrieved July 15, 2011, from http://wwwp.oakland.edu/ais/publications.

Newell, W. H. (2012). Conclusion. In A. F. Repko, W. H. Newell, & R. Szostak (Eds.), *Case studies in interdisciplinary research* (pp. 299–314). Thousand Oaks, CA: Sage.

Newell, W. H., & Green, W. J. (1982). Defining and teaching interdisciplinary studies. *Improving College and University Teaching, 30*(1), 23–30.

Newman, K., Fox, S. C., Harding, D. J., Mehta, J., & Roth, W. (2004). *Rampage: The social roots of school shootings.* New York, NY: Basic Books.

Nicholas, D., Huntington, P., & Jamali, H. R. (2007). The use, users, and role of abstracts in the digital scholarly environment. *The Journal of Academic Librarianship, 33*(4), 446–453.

Nicholas, D., Huntington, P., Williams, P., & Dobrowolski, T. (2004). Re-appraising information seeking behavior in a digital environment: Bouncers, checkers, returnees and the like. *Journal of Documentation, 60*(1), 24–43.

Nikitina, S. (2005). Pathways of interdisciplinary cognition. *Cognition and Instruction, 23*(3), 389–425.

Nikitina, S. (2006). Three strategies for interdisciplinary teaching: Contextualization, conceptualization, and problem-centering. *Journal of Curriculum Studies, 38*(3), 251–271.

Novak, J. D. (1998). *Learning, creating, and using knowledge: Concept maps as facilitative tools in schools and corporations.* Mahwah, NJ: Lawrence Erlbaum Associates.

Novick, P. (1998). *That noble dream: The "objectivity question" and the American historical profession.* New York, NY: Cambridge University Press.

O'Rourke, M., Crowley, S., Eigenbrode, S. D., & Wulfhorst, J. D. (Eds.). (2014). *Enhancing communication and collaboration in interdisciplinary research.* Thousand Oaks, CA: Sage.

O'Rourke, M., Crowley, S., & Gonnerman, C. (2016). On the nature of cross-disciplinary integration: A philosophical framework. *Studies in History and Philosophy of Biological and Biomedical Sciences, 56,* 62–70.

Palmer, C. L. (2010). Information research on interdisciplinarity. In R. Frodeman, J. T. Klein, C. Mitcham, & J. B. Holbrook (Eds.), *The Oxford handbook of interdisciplinarity* (pp. 174–188). New York, NY: Oxford University Press.

Palmer, C. L., Teffeau, L. C., & Pirmann, C. M. (2009). *Scholarly information practices in an online environment: Themes from the literature and implications for library service development.* Dublin, OH: Online Computer Literacy Center.

Palys, T. (1997). *Research decisions: Quantitative and qualitative perspectives.* Toronto, Canada: Harcourt Brace.

Paternoster, R., & Bachman, R. (Eds.). (2001). *Explaining criminals and crime.* Los Angeles, CA: Roxbury.

Petrie, H. (1976). Do you see what I see? The epistemology of interdisciplinary inquiry. *Journal of American Education, 10,* 29–43.

Pieters, R., & Baumgartner, H. (2002). "Who Talks to Whom? Intra- and Interdisciplinary Communication of Economics Journals." *Journal of Economic Literature, 40*(2), 483–509.

Piso, Z. (2016). Language games of "Language Games." *Issues in Interdisciplinary Studies, 34,* 213–216.

Piso, Z., O'Rourke, M., & Weathers, K. (2016) Out of the fog: Catalyzing integrative capacity in interdisciplinary research. *Studies in History and Philosophy of Science, 56,* 84–94.

Pohl, C., van Kerkhoff, L., Hadorn, G. H., & Bammer, G. (2008). Integration. In G. H. Harorn, H. Hoffman-Riem, S. Biber-Klemm, W. Grossenbacher-Mansuy, D. Joy, C. Pohl, U. Wiesmann, & E. Zemp (Eds.), *Handbook of transdisciplinary research* (pp. 411–426). Berlin Heidelberg, Germany: Springer.

Polkinghorne, J. (1996). *Beyond science: The wider human context.* Cambridge, UK: Cambridge University Press.

Popham, W. J. (1997, October). What's wrong—and what's right—with rubrics. *Educational Leadership,* pp. 72–75.

Post, G. M. (1998). Terrorist psycho-logic: Terrorist behavior as a product of psychological forces. In W. Reich (Ed.), *Origins of terrorism: Psychologies, ideologies, theologies, states of mind* (pp. 25–40). Washington, DC: Woodrow Wilson Center Press.

Preziosi, D. (1989). *Rethinking art history: Meditations on a coy science.* New Haven, CT: Yale University Press.

Quinn, G. P., & Keough, M. J. (2002). *Experimental design and data analysis for biologists.* Cambridge, UK: Cambridge University Press.

Rapoport, D. C. (1998). Sacred terror: A contemporary example from Islam. In W. Reich (Ed.), *Origins of terrorism: Psychologies, ideologies, theologies, states of mind* (pp. 103–130). Washington, DC: Woodrow Wilson Center Press.

Reisberg, D. (2006). *Cognition: Exploring the science of the mind* (3rd ed.). New York, NY: Norton.

Remler, D. K., & van Ryzin, G. G. (2011). *Research methods in practice: Strategies for description and causation.* Thousand Oaks, CA: Sage.

Repko, A. F. (2006). Disciplining interdisciplinarity: The case for textbooks. *Issues in Integrative Studies, 24,* 112–142.

Repko, A. F. (2007) Integrating interdisciplinarity: How the theories of common ground and cognitive interdisciplinarity are informing the debate on interdisciplinary integration. *Issues in Integrative Studies, 25,* pp. 1–31.

Repko, A. F. (2012). Integrating theory-based insights on the causes of suicide terrorism. In A. F. Repko, W. H. Newell, & R. Szostak (Eds.), *Case studies in interdisciplinary research* (pp. 125–157). Thousand Oaks, CA: Sage.

Repko, A. F., Newell, W. H., & Szostak, R. (Eds.). (2012). *Case studies in interdisciplinary research.* Thousand Oaks, CA: Sage.

Repko, A. F., Szostak, R., & Buchberger, M. (2020). *Introduction to interdisciplinary studies* (3rd ed). Thousand Oaks, CA: Sage.

Reshef, N. (2008). Writing research reports. In C. Frankfort-Nachmias & D. Nachmias (Eds.), *Research methods in the social sciences* (7th ed.). New York, NY: Worth.

Rhoten, D., Boix Mansilla, V., Chun, M., & Klein, J. T. (2006). *Interdisciplinary education at liberal arts institutions.* Brooklyn, NY: Social Science Research Council. Retrieved September 22, 2019, from http://www.teaglefoundation.org/Teagle/media/GlobalMediaLibrary/documents/resources/Interdisciplinary_Education.pdf?ext=.pdf.

Richards, D. G. (1996). The meaning and relevance of "synthesis" in interdisciplinary studies. *The Journal of Education, 45*(2), 114–128.

Rogers, Y., Scaife, M., & Rizzo, A. (2005). Interdisciplinarity: An emergent or engineered process? In S. J. Derry, C. D. Schunn, & M. A. Gernsbacher (Eds.), *Interdisciplinary collaboration: An emerging cognitive science* (pp. 265–285). Mahwah, NJ: Lawrence Erlbaum Associates.

Rosenau, P. M. (1992). *Post-modernism and the social sciences: Insights, inroads, and intrusions.* Princeton, NJ: Princeton University Press.

Rosenberg, A. (2000). *Philosophy of science* (2nd ed.). New York, NY: Routledge.

Rosenfeld, P. L. (1992, December). The potential of transdisciplinary research for sustaining and extending linkages between the health and social sciences. *Social Science and Medicine, 35*(11), 1343–1357.

Salmon, M. H. (1997). Ethical considerations in anthropology and archeology: Or, relativism and justice for all. *Journal of Anthropological Research, 53,* 47–63.

Salter, L., & Hearn, A. (1996). Introduction. In L. Salter & A. Hearn (Eds.), *Outside the lines: Issues in interdisciplinary research* (pp. 3–15). Montreal, Canada: McGill-Queen's University Press.

Saxe, J. G. (1963). *The blind men and the elephant.* New York, NY: McGraw-Hill.

Scheurich, J. J. (1997). *Research method in the postmodern.* Washington, DC: The Falmer Press.

Schneider, C. G. (2010). Foreword. In J. T. Kline (Ed.), *Creating interdisciplinary campus cultures: A model for strength and sustainability* (pp. xiii–xvii). San Francisco, CA: John Wiley & Sons.

Schoenfeld, K. (2005). *Customer service: The ultimate return policy.* Unpublished manuscript, Miami of Ohio University.

Seabury, M. B. (2002). Writing in interdisciplinary courses: Coaching integrative thinking. In C. Haynes (Ed.), *Innovations in interdisciplinary teaching* (pp. 38–64). Westport, CT: Oryx Press.

Searing, S. E. (1992). How libraries cope with interdisciplinarity: The case of women's studies. *Issues in Integrative Studies, 10,* 7–25.

Seipel, M. (2002). *Interdisciplinarity: An introduction.* Retrieved July 15, 2011, from http://www2.truman.edu/~mseipel/.

Shoemaker, D. J. (1996). *Theories of delinquency: An examination of explanations of delinquent behavior* (3rd ed.). New York, NY: Oxford University Press.

Shulman, L. (1998). Course anatomy: The dissection and analysis of knowledge through teaching. In P. Hutchings (Ed.), *The course portfolio: How faculty can examine their teaching to advance practice and improve student learning* (pp. 5–12). Washington, DC: American Association for Higher Education.

Silberberg, M. S. (2006). *Chemistry: The molecular nature of matter and change* (4th ed.). Boston, MA: McGraw-Hill.

Sill, D. (1996). Integrative thinking, synthesis, and creativity in interdisciplinary studies. *Journal of General Education, 45*(2), 129–151.

Silver, L. (2005). *Composing race and gender: The appropriation of social identity in fiction.* Unpublished manuscript, Miami of Ohio University.

Silverman, D. (2000). *Doing qualitative research: A practical handbook.* London, UK: Sage.

Simpson, D. (1999). How Marxism reads "The Rime of the Ancient Mariner." In P. H. Fry (Ed.), *Samuel Taylor Coleridge: The rime of the ancient mariner* (pp. 148–167). Boston, MA: Bedford/St. Martin's.

Smolinski, W. J. (2005). *Freshwater scarcity in Texas.* Unpublished manuscript, University of Texas at Arlington.

Sokolowski, R. (1998). The method of philosophy: Making distinctions. *The Review of Metaphysics, 51*(3), 1–11.

Somit, A., & Tanenhaus, J. (1967). *The development of American political science.* Boston, MA: Allyn & Bacon.

Stanford Encyclopedia of Philosophy. (2010). "Implicature." Retrieved July 15, 2011, from http://plato.stanford.edu/entries/implicature/.

Stember, M. (1991). Advancing the social sciences through the interdisciplinary enterprise. *The Social Science Journal*, *28*(1), 1–14.

Stoll, C. (1999). *High-tech heretic: Why computers don't belong in the classroom and other reflections by a computer contrarian.* New York, NY: Doubleday.

Stone, J. R. (1998). Introduction. In J. R. Stone (Ed.), *The craft of religious studies* (pp. 1–17). New York, NY: St. Martin's Press.

Sturgeon, S., Martin, M. G. F., & Grayling, A. C. (1995). Epistemology. In A. C. Grayling (Ed.), *Philosophy 1: A guide through the subject* (pp. 7–60). New York, NY: Oxford University Press.

Szostak, R. (2002). How to do interdisciplinarity: Integrating the debate. *Issues in Integrative Studies*, *20*, 103–137.

Szostak, R. (2004). *Classifying science: Phenomena, data, theory, method, practice.* Dordrecht, The Netherlands: Springer.

Szostak, R. (2007a). Modernism, postmodernism, and interdisciplinarity. *Issues in Integrative Studies 26*, 32–83.

Szostak, R. (2007b). How and why to teach interdisciplinary research practice. *Journal of Research Practice*, *3*(2), Article M17. Retrieved February 25, 2011, from http://jrp.icaap.org/index.php/jrp/article/view/912/89.

Szostak, R. (2009). *The causes of economic growth: Interdisciplinary perspectives.* Berlin, Germany: Springer.

Szostak, R. (2012). An interdisciplinary analysis of the causes of economic growth. In A. F. Repko, W. H. Newell, & R. Szostak (Eds.), *Case studies in interdisciplinary research* (pp. 159–189). Thousand Oaks, CA: Sage.

Szostak, R. (2013). Communicating complex concepts. In M. O'Rourke, S. Crowley, S. D. Eigenbrode, & J. D. Wulfhorst (Eds.), *Enhancing communicating and collaboration in interdisciplinary research* (pp. 34–55). Thousand Oaks, CA: Sage.

Szostak, R. (2015a). Interdisciplinary and transdisciplinary approaches to multimethod and mixed method research. In S. N. Hesse-Biber & R. B. Johnson (Eds.), *The Oxford handbook of multimethod and mixed methods research inquiry* (pp. 128–143). Oxford, UK: Oxford University Press.

Szostak, R. (2015b). Extensional definition of interdisciplinarity. *Issues in Interdisciplinary Studies*, *33*, 94–117.

Szostak, R. (2016). What is lost? *Issues in Interdisciplinary Studies*, *34*, 208–213.

Szostak, R. (2017a). Interdisciplinary research as a creative design process. In F. Darbellay, Z. Moody, & T. Lubart, (Eds.), *Creative design thinking from an interdisciplinary perspective.* Berlin, Germany: Springer.

Szostak, R. (2017b). Stability, instability, and interdisciplinarity. *Issues in Interdisciplinary Studies*, *35*, 65–87.

Szostak, R. (2019). *Manifesto of interdisciplinarity.* Retrieved from https://sites.google.com/a/ualberta.ca/manifesto-of-interdisciplinarity/manifesto-of-interdisciplinarity.

Szostak, R., Gnoli, C., & Lopez-Huertas, M. (2016). *Interdisciplinary knowledge organization.* Berlin, Germany: Springer.

Taffel, A. (1992). *Physics: Its methods and meanings* (6th ed.). Upper Saddle River, NJ: Prentice Hall.

Taper, M. L., & Lele, S. R. (2004). The nature of scientific evidence: A forward-looking synthesis. In M. L. Taper & S. R. Lele (Eds.), *The nature of scientific evidence: Statistical, philosophical, and empirical considerations* (pp. 527–551). Chicago, IL: University of Chicago Press.

Tashakkori, A., & Teddlie, C. (1998). *Mixed methodology: Combining qualitative and quantitative approaches.* Thousand Oaks, CA: Sage.

Tayler, M. R. (2012). Jewish marriage as an expression of Israel's conflicted identity. In A. F. Repko, W. H. Newell, & R. Szostak (Eds.), *Case studies in interdisciplinary research* (pp. 23–51). Thousand Oaks, CA: Sage.

Tenopir, C., King, D. W., Boyce, P., Grayson, M., & Paulson, K. L. (2005). Relying on electronic journals: Reading patterns of astronomers. *Journal of the American Society for Information Science and Technology*, *56*(8), 786–802.

Terpstra, J. L., Best, A., Abrams, D., & Moor, G. (2010). Interdisciplinary health sciences and health systems. In J. T. Klein & C. Mitcham (Eds.), *The Oxford handbook of interdisciplinarity.* Oxford University Press, UK: Oxford.

Toynton, R. (2005). Degrees of disciplinarity in equipping students in higher education for engagement and success in lifelong learning. *Active Learning in Higher Education*, *6*(2), 106–117.

Tress, B., Tress, G., & Fry, G. (2006). Defining concepts and the process of knowledge production in integrative research. In B. Tress, G. Tress, G. Fry, & P. Opdam (Eds.), *From landscape research to landscape planning: Aspects of integration, education, and application* (pp. 13–25). Dordrecht, The Netherlands: Springer.

Tress, B., Tress, G., Fry, G., & Opdam, P. (Eds.). (2006). *From landscape research to landscape planning: Aspects of integration, education, and application.* Dordrecht, The Netherlands: Springer.

Turner, M. (2001). *Cognitive dimensions of social science.* New York, NY: Offord University Press.

van der Lecq, R. (2012). Why we talk: An interdisciplinary approach to the evolutionary origin of language. In A. F. Repko, W. H. Newell, & R. Szostak (Eds.), *Case studies in interdisciplinary research* (pp. 191–223). Thousand Oaks, CA: Sage.

Vess, D., & Linkon, S. (2002). Navigating the interdisciplinary archipelago: The scholarship of interdisciplinary teaching and learning. In M. Taylor Huber & S. P. Morreale (Eds.), *Disciplinary styles in the scholarship of teaching and learning: Exploring common ground* (pp. 87–106). Washington, DC: American Association for Higher Education and the Carnegie Foundation for the Advancement of Teaching.

Vickers, J. (1998). "[U]framed in open, unmapped fields": Teaching the practice of interdisciplinarity. *Arachne: An Interdisciplinary Journal of the Humanities*, *4*(2), 11–42.

von Wehrden, H., Guimarães, M. H., Bina, O., Varanda, M., Lang, D. J., John, B., . . . Lawrence, R. J. (2019). Interdisciplinary and transdisciplinary research: finding the common ground of multi-faceted concepts. *Sustainability Science, 14*, 875. Retrieved from https://doi.org/10.1007/s11625-018-0594-x.

Wallace, R. A., & Wolf, A. (2006). *Contemporary sociological theory: Expanding the classical tradition* (6th ed.). Upper Saddle River, NJ: Pearson.

Wallis, S. E. (2014). Existing and emerging methods for integrating theories within and between disciplines. *Journal of Organisational Transformation & Social Change, 11*, 3–24.

Walvoord, B. E. F., & Anderson, V. J. (1998). *Effective grading: A tool for learning and assessment.* San Francisco, CA: Jossey-Bass.

Weingast, B. (1998). Political institutions: Rational choice perspectives. In R. Goodin & H. Klingerman (Eds.), *A new handbook of political science* (pp. 167–190). Oxford, UK: Oxford University Press.

Welch, J., IV. (2003). Future directions for interdisciplinarity effectiveness in higher education: A Delphi study. *Issues in Integrative Studies, 21*, 3, 5–6, 170–203.

Welch, J., IV. (2007). The role of intuition in interdisciplinary insight. *Issues in Integrative Studies, 25*, 131–155.

Welch, J., IV (2011). The emergence of interdisciplinarity from epistemological thought. *Issues in Integrative Studies, 29, 1–39.*

Welch, J., IV. (2017). All too human: Conflict and common ground. *Issues in Interdisciplinary Studies, 35*, 88–112.

Welch, J., IV. (2018). The impact of Newell's "A Theory of Interdisciplinary Studies": Reflection and analysis. *Issues in Interdisciplinary Studies, 36*(2), 193–211.

Werder, K. P., Nothhaft, H., Verčič, D., & Zerfass, A. (2018) strategic communication as an emerging interdisciplinary paradigm. *International Journal of Strategic Communication, 12*(4), 333–351, DOI: 10.1080/1553118X.2018.1494181

Wheeler, L., & Miller, E. (1970, October). *Multidisciplinary approach to planning.* Paper presented at Council of Education Facilities Planners 47th Annual Conference in Oklahoma City, OK, October 6, 1976. (ERIC Document Reproduction Service No. ED044814)

Whitaker, M. P. (1996). Relativism. In A. Barnard & J. Spencer (Eds.), *Encyclopedia of social and cultural anthropology* (pp. 478–482). New York, NY: Routledge.

Wiersma, W., & Jurs, S. G. (2005). *Research methods in education: An introduction.* Boston, MA: Allyn & Bacon.

Wolfe, C., & Haynes, C. (2003). Interdisciplinary writing assessment profiles. *Issues in Integrative Studies, 21*, 126–169.

Worden, R. (1998). The evolution of language from social intelligence. In J. R. Hurford, M. Studdert-Kennedy, & C. Knight (Eds.), *Approaches to the evolution of language: Social and cognitive bases* (pp. 148–168). Cambridge, UK: Cambridge University Press.

Xio, H. (2005). *Research methods for English studies.* Edinburgh, Scotland: Edinburgh University Press.

Zajonc, A. (1993). *Catching the light: The entwined history of light and mind.* New York, NY: Bantam.

INDEX